Terahertz Technology in Microwave and Photonics for Effective Communications

This book explores the utilization of terahertz technology within the realms of microwave and photonics to advance the development of next-generation communication systems.

Terahertz Technology in Microwave and Photonics for Effective Communications offers a fundamental comprehension of terahertz technology and its significance in reshaping research in microwave and photonics. It delves into various cutting-edge technologies such as the application of metamaterials to harness terahertz capabilities, inter-satellite communication, next-generation optical fiber communication, and the utilization of metamaterial absorbers for secure communications. This book also deals with the advancements in multi-beam and beamforming terahertz antenna technologies, incorporating multi-user multiple-input multiple-output (MU-MIMO) features. Additionally, it elucidates the reconfigurable capabilities and the impact of artificial intelligence, machine learning, and deep learning technologies on the evolution of the next generation of communication. This book also highlights recent attempts by researchers to utilize terahertz technology in the optical domain and discusses terahertz optical sources and detectors, optical communication, leveraging quantum technology for communication technologies, and inter-satellite optical communication.

This book is intended for experts and professionals in the fields of advanced communications, computer science, networking, telecommunications, and technology policy.

Terahertz Technology in Microwave and Photonics for Effective Communications

Edited by
Prakash Pareek, Naveen Mishra,
Jitendra K. Mishra, Indrasen Singh,
and Abhinav Gautam

CRC Press
Taylor & Francis Group
Boca Raton London New York

CRC Press is an imprint of the
Taylor & Francis Group, an **informa** business

Designed cover image: Getty Images ©

First edition published 2026
by CRC Press
2385 NW Executive Center Drive, Suite 320, Boca Raton FL 33431

and by CRC Press
4 Park Square, Milton Park, Abingdon, Oxon, OX14 4RN

CRC Press is an imprint of Taylor & Francis Group, LLC

ISBN: 978-1-032-98522-0 (hbk)
ISBN: 978-1-032-98523-7 (pbk)
ISBN: 978-1-003-59911-1 (ebk)

DOI: 10.1201/9781003599111

Typeset in Times
by codeMantra

Contents

About the Editors

Dr. Prakash Pareek earned his Ph.D. degree from the Indian Institute of Technology Dhanbad, India, in 2017. He is currently working as an Associate Professor in the Department of Electronics and Communication Engineering at Vishnu Institute of Technology, Bhimavaram, Andhra Pradesh, India. His research interests include optical communications, Group IV optoelectronic devices, plasmonics and integrated optics, and THz photonics.

Dr. Naveen Mishra is a Senior Assistant Professor in the School of Electronics Engineering at Vellore Institute of Technology, VIT University, Vellore. He joined the VIT University in May 2019. He joined the Department of Electronics Engineering, Indian Institute of Technology (Indian School of Mines) Dhanbad, in January 2014, as a Junior Research Fellow, where he worked as a Senior Research Fellow from January 2016 to May 2019. His current research interests include metamaterial structures and their applications.

Dr. Jitendra K. Mishra earned his Ph.D. degree in Electronics and Communication Engineering from the Indian Institute of Technology (ISM) Dhanbad, India, in 2016. Currently, he is with the Department of Electronics and Communication Engineering, Indian Institute of Information Technology Ranchi, Jharkhand, India. His research interests include optical fiber communication, optical wireless communication, optical quantum communication, and the application of machine learning in information and communication technology.

Dr. Indrasen Singh is an Assistant Professor in the School of Electronics Engineering at Vellore Institute of Technology (VIT), Vellore, Tamil Nadu, India. He earned his B.Tech. and M.Tech. degrees in Electronics and Communication Engineering from Uttar Pradesh Technical University, Lucknow, India, in 2006 and 2010, respectively. He earned his Ph.D. degree in Electronics and Communication Engineering from the National Institute of Technology Kurukshetra, Haryana, India, in 2019. His research interests include cooperative communication, stochastic geometry, modeling of wireless networks, heterogeneous networks, millimeter-wave communications, device-to-device communication, and 5G/6G communication.

Dr. Abhinav Gautam is a Marie Curie Postdoctoral Fellow in the Photonic Systems Group at Tyndall National Institute, Cork, Ireland. He earned his Ph.D. degree. in Electronics and Communication Engineering from the Indian Institute of Technology (ISM) in 2022, where he also earned his master's degree in 2016. He received his bachelor's degree in Electronics and Communication Engineering from Rajiv Gandhi Technical University in 2013. His research interests encompass fiber optic sensing, optical communication, photonic systems, and microwave photonics.

Contributors

Praveena A
Department of ECE
PSG College of Technology
Coimbatore, Tamil Nadu, India

Saif Ahmad
Department of ECE
Integral University
Lucknow, Uttar Pradesh, India

Amit Bage
Department of Electronics and Communication Engineering
National Institute of Technology
Hamirpur, Himachal Pradesh, India

Venkata Naga Rani Bandaru
Department of Information Technology
Vishnu Institute of Technology
Bhimavaram, Andhra Pradesh, India

Rakesh Chowdhury
Department of EEE
ABV-Indian Institute of Information Technology and Management
Gwalior, Madhya Pradesh, India

Pradeep Dhara
Department of ECE
SR University
Warangal, Telangana, India

Vipul Dixit
School of Computer Science Engineering and Technology
Bennett University
Greater Noida, Uttar Pradesh, India

Prashant Dwivedy
PDPM-Indian Institute of Information Technology, Design and Manufacturing
Jabalpur, Madhya Pradesh, India

Zeenat Fatima
Department of ME
University of Polytechnic, Integral University
Lucknow, Uttar Pradesh, India

M Harikrishnan
Department of Electronics and Communication Engineering
Sri Manakula Vinayagar Engineering College
Madagadipet, Puducherry, India

Rachit Jain
Department of IT
Prestige Institute of Management and Research
Gwalior, Madhya Pradesh, India

Raja Rajeshwari Kamaja
Department of Computer Science and Technology
SRKR Engineering College
Bhimavaram, Andhra Pradesh, India

Madurakavi Karthikeyan
School of Electronics Engineering
Vellore Institute of Technology
Vellore, Tamil Nadu, India

Mohd Javed Khan
Department of ECE
Integral University
Lucknow, Uttar Pradesh, India

Atul Kumar
Department of Electronics and Communication Engineering
Maulana Azad National Institute of Technology
Bhopal, Madhya Pradesh, India

Saurabh Kumar
Department of Electronics and Communication Engineering
National Institute of Technology Hamirpur
Hamirpur, Himachal Pradesh, India

Anoushka Kumari
Department of Electronics and Communication Engineering
National Institute of Technology Hamirpur
Hamirpur, Himachal Pradesh, India

P. Krishna Madhuri
Department of Computer Science and Engineering
Vishnu Institute of Technology
Bhimavaram, Andhra Pradesh, India

Sumalatha Madugula
Department of Electronics and Communication Engineering
Shri Vishnu Engineering College for Women
Bhimavaram, Andhra Pradesh, India

Naveen Kumar Maurya
Department of ECE
Vishnu Institute of Technology
Bhimavaram, Andhra Pradesh, India

Rajesh Natarajan
School of Electronics Engineering
Vellore Institute of Technology
Vellore, Tamil Nadu, India

Swati Nema
Vellore Institute of Technology
Vellore, Tamil Nadu, India

Durai Pandurangan
Department of Communication Engineering
School of Electronics Engineering
Vellore Institute of Technology
Vellore, Tamil Nadu, India

SD Nageena Parveen
Department of ECE
SR University
Warangal, Telangana, India

J Pradeep
Department of Electronics and Communication Engineering
Sri Manakula Vinayagar Engineering College
Madagadipet, Puducherry, India

Munjala Praharsha
Department of ECE
SR University
Warangal, Telangana, India

Jayant Kumar Rai
Department of EEE
ABV-Indian Institute of Information Technology and Management
Gwalior, Madhya Pradesh, India

Natarajan Rajesh
School of Electronics Engineering
Vellore Institute of Technology
Vellore, Tamil Nadu, India

G. Challa Ram
Department of Electronics and Communication Engineering
Shri Vishnu Engineering College for Women
Bhimavaram, Andhra Pradesh, India

Pinku Ranjan
Department of EEE
ABV-Indian Institute of Information Technology and Management
Gwalior, Madhya Pradesh, India

Sureshkumar S
Department of CSE
Shiv Nadar University
Chennai, Tamil Nadu, India

Anusha Sanapathi
Department of Computer Science and Engineering
Vishnu Institute of Technology
Bhimavaram, Andhra Pradesh, India

Venkata Ashok Durga Prasad Sanapathi
Department of Computer Science and Engineering
Vishnu Institute of Technology
Bhimavaram, Andhra Pradesh, India

S. Sangeetha
College of Engineering, Guindy
Anna University
Chennai, Tamil Nadu, India

Ramprabhu Sivasamy
Department of Electronics Engineering
Madras Institute of Technology
Anna University
Chennai, Tamil Nadu, India

Deepika Rani Sona
Vellore Institute of Technology
Vellore, Tamil Nadu, India

Vegesna S M Srinivasavarma
Department of CSE
Shiv Nadar University
Chennai, Tamil Nadu, India

Vineetha Paul Srungarapati
Department of ECE
SR University
Warangal, Telangana, India

Sk Thaherbasha
Department of ECE
SR University
Warangal, Telangana, India

Shweta B. Thomas
Vellore Institute of Technology
Vellore, Tamil Nadu, India

Poonam Tiwari
Department of ECE
Indian Institute of Technology
Indore, Madhya Pradesh, India

Sukumar Vidhya
School of Electronics Engineering
Vellore Institute of Technology
Vellore, Tamil Nadu, India

Kanniyappan Vinayagam
School of Electronics Engineering
Vellore Institute of Technology
Vellore, Tamil Nadu, India

Aluvoju Vivek
Department of ECE
SR University
Warangal, Telangana, India

S. Yuvaraj
Department of Electronics and Communication Engineering
Indian Institute of Information Technology, Design and Manufacturing
Chennai, Tamil Nadu, India

1 Advancements in Terahertz Antenna for 5G and Beyond

Anoushka Kumari, Saurabh Kumar, and Amit Bage

1.1 INTRODUCTION

Wireless communication technologies have transformed modern lifestyles, shifting services such as classes, meetings, conferences, and payments to online platforms. The transformation has triggered an expansion in wireless data traffic, pushing existing systems to their limits and creating the need for innovative technologies capable of supporting higher data rates and lower latencies. Projections estimate that by 2025, approximately 18.22 billion mobile-connected devices will be in use [1]. Such an explosion in demand has motivated researchers to explore the spectrum, which can potentially meet the growing requirements for data capacity and transmission speed.

Wireless communication has progressed through multiple generations, addressing different technical challenges. In the 1980s, the first generation (1G) was developed using analog systems that had the Advanced Mobile Phone Systems (AMPS) and the Nordic Mobile Telephone (NMT) as the standards. The shift to second-generation (2G) technologies introduced digital systems like IS-95/CDMA and GSM, enabling voice communication and basic data services. Further enhancements were made in 2G, which led to the development of Enhanced Data Rates for GSM Evolution (EDGE) as the 2.5G standard that can support low-speed internet. For higher data rates, the third generation (3G) was introduced in which the speed can reach 2 Mbps. 3G enabled various applications such as web browsing, video calling, and multimedia messaging through standards like CDMA-2000 and Wideband-CDMA [2–4]. The fourth generation (4G) significantly improved communication efficiency with technologies like Orthogonal Frequency Division Multiple Access (OFDMA), delivering more than 20 Mbps speed. HD video streaming, online gaming, and mobile broadband services are the services supported by 4G [5,6]. Despite these advancements, 4G networks could not meet the demands for low-latency communication and massive connectivity required by smart cities, industrial automation, and device-to-device communication. In response to this, fifth-generation (5G) technologies are being developed, which provide faster data rates and reduced latency to satisfy the demands of applications of the future. The sixth generation (6G) aims to surpass 5G capabilities by having ultra-low latency and delivering high data rates [7–9]. The 6G mobile network will require wireless transmission speeds of up to 1 Tbps, for which the THz spectrum is desirable, along with highly directive antennas [10,11].

DOI: 10.1201/9781003599111-1

This chapter delves into terahertz (THz) communication while addressing the challenges associated. Key antenna technologies are discussed, including metamaterial-based designs for improved performance, FSS-based antennas for frequency control, plasmonic antennas for compactness and efficiency, Dielectric Resonator Antennas (DRAs) for low-loss integration, and ultra-massive multiple-input multiple-output (UM-MIMO) antennas for enhanced spectral efficiency. The chapter highlights advancements in antenna technologies and their importance in addressing the challenges of 5G and 6G systems.

1.2 TERAHERTZ COMMUNICATION

The THz spectrum, as shown in Figure 1.1, has frequency ranging from 0.1 to 10 THz and wavelengths spanning from 3 mm to 30 μm. The THz region is positioned between the infrared region on the higher frequency side and the microwave region on the lower frequency side of the electromagnetic spectrum. Known as the "THz Gap", this region was historically underutilized due to challenges in generating, detecting, and manipulating THz signals [12]. However, recent developments have made it feasible to utilize this spectrum's unique characteristics for sensing and high-speed communication applications. With the maximum licensing limit of 3 THz, the Federal Communications Commission (FCC) has reserved frequencies above 0.095 THz for experimental research, which includes the 0.116–0.123, 0.1748–0.182, and 0.244–0.246 THz bands [9]. These allocations aim to exploit the high bandwidth available in the THz spectrum, which enables data rates exceeding 1 Tbps and addresses the limitations of lower-frequency systems.

The THz spectrum exhibits several distinctive properties due to which it can be implemented for real-time applications, as shown in Figure 1.2. Its non-ionizing radiation, with photon energy levels between 0.4 and 40 meV, makes it safe for biomedical imaging applications [13]. The minimal scattering associated with sub-millimeter wavelengths ensures high penetration depths, enabling THz waves to detect concealed objects in security systems at airports and metro stations. Furthermore, shorter wavelengths offer high spatial resolution, allowing THz waves to identify molecular vibrations for hazardous chemical detection [14]. In the future, THz communication is expected to realize the sixth generation (6G). THz has a shorter wavelength

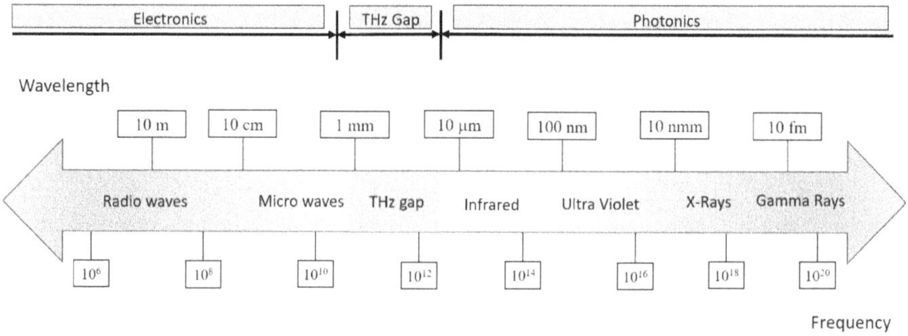

FIGURE 1.1 The THz gap in the electromagnetic spectrum.

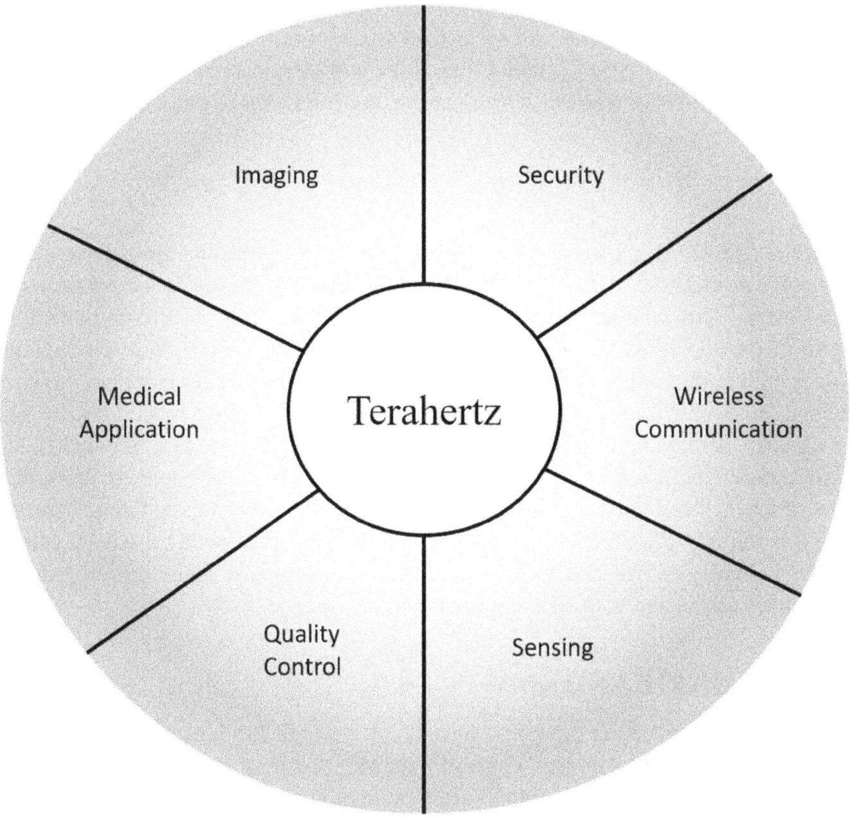

FIGURE 1.2 Various applications of THz spectrum.

resulting in smaller footprints as compared to mmWave communication. Hence, it can have higher directionality, less sensitivity to inter-antennas interferences, and high-capacity links for applications like vehicular networks and drone-based communication systems [15–17]. For example, 50 Mbps and 50 ms of data rate and latency are required respectively for the vehicle's surround view system. Also, an automatic takeover requires a latency of less than 10 ms. Hence, vehicle platooning, remote driving, and other safety measures are possible with the THz spectrum. Similarly, THz frequency enhances the localization accuracy of the drones, making them suitable for path planning and tracking applications [18,19].

1.3 TERAHERTZ COMMUNICATION CHALLENGES

Despite the immense potential, THz communication still has many issues that need to be resolved before it can be fully utilized. One of the primary issues is propagation loss, which arises because of the spreading (A_s) and absorption (A_{ma}) of THz signals [20]. Compared to the microwave band, the loss due to spreading in the THz band is approximately 60 dB more, which significantly reduces receiving power

[21]. Additionally, molecular absorption introduces high attenuation levels, further limiting signal transmission. A key factor contributing to propagation losses is the free space path loss (FSPL), which increases with the increase in frequency [22]. According to the Friis transmission equation, the FSPL can be calculated as:

$$\text{FSPL}(\text{in dB}) = 20\log\left(\frac{4\pi d}{\lambda}\right) \tag{1.1}$$

where d denotes the distance and λ represents the wavelength. Since path loss is inversely proportional to wavelength, higher frequencies result in larger propagation losses. This characteristic limits the effective range of THz communication to short distances, typically within a few tens of meters. Apart from propagation losses, multipath scattering presents another challenge, especially in indoor environments. Surfaces with micron-scale roughness can cause diffuse reflections, leading to signals reaching the receiver through multiple paths, thereby causing interference. Unlike lower-frequency systems, reflections in the THz band are strongly influenced by surface textures, making non-line-of-sight (NLoS) communication more complex and less reliable [23–25]. Given these limitations, there is a need for advancements in the technologies that can have reliable signal transmission while minimizing the challenges associated with THz communications.

1.4 TERAHERTZ ANTENNAS

A basic communication system consists of several fundamental components, including a source, transmitter, channel, receiver, and destination. The information to be sent is generated by the source after which it is encoded and modulated by the transmitter so that it can go over the channel. The channel such as optical fiber or free space acts as the medium through which the signal can travel. The information is then demodulated and decoded by the receiver so that the original information can be extracted and delivered to the intended destination. Among these components, the antenna is a crucial element used at both the transmitting and receiving ends. Its performance directly impacts the overall efficiency, range, and reliability of the communication system. Given its critical function, the design and optimization of antennas become even more significant, especially in THz communication systems. Hence, this section focuses on advancements in THz antenna technologies, exploring various designs and techniques aimed at addressing these challenges to have reliability and efficiency for 5G and beyond communication.

1.4.1 Metamaterial-Based Terahertz Antennas

The increasing demand for advanced and efficient materials in emerging technologies has driven the development of innovative artificial substances known as metamaterials. This breakthrough has sparked considerable interest and extensive research due to the unique characteristics and advantages offered by these materials, distinguishing them from naturally occurring substances. Metamaterials are artificially engineered materials composed of periodically arranged microstructures, commonly known as unit cells. Unit cell's symmetric or asymmetrical arrangements allow for precise control over their electromagnetic characteristics. The size, shape, and arrangement of these unit cells play a crucial role in determining key parameters such as resonant frequency, permittivity, and permeability [26,27]. Two primary parameters can define the electromagnetic behavior of metamaterials, i.e., permittivity and permeability. Notably, these properties can exhibit values that are either positive or negative, which is rarely observed in natural materials. Metamaterials can be classified into four main categories based on their permittivity and permeability characteristics. Double-negative materials, which exhibit both negative permittivity and permeability, are entirely synthetic and not found in nature. These materials are often utilized in applications requiring negative refraction. On the other hand, materials that have a positive permittivity are called ε-positive materials, whereas materials that have positive permittivity and permeability values are called as double-positive materials. Such double-positive properties are often found in natural materials, including certain types of plasma. Additionally, μ-negative materials are the materials that have negative permeability [28–31]. The performance of antennas and overall systems can be improved through the incorporation of metamaterials, as Figure 1.3 illustrates. For maximizing the resolution, detection range, and long-distance data transmission between remote sites, buildings, stations, or across geographical barriers, an important parameter of antennas plays a crucial role, i.e., gain. Despite being used extensively because of their small size and ease of fabrication, microstrip patch antennas often lack the gain needed for 5G and beyond applications. To address this limitation, metamaterials integrated with materials such as graphene, indium antimonide (InSb), and silicon dioxide (SiO_2) have been explored. For instance, the application of near-zero epsilon metamaterials, such as InSb and SiO_2, has improved antenna gain from 5.3 to 7.8 dBi [32]. Similarly, a 100% gain improvement has been achieved by integrating a multi-layer THz antenna with a graphene-loaded layer [33]. In addition to gain, radiation efficiency is another crucial factor for achieving optimal antenna performance. Antennas with high radiation efficiency reduce energy losses by allowing more power to be transferred from the source to the receiver. Energy efficiency is a critical requirement in ultra-low power applications. THz antennas can have their radiation efficiency increased by using metamaterials. Another key parameter

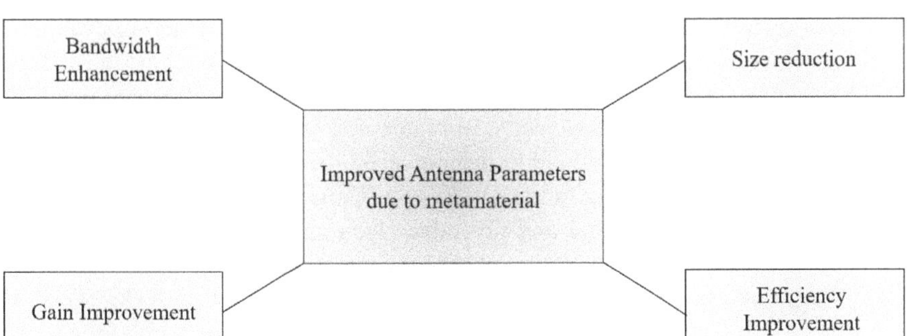

FIGURE 1.3 Various enhancements in antenna performance parameters through metamaterials.

is bandwidth, which determines the antenna's capacity to handle high data rates and accommodate multiple connected devices simultaneously. Enhanced bandwidth is vital for 5G and 6G systems to support ultra-fast data transmission and seamless connectivity. Metamaterials are utilized for enhancing the bandwidth of an antenna; the underlying principle is that there is zero propagation for the zero-order antennas [34]. Moreover, metamaterials contribute to size reduction, allowing antennas to be miniaturized. This miniaturization is essential for integrating antennas into compact and portable devices, making them suitable for modern wireless systems.

This miniaturization is essential for integrating antennas into compact and portable devices, making them suitable for modern wireless systems. The basic structure usually deployed in metamaterials is the Split Ring Resonator (SRR) or the Complementary Split Ring Resonator (CSRR). Scattering parameters are used for analyzing the antenna structure's metamaterial properties. The following formulas can be used to get the refractive index and impedance from the reflectance (S_{11}) and transmittance (S_{21}) [35]:

$$n = \frac{1}{kd}\cos^{-1}\left[\frac{1}{2S_{21}}\left(1 - S_{11}^2 + S_{21}^2\right)\right] \tag{1.2}$$

$$z = \sqrt{\frac{\left(1 + S_{11}^2\right) - S_{21}^2}{\left(1 - S_{11}^2\right) - S_{21}^2}} \tag{1.3}$$

$$\mu = nz \tag{1.4}$$

$$\epsilon = \frac{n}{z} \tag{1.5}$$

where n represents the refractive index, k denotes the wave vector, d represents the height, z denotes wave impedance, ϵ is the permittivity, and μ represents the permeability. These factors are essential in determining the electromagnetic response of metamaterial-based antennas, especially for THz applications where 6G demands high performance. A significant amount of literature has been published in the context of metamaterial-based THz antenna. One of them was done by Jeyakumar et al. who designed and optimized metamaterial-based THz antennas to address the increasing requirements of sixth-generation (6G) communication systems [36]. The study utilized CSRRs, as depicted in Figure 1.4, which are incorporated in the ground plane to enhance bandwidth, minimize return loss, and enhance radiation performance. The antenna was fabricated using a Flame Retardant 4 substrate, which has a dielectric constant of 4.4 and favorable electromagnetic characteristics. The substrate thickness was set to 10 μm, ensuring compatibility with THz frequencies. The initial dimensions of the antenna included a width and a length of 30 and 18 μm respectively. The antenna is microstrip fed having length and width 26 and 2.5 μm respectively. Despite these features, the initial design exhibited certain limitations, including a return loss and bandwidth of 18.5 dB and 1.7 THz respectively,

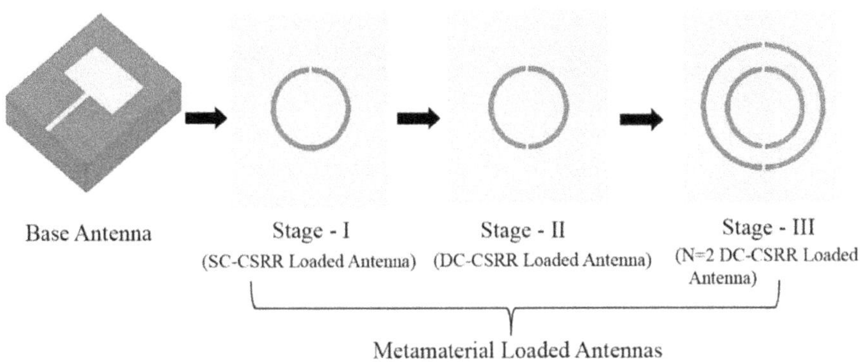

FIGURE 1.4 Evolution of the proposed metamaterial antenna [36]. (Copyright permission taken.)

which required further enhancements to meet 6G standards. To address these issues, a single-cut complementary split-ring resonator (SC-CSRR) was incorporated into the design. The SC-CSRR consisted of a circular ring structure with an inner radius and outer radius of 11 and 12.5 μm respectively, and a single cut with a 1 μm gap width. This configuration formed an LC resonant circuit. The resonating frequency is determined through [37]:

$$f_{\text{CSRR}} = \frac{1}{2\pi\sqrt{\text{LC}}} \tag{1.6}$$

The antenna's resonance frequency shifted from 2 to 2.2 THz by the SC-CSRR, which also decreased the return loss and expanded the bandwidth to −32.6 dB and 2.32 THz respectively. These improvements were attributed to better impedance matching and effective charge redistribution on the ground plane. Further improvements were achieved with DC-CSRR, i.e., dual-cut CSRR and N = 2 DC-CSRR designs. DC-CSRR design added a second cut in the ring structure, which increased its capacitive properties. Each cut had a 1 μm gap width, leading to a return loss and bandwidth of −49.9 dB and 2.35 THz respectively. The N = 2 DC-CSRR design introduced an additional concentric ring with dua, having 18 and 19.5 μm as the inner and outer radius. This configuration further optimized the electromagnetic performance, achieving −57 dB return loss and 2.36 THz bandwidth. Parametric analysis using the Nicolson Ross Weir (NRW) method validated the performance of the antenna. The analysis revealed sharp dual-negative permittivity and permeability values across the THz range. The proposed designs effectively addressed the challenges posed by 6G communication systems, demonstrating significant improvements in bandwidth and return loss. The N = 2 DC-CSRR loaded antenna is compact and high-performing, making it well-suited for applications such as ultra-reliable low latency communications (URLLC) and the Internet of Everything (IoE). The findings highlight the role of metamaterials in advancing THz antenna technologies and emphasize their importance in fulfilling the needs of next-generation communication systems.

1.4.2 Frequency Selective Surface-Based Terahertz Antennas

With advancements in metamaterials and communication technologies, Frequency Selective Surfaces (FSS) have emerged as a viable approach for 6G networks. Initially introduced in 1949, FSS gradually found applications in antenna design, beamforming, and other communication fields. To improve the antenna performance, FSS collaborated with antenna designers [38,39]. As 6G has various strict quality of service requirements and involves more available spectrum resources, the FSS technology can serve as a vital solution. FSS has a periodic array structure of either conducting wire (radiating element) or slot (non-radiating element). These structures can behave as bandstop and bandpass filters, respectively [40,41]. FSS's typical frequency response is shown in Figure 1.5 in which the transmission coefficient, $T=S_{21}$, is the response due to the array of conducting wires, and the reflection coefficient, $\Gamma=S_{11}$, is due to the array of slots. Also, here f_t is the starting of the bandpass and f_r is the bandstop center frequency. Bandstop FSS has a series connection of capacitance and inductance, whereas in bandpass FSS, there is a parallel connection between capacitance and inductance [42]. Through proper optimization of the FSS structure, the transmission characteristics such as reflection and transmission coefficient can be adjusted and controlled. Since an FSS's unit cells are stimulated by the electromagnetic field, they can be replaced by an electrically equivalent resonating circuit. The resonance frequency (f_{EM}) of an electromagnetic wave can be computed as follows [39]:

$$f_{EM} = \frac{1}{2\pi\sqrt{LC}} \tag{1.7}$$

In the above formula, L denotes the equivalent inductance and C denotes the equivalent capacitance of an FSS unit. Further, for simple FSS units, the value of L and C can be calculated as follows:

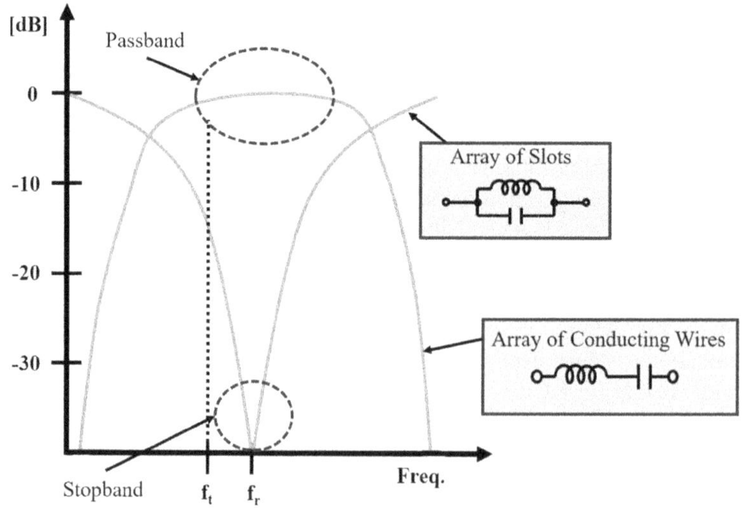

FIGURE 1.5 FSS's frequency response indicating the passband and stopband.

$$L = \frac{\left(\mu N^2 A\right)}{d_{\text{pi}}} \quad (1.8)$$

$$C = \frac{(\varepsilon A)}{d_{\text{pl}}} \quad (1.9)$$

Here, ε represents the dielectric constant, μ is the equivalent permeability, N denotes the number of turns, A is the area between metal plates, and d_{pi} and d_{pl} represent the distance between metal pieces and metal plates respectively. Structures with complex shapes and multiple metal plates have no generalized expression for calculating capacitance and inductance. Therefore, to calculate the capacitance and inductance value, the finite element method or electromagnetic field stimulation method can be utilized.

Many researches have been reported in which THz antennas are designed using FSS. In [43], a method was proposed by Aqlan et al. to improve the direct gain of the THz optical mixer antenna by incorporating it with structure. Nguyen et al. [44] combine the FSS with the leaky wave slit antenna to optimize signal transmission. Hence, by utilizing FSS structures to design a THz antenna, one can get a high value for gain and efficiency, which makes it appropriate for 5G and beyond applications. In addition to improving gain, FSS-based antenna systems can effectively mitigate angular interference during signal transmission and reception, which is a common issue in traditional phased-array systems [45,46]. FSS can further improve performance by reducing mutual interaction between closely spaced antenna elements. High spectral efficiency in 5G and beyond can be attained by using multiple-input multiple-output (MIMO) antennas, which are now widely utilized. However, MIMO systems often face challenges related to frequency selection and interference, which can degrade signal quality due to mutual coupling effects. FSS-assisted MIMO antenna systems offer a promising solution to these issues by improving stability and reliability through reduced multipath interference. Sharma et al. [47] proposed a THz MIMO antenna, as shown in Figure 1.6, which has one bandwidth from 2.3 to 2.7 THz and another from 5.5 to 8.9 THz. To enhance the two-port antenna's performance, FSS was positioned between the orthogonal MIMO antenna's radiators. As illustrated in Figure 1.7, FSS suppresses mutual coupling of greater than −10 dB for the frequency range of 2–3.5 THz and more than −5 dB for the frequency range of 3.5–10 THz. The average coupling coefficient without FSS is −30 dB, but with FSS, it exceeds −35 dB across the entire frequency that ranges from 2 to 10 THz. As shown in Figure 1.8, the gain of an antenna with or without FSS is analyzed as well. The antenna gain increases to more than 7 DBI at 9 THz because the FSS surface serves as a reflector. This improvement underscores the potential of FSS for enhancing antenna performance in high-frequency applications. While these developments highlight the potential of FSS in improving MIMO antenna systems, certain challenges remain. The multi-antenna configuration and beam assignment of the MIMO system may require management and proper allocation, and FSS needs to consider attenuation and power transmission to maximize system performance. Hence, power consumption and balancing the performance between MIMO and FSS systems are challenging tasks.

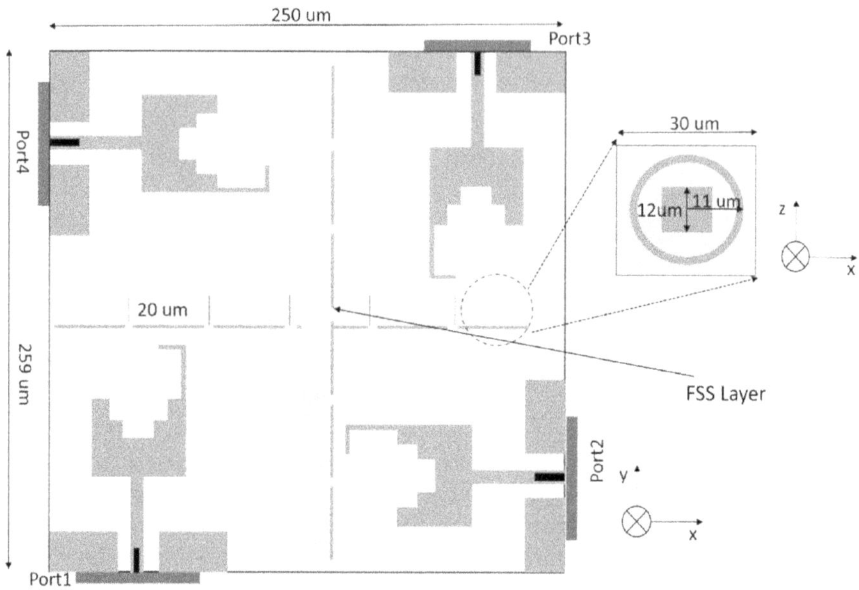

FIGURE 1.6 Structure of the proposed 4-port FSS MIMO antenna [47]. (Copyright permission taken.)

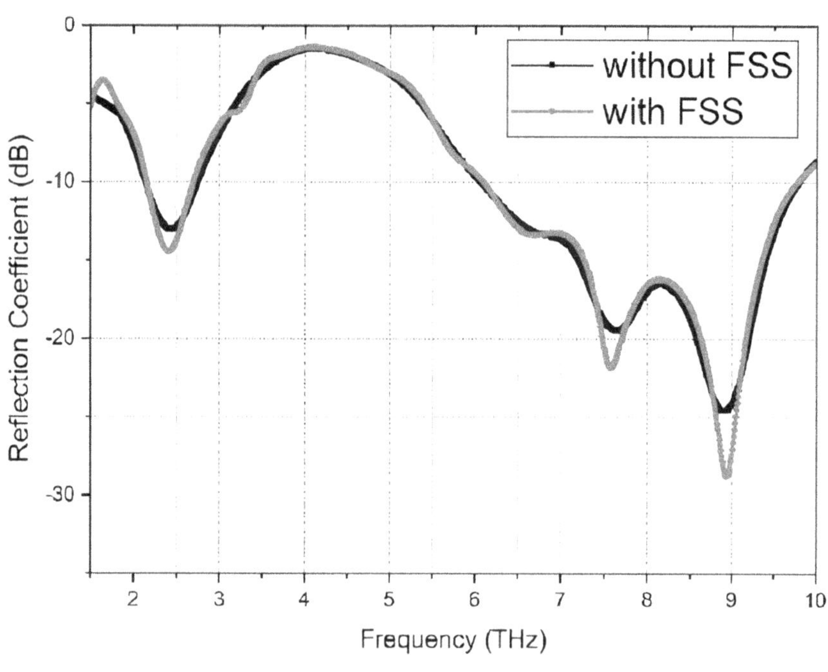

FIGURE 1.7 Reflection coefficient of the 2-port orthogonal antenna with and without FSS in dB [47]. (Copyright permission taken.)

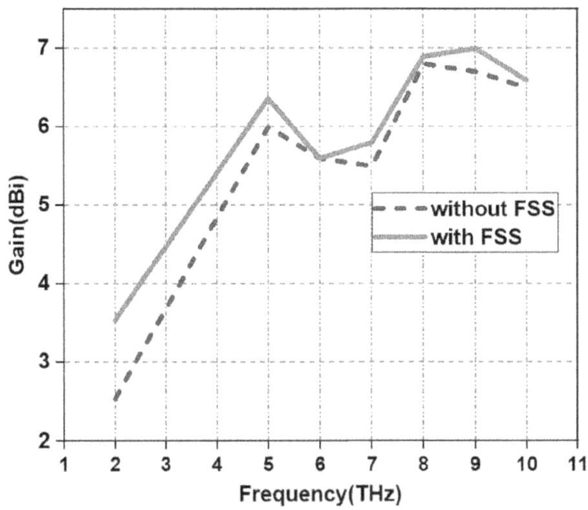

FIGURE 1.8 Gain of 2-port orthogonal antenna with and without FSS in dBi [47]. (Copyright permission taken).

1.4.3 Plasmonic Terahertz Antennas

Plasmonic antennas offer exceptionally high bandwidth and speed, making them suitable for enabling THz wireless communication at the nanoscale. Among the various materials considered for plasmonic applications, graphene stands out as a promising candidate. As seen in Figure 1.9, graphene is a carbon allotrope made up of sp_2-bonded carbon atoms organized in a honeycomb lattice. Its distinct mechanical, electrical, thermal, and electromagnetic characteristics have attracted a lot of attention lately. A key feature of graphene is the presence of a Dirac point, where the conduction and valence bands meet. This property results in tunable carrier densities, enabling gate-controlled electron–hole pair generation and providing ultra-wideband accessibility across different wavelengths. A major merit of graphene is its ability to adjust conductivity through external biasing via a gate voltage or chemical doping. The conductivity of graphene is influenced by factors such as relaxation time, chemical potential, electron mobility, doping level, operating frequency, and environmental temperature [48–52].

At THz frequencies, graphene conductivity primarily follows the intraband mechanism, as described by Kubo's model. Equation (1.10) is valid for the range of frequency of 0.1–10 THz if $\hbar\omega < 2|\mu_c|$. The equation for the graphene's conductivity is as follows [53]:

$$\sigma = -j\frac{e^2 k_B T}{\pi\hbar\left(\omega - j\tau^{-1}\right)}\ln\left\{2\left[1+\cosh\left(\frac{\mu_c}{k_B T}\right)\right]\right\} \tag{1.10}$$

where k_B denotes the Boltzmann constant, $\hbar$ denotes the reduced Planck's constant, μ_c represents the chemical potential of graphene, e is the electron charge, and τ represents the relaxation time of graphene electron. ω is the angular frequency and

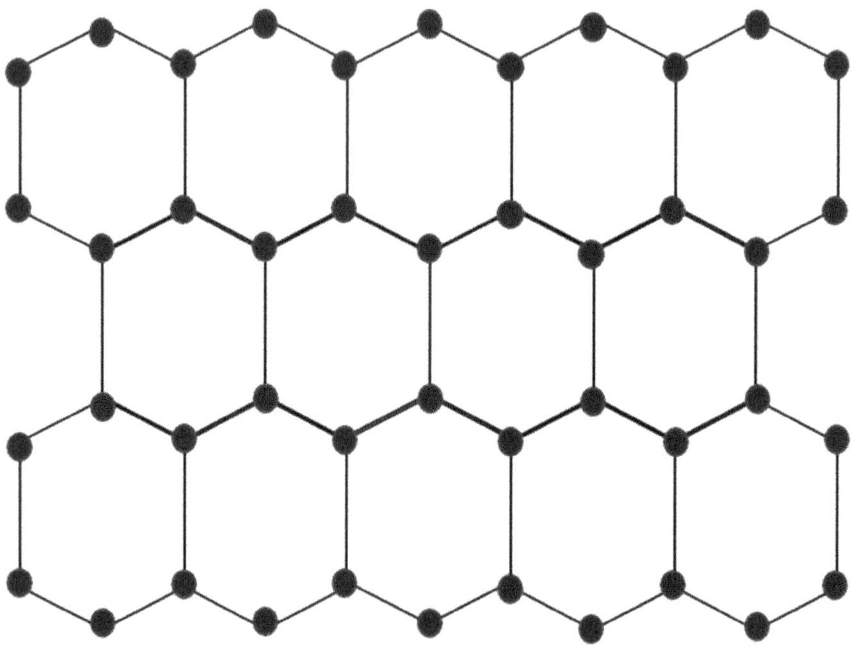

FIGURE 1.9 Structure of carbon allotrope—graphene.

T represents the temperature. Graphene also supports a unique phenomenon known as graphene plasmons, which are collective oscillations of charge carriers induced by strong light–matter interactions. The characteristic impedance and dispersion relation of the supported Transverse Magnetic Plasmon in a free space standing graphene sheet with σ conductivity can be computed as follows [54,55]:

$$k_{\text{SPP}} = k_0 \sqrt{1 - \left(\frac{2}{\eta\sigma}\right)^2} \tag{1.11}$$

$$Z_c = \frac{k_{\text{SPP}}}{\omega\varepsilon_0\varepsilon_{\text{eff}}} \tag{1.12}$$

Here, ε_0 denotes permittivity of vacuum and ε_{eff} represents the effective permittivity. While η is the impedance and k_0 represents the free space wave number. In the context of 5G and beyond, where efficiency and adaptability are crucial, graphene's ability to dynamically modulate its chemical potential by applying an external bias field orthogonal to its surface offers a significant advantage. Graphene-based THz array antennas can achieve high gain and excellent radiation characteristics. By altering the gate potential of individual antenna elements, the radiation pattern of an array can be dynamically reconfigured, eliminating the need for conventional switches and phase shifters for beam steering [56–59]. Numerous studies have been reported on graphene-based plasmonic THz antennas. The high-performance multiband plasmonic antenna, which utilizes a graphene patch on a SiO_3N_4 substrate depicted in

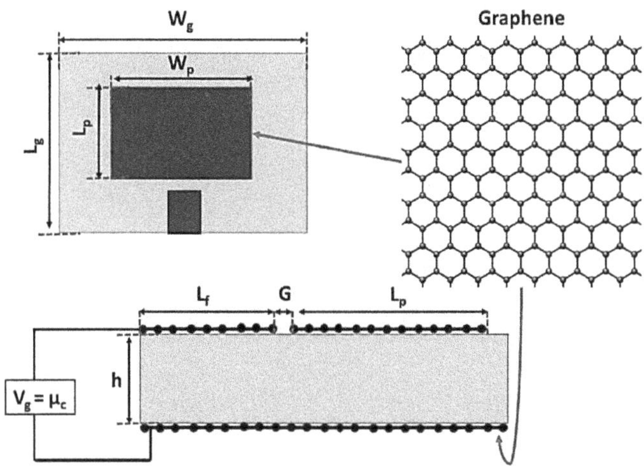

FIGURE 1.10 The proposed graphene plasmonic antenna top and cross-sectional view [60]. (Copyright permission taken.)

Figure 1.10, demonstrates gains of 3.06, 7.69, and 6.53 dB at 0.766, 3.285, and 4.510 THz respectively [60]. A reconfigurable square spiral antenna, composed of two graphene sheets with varying chemical potentials, operating across a range of 0.1–2 THz was proposed in [61]. It achieves a high radiation efficiency of 85.06%, with tunable frequency and beam angle, making it suitable for advanced THz communication technologies. In [62], the dual/tri-band graphene patch antenna using different substrates such as quartz, GaAs, and SiO_3N_4 was simulated from 0.1 to 5 THz, delivering gains of 2.510, 2.730, and 6.560 dB corresponding to 0.766, 3.285, and 4.510 THz in the case of GaAs. Moreover, the MIMO plasmonic antenna array designed as a series-fed graphene MIMO configuration operates at frequencies of 3.4, 3.6, 4.96, and 5.6 THz. It achieves gains exceeding 6.0 dB and supports a bandwidth greater than 120 GHz across all bands. Its excellent isolation (>20 dB) and diversity gain, combined with scalability for massive MIMO systems, make it an ideal candidate for 6G applications [63]. These investigations demonstrate that integrating graphene with THz antennas can provide highly reconfigurable, energy-efficient, and compact solutions. Such advancements address key challenges in developing 6G networks, paving the way for next-generation communication technologies.

1.4.4 DRA-Based Terahertz Antennas

In 1939, the concept of the DRA was introduced by Robert Richtmyer [64]. One of the main advantages of DRAs is that they do not suffer from conductor losses, resulting in high radiation efficiency. This property is particularly significant for high frequencies, such as THz, where metals tend to exhibit high losses. For the DRA, the dimension is in the order of $\lambda_0 / \sqrt{\epsilon_r}$, where ϵ_r is the dielectric constant and λ_0 represents the wavelength at the resonant frequency. As the dielectric constant and antenna size are inversely proportional to each other, materials with higher dielectric constants can

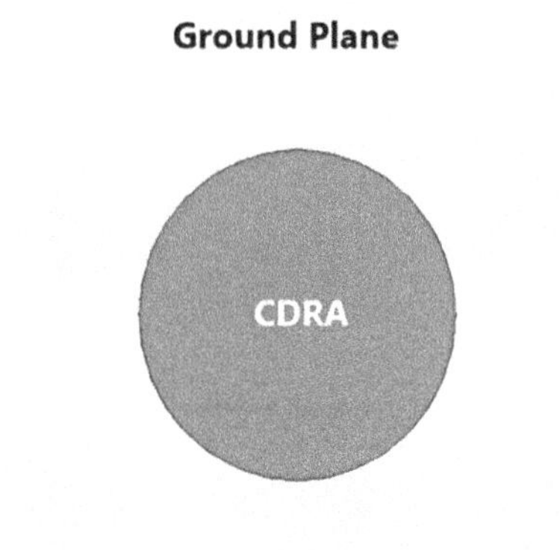

FIGURE 1.11 The top view of the proposed cylindrical Dielectric Resonator Antenna [70]. (Copyright permission taken.)

be utilized to make the antenna smaller and more compact for advanced applications [65,66]. Further, DRAs can provide wide bandwidth and design flexibility. While maintaining the desired resonant frequency, they can be designed into various geometric shapes, including rectangular, conical, and cylindrical. This flexibility allows DRAs to be adapted for different frequency bands and performance requirements. An experimental demonstration of a magnetic mirror with reflectivity near unity was presented in [67], with 0.8 THz as the central frequency. This mirror utilized cylindrical DRAs that were etched from silicon wafers having a permittivity of 11.8 and were bonded onto a ground plane made of gold. Similarly, in [68], a conical DRA operating at 10 THz was designed with 4.91 dBi gain and a boresight radiation pattern. In another study [69], a T-shaped DRA structure mounted on a polyimide substrate and excited with a coaxial probe exhibited a tri-band response, delivering gains of 7.81, 8.27, and 7.43 dBi at respective resonant frequencies. Furthermore, in [70], a cylindrical DRA combined with a rectangular dielectric resonator slab made up of silicon demonstrates a quad-band response, as shown in Figure 1.11, with a maximum gain and radiation efficiency of 6.67 dBi and 89% respectively. Further, for the tunability of the antenna, graphene is added on the top layer of cylindrical DRA. Therefore, DRAs are suitable for applications in 5G and beyond.

1.4.5 Ultra-massive MIMO Antennas

The integration of THz multi-beam UM-MIMO technology presents a viable approach to overcome capacity limitations in future 6G wireless communication networks. Combining THz frequencies with UM-MIMO systems offers practical approaches to fulfill the growing demand for high data rates and network capacity. UM-MIMO represents an advanced evolution of current MIMO systems used in 5G networks, accommodating 100–1,000 antennas at the base station while supporting a lot of users concurrently. The fundamental idea is to combine the signals

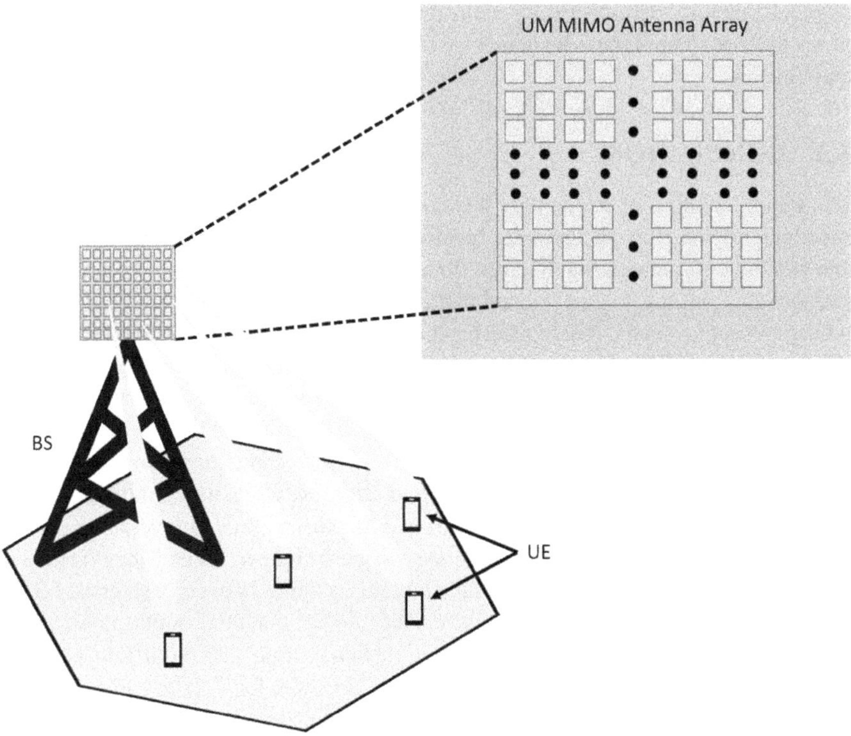

FIGURE 1.12 Representation of THz UM-MIMO approach at cell site.

from every antenna to improve the overall gain and performance of the system [71]. To improve capacity and alleviate distance-related challenges in THz-band 6G wireless links, UM-MIMO leverages multipath signal propagation and optimizes spectral efficiency under a fixed total transmit power. Multi-beam UM-MIMO antennas not only amplify signal coverage but also boost data rates through enhanced directivity. This is achieved using highly directional antennas or transmitting extensive data streams, thereby increasing spectral efficiency and communication reliability [71–73]. Consequently, UM-MIMO is anticipated to be a pivotal technique in 6G systems, enabling massive antenna arrays to deliver high-capacity data streams within compact cells. The schematic representation of this approach is illustrated in Figure 1.12. The inherently shorter wavelengths of THz signals make it feasible to integrate an exceptionally large number of antenna elements into compact transceiver layouts, resulting in multi-beam UM-MIMO arrays [58]. These arrays are meticulously engineered to achieve high directional gains, typically ranging from 30 to 50 dB [74]. Such gains are instrumental in overcoming the significant absorption and propagation losses that characterize THz communication channels. Chong et al. [75] further explored the potential of graphene-based plasmonic nano-antenna arrays for the UM-MIMO THz system. Their research investigated the properties of 1,024 × 1,024 nano-THz antenna arrays embedded in nanoscale footprints. The study assessed key performance metrics such as array factor, path gain, and channel capacity, demonstrating multi-terabit per second data rates over distances of up to 20 m. Thus, the advantages of UM-MIMO make them well-suited for 5G and beyond.

1.5 CONCLUSION

The rapid evolution of technology has created a significant demand for higher data speeds, lower latency, and broader bandwidth, which are essential for realizing 5G and beyond communication systems. To address these needs, advanced components and technologies must be developed. Antennas, as a core component of communication systems, play a vital role in meeting these demands. This chapter explores various advancements in THz antennas, focusing on how innovations can enhance performance, efficiency, and adaptability in next-generation wireless networks. One key development discussed is the integration of metamaterials with THz antennas. For high-performance systems, these engineered materials can enhance antenna properties like gain, bandwidth, size reduction, and radiation efficiency. Additionally, FSS are examined for their role in spatial filtering. FSS can be used to implement bandpass and bandstop mechanisms, enhancing system performance. When integrated with MIMO THz antennas, FSS helps reduce mutual coupling between antennas, further optimizing efficiency. The chapter also delves into the potential of plasmonic antennas, with a particular focus on graphene. Graphene's ability to modify its chemical potential through electric bias offers substantial benefits for THz antennas, improving parameters such as adaptability and efficiency. Furthermore, DRAs are explored for their advantages in THz applications, including low conductor losses, wide bandwidth, and compact design. Finally, the combination of UM-MIMO technology with THz antennas is discussed as a promising solution for enhancing data capacity and spectral efficiency in future wireless networks. Therefore, these advancements will contribute significantly to the evolution of global wireless networks, ensuring their ability to fulfill the demands of 5G and beyond systems.

REFERENCES

1. Statista, Forecast number of mobile devices worldwide from 2020 to 2025 (in billions)*, [Online]. Available: https://bit.ly/3pGyOZ3
2. W. C. Lee, *Mobile Communications Design Fundamentals*. John Wiley & Sons, 2010.
3. A. K. Jagannatham, *Principles of Modern Wireless Communication Systems*. McGraw-Hill Education, 2015.
4. Rappaport, *Wireless Communications: Principles and Practice, 2e*. Pearson Education India, 1996.
5. K. R. Santhi, V. K. Srivastava, G. SenthilKumaran and A. Butare, Goals of true broad band's wireless next wave (4G-5G). *2003 IEEE 58th Vehicular Technology Conference. VTC 2003-Fall (IEEE Cat. No.03CH37484)*, Orlando, FL, USA, 2003, vol. 4, pp. 2317–2321.
6. E. Ezhilarasan and M. Dinakaran. A review on mobile technologies: 3G, 4G and 5G. *2017 Second International Conference on Recent Trends and Challenges in Computational Models (ICRTCCM)*, Tindivanam, India, 2017, pp. 369–373.
7. J. G. Andrews, S. Buzzi, W. Choi, S. V. Hanly, A. Lozano and A. C. K. Soong, What will 5G be? *IEEE Journal on Selected Areas in Communications*, vol. 32, no. 6, pp. 1065–1082, June 2014. https://doi.org/10.1109/JSAC.2014.2328098

8. W. Jiang, B. Han, M. A. Habibi and H. D. Schotten, The road towards 6G: A comprehensive survey, *IEEE Open Journal of the Communications Society*, vol. 2, pp. 334–366, 2021. https://doi.org/10.1109/OJCOMS.2021.3057679
9. T. S. Rappaport, Y. Xing, O. Kanhere, S. Ju, A. Madanayake, and S. Mandal, Wireless communications and applications above 100GHz: Opportunities and challenges for 6G and beyond, *IEEE Access*, vol. 7, pp. 78729–78757, 2019. https://doi.org/10.1109/ACCESS.2019.2921522
10. S. Kumar and S. Kumar, A circularly polarized sub-THz antenna for 6G wireless communication, *2023 IEEE Microwaves, Antennas, and Propagation Conference (MAPCON)*, Ahmedabad, India, 2023, pp. 1–5. https://doi.org/10.1109/MAPCON58678.2023.10463839
11. S. Kumar and S. Kumar, Sub-THz pyramidal horn antenna for 6G wireless communication, *Frequenz*, vol. 78, no. 3–4, pp. 159–167, 2024. https://doi.org/10.1515/freq-2023-0114
12. C. Chaccour, M. N. Soorki, W. Saad, M. Bennis, P. Popovski and M. Debbah, Seven defining features of terahertz (THz) wireless systems: A fellowship of communication and sensing, *IEEE Communications Surveys & Tutorials*, vol. 24, no. 2, pp. 967–993, 2022. https://doi.org/10.1109/COMST.2022.3143454
13. M. Wan, J. J. Healy and J. T. Sheridan, Terahertz phase imaging and biomedical applications, Optics & Laser Technology, vol. 122, p. 105859, 2020. https://doi.org/10.1016/j.optlastec.2019.105859
14. A. Y. Pawar, D. D. Sonawane, K. B. Erande and D. V. Derle, Terahertz technology and its applications, *Drug Invention Today*, vol. 5, no. 2, pp. 157–163, 2013. https://doi.org/10.1016/j.dit.2013.03.009
15. K. M. S. Huq, S. A. Busari, J. Rodriguez, V. Frascolla, W. Bazzi and D. C. Sicker, Terahertz-enabled wireless system for beyond-5G ultra-fast networks: A brief survey, *IEEE Network*, vol. 33, no. 4, pp. 89–95, July/August 2019. https://doi.org/10.1109/MNET.2019.1800430
16. W. Saad, M. Bennis and M. Chen, A vision of 6G wireless systems: Applications, trends, technologies, and open research problems, *IEEE Network*, vol. 34, no. 3, pp. 134–142, May/June 2020. https://doi.org/10.1109/MNET.001.1900287
17. N. Rajatheva, I. Atzeni, S. Bicais, E. Bjornson, A. Bourdoux, S. Buzzi, et al., Scoring the terabit/s goal: Broadband connectivity in 6G. *arXiv preprint* arXiv:2008.07220, 2020.
18. J. Choi, V. Va, N. Gonzalez-Prelcic, R. Daniels, C. R. Bhat, and R. W. Heath, Millimeter-wave vehicular communication to support massive automotive sensing. *IEEE Communications Magazine*, vol. 54, no. 12, pp. 160–167, 2016.
19. R. Mendrzik, D. Cabric and G. Bauch, Error bounds for terahertz MIMO positioning of swarm UAVs for distributed sensing, *2018 IEEE International Conference on Communications Workshops (ICC Workshops)*, Kansas City, MO, USA, 2018, pp. 1–6. https://doi.org/10.1109/ICCW.2018.8403688
20. A. Tamminen, S. Makela, J. Ala-Laurinaho, J. Hakli, P. Koivisto, P. Rantakari, J. Saily, A. Luukanen and A. V. Raisanen, Reflectarray design for 120-GHz radar application: Measurement results, *IEEE Transactions on Antennas and Propagation*, vol. 61, no. 10, pp. 5036–5047, 2013.
21. Y. Huang, Y. Shen and J. Wang, From terahertz imaging to terahertz wireless communications, *Engineering*, vol. 22, pp. 106–124, 2023. https://doi.org/10.1016/j.eng.2022.06.023
22. J. M. Jornet and I. F. Akyildiz, Channel modeling and capacity analysis for EM wireless nanonetworks in the terahertz band, *IEEE Transactions on Wireless Communications*, vol. 10, no. 10, pp. 3211–3221, Oct. 2011.
23. C. Jansen, S. Priebe, C. Moller, M. Jacob, H. Dierke, M. Koch and T. Kurner, Diffuse scattering from rough surfaces in THz communication channels, *IEEE Transactions on Terahertz Science and Technology*, vol. 1, no. 2, pp. 462–472, Nov. 2011.

24. R. Piesiewicz, C. Jansen, D. Mittleman, T. K. Ostmann, M. Koch and T. Kurner, Scattering analysis for the modeling of THz communication systems, *IEEE Transactions on Antennas and Propagation*, vol. 55, no. 11, pp. 3002–3009, Nov. 2007.
25. P. Beckmann and A. Spizzichino, *The Scattering of Electromagnetic Waves from Rough Surfaces*. Pergamon, 1963.
26. Y. Liu and X. Zhang, Metamaterials: A new frontier of science and technology. *Chemical Society Reviews*, vol. 40, no. 5, pp. 2494–2507. https://doi.org/10.1039/c0cs00184h. Epub 2011 Jan 14.
27. P. D. Tung, P. H. Lam and N. T. Q. Hoa, A miniaturization of microstrip antenna using negative permittivity metamaterial based on CSRR-loaded ground for WLAN applications. *Vietnam Journal of Science and Technology*, vol. 54, no. 6, pp. 689–697, 2016.
28. B. P. Smyth, S. Barth and A. K. Iyer, Dual-band microstrip patch antenna using integrated uniplanar metamaterial-based EBGs. *IEEE Transactions on Antennas and Propagation*, vol. 64, pp. 5046–5053, 2016.
29. A. H. Naqvi and S. Lim, A beam-steering antenna with a fluidically programmable metasurface. *IEEE Transactions on Antennas and Propagation*, vol. 67, pp. 3704–3711, 2019.
30. J. Zhu and G. Eleftheriades, Dual-band metamaterial-inspired small monopole antenna for WiFi applications. *Electronics Letters*, vol. 45, pp. 1104–1106, 2009.
31. T. Ali, S. Pathan and R. C. Biradar, Multiband, frequency reconfigurable, and metamaterial antennas design techniques: Present and future research directions. *Internet Technology Letters*, vol. 1, p. e19, 2018.
32. C. Cheng, Y. Lu, D. Zhang, F. Ruan and G. Li, Gain enhancement of terahertz patch antennas by coating epsilon-near-zero metamaterials. *Superlattices Microstructures*, vol. 139, p. 106390, 2020.
33. M. M. Seyedsharbaty and R. A. Sadeghzadeh, Antenna gain enhancement by using metamaterial radome at THz band with reconfigurable characteristics based on graphene load. *Optical and Quantum Electronics*, vol. 49, pp. 1–13, 2017.
34. V. V. Yem and N. N. Lan, Gain and bandwidth enhancement of array antenna using novel metamaterial structure. *Journal of Communications*, vol. 13, pp. 101–107, 2018.
35. A. H. M. Almawgani, S. K. Patel, T. K. Nguyen and A. A. H. Al-Athwary, Design and analysis of split ring resonator engraved metamaterial broadband and high gain patch antenna for THz applications, *Optik*, vol. 287, p. 171054, 2023. https://doi.org/10.1016/j.ijleo.2023.171054
36. P. Jeyakumar, R. Pandeeswari, S. Saranya and B. Neeththi Aadithiya, Broadband complementary ring-resonator based terahertz antenna for 6G application. *Applied Physics A*, 130, p. 604, 2024. https://doi.org/10.1007/s00339-024-07763-6
37. K. Ren, P. Zhu, T. Sun, J. Wang, D. Wang, J. Liu and W. Zhao, A complementary split-ring resonator (CSRR)-based 2D displacement sensor. *Symmetry*, vol. 14, no. 6, p. 1116, 2022. https://doi.org/10.3390/sym14061116
38. G. H. H. Sung, K. W. Sowerby and A. G. Williamson, The impact of frequency selective surfaces applied to standard wall construction materials, *IEEE Antennas and Propagation Society Symposium*, Monterey, CA, USA, 2004, vol. 2, pp. 2187–2190.
39. X. Chen, J. Tan, L. Kang, F. Tang, M. Zhao and N. Kato, Frequency selective surface toward 6G communication systems: A contemporary survey, *IEEE Communications Surveys & Tutorials*, vol. 26, no. 3, pp. 1635–1675, 2024, https://doi.org/10.1109/COMST.2024.3369250
40. B. A. Munk, *Frequency Selective Surfaces*. John Wiley & Sons, 2005.
41. F. Capolino, *Theory and Phenomena of Metamaterials*. CRC Press, 2009.
42. U. Nissanov and G. Singh (Eds.), Terahertz microstrip antenna with frequency selective surface for 6G communication. In: Antenna Technology for Terahertz Wireless Communication. Springer, Cham, 2023. https://doi.org/10.1007/978-3-031-35900-2_6

43. B. Aqlan, M. Himdi, H. Vettikalladi and L. Le-Coq, A circularly polarized sub-terahertz antenna with low-profile and high-gain for 6G wireless communication systems, *IEEE Access*, vol. 9, pp. 122607–122617, 2021.
44. T. K. Nguyen, F. Rotermund and P. Ikmo, Leaky-wave slit antenna incorporated with a frequency selective surface for terahertz application, *Proceedings of the USNC-URSI Radio Science Meeting (Joint AP-S Symposium)*, Memphis, TN, USA, 2014, pp. 317–317.
45. H. Tang, B. Zheng, S. An, H. Li, M. Haerinia and C. Fowler, A broadband high-efficiency dipole array based on frequency selective surface and integrated feeding structure, *IEEE Open Journal Antennas and Propagation*, vol. 2, pp. 1087–1097, 2021.
46. L.-M. Zhang, X. Ding and W. Shao, A phase compensation beam switching antenna based on frequency selective surface, *IEEE Antennas Wireless Propagation Letters*, vol. 20, pp. 1741–1744, 2021.
47. M. K. Sharma, A. Sharma and R. Kumari, A CPW THz-MIMO antenna with reduced mutual coupling using frequency selective surface for future wireless applications, *Optical Materials*, vol. 148, p. 114929, 2024. https://doi.org/10.1016/j.optmat.2024.114929
48. E. N. Lallas, Key roles of plasmonics in wireless THz nanocommunications: A survey. *Applied Sciences*, vol. 9, p. 5488, 2019. https://doi.org/10.3390/app9245488
49. I. F. Akyildiz, J. M. Jornet and C. Han, TeraNets: Ultra-broadband communication networks in the terahertz band, *IEEE Wireless Communications*, vol. 21, no. 4, pp. 130–135, Aug. 2014.
50. I. Laser, C. Kremers, D. N. Chigrin, J. M. Jornet, M. C. Lemme, A. Cabellos-Aparicio, and E. Alarcon, Characterization of graphene-established nano-antennas in the terahertz band, *Proceedings of IEEE 6th European Conference on Antennas and Propagation (EUCAP)*, Prague, Czech Republic, 26–30 March 2012, pp. 1–5.
51. Y. Luo, Q. Zeng, X. Yan, T. Jiang, R. Yang, J. Wang, Y. Wu, Q. Lu and X. Zhang, A graphene-established tunable negative refractive index metamaterial and its application in dynamic beam-tilting terahertz antenna, *Microwave and Optical Technology Letters*, vol. 61, no. 12, pp. 2266–2672, Dec. 2019.
52. S. Kavitha, K. V. S. S. S. S. Sairam and A. Singh, Graphene plasmonic nano-antenna for terahertz communication, *S N Applied Science*, vol. 4, p. 114, Mar. 2022.
53. G. W. Hanson, Dyadic Green's functions and guided surface waves for a surface conductivity model of graphene, *Journal of Applied Physics*, vol. 103, no. 06430, pp. 0643021–0643028, Jan. 2008.
54. L. Cui, J. Wang and M. Sun, Graphene plasmon for optoelectronics, *Reviews in Physics*, vol. 6, p. 100054, June 2021.
55. C. L. Wang, Y. Q. Wang, H. Hu, D. J. Liu, D. L. Gao and L. Gao, Reconfigurable sensor and nano-antenna by graphene-tuned fano resonance, *Optics Express*, vol. 27, no. 24/25, pp. 35925–35935, Nov. 2019.
56. X. Fu, F. Yang, C. Liu, X. Wu and T. J. Cui, Terahertz beam steering technologies: from phased arrays to field-programmable metasurfaces, *Advanced Optical Material*, vol. 8, no. 3, p. 1900628, 2020.
57. M. H. Mubarak, S. Hara, I. Watanabe and A. Kasamatsu, Beam steering in graphene-based sub-THz dipole phased array antenna, *Proceedings of IEEE International Symposium on Radio-Frequency Integration Technology (RFIT)*, Hiroshima, Japan, 2–4 Sep. 2020, pp. 154–156.
58. A. Singh, M. Andrello, N. Thawdar and J. M. Jornet, Design and operation of a graphene based plasmonic nano-antenna array for communication in the terahertz band, *IEEE Journal on Selected Areas in Communications*, vol. 38, no. 9, pp. 2104–2117, Sep. 2020.
59. D. Yang, Analysis on the conductivity of graphene-based composite material, *IOP Publishing Journal of Physics: Conference Series*, vol. 2011, no. 012002, pp. 1–8, July 2021.

60. S. O. Hasan, S. K. Ezzulddin, H. A. Khizir, M. A. Hamad and B. A. Rahman, Design of graphene-based tunable plasmonic antenna for multiband terahertz application systems, *Plasmonics*, vol. 19, pp. 2107–2118, 2024. https://doi.org/10.1007/s11468-023-02153-9
61. X. Zhang, R. Cunjun, J. Dai, Y. Ding, S. Ullah and A. Kosar Fahad, Design of a reconfigurable antenna based on graphene for terahertz communication, *International Journal of Numerical Modelling: Electronic Networks, Devices and Fields*, vol. 34, no. 5, p. e2911.
62. S. K. Ezzulddin, S. O. Hasan, and M. M. Ameen, Performance analysis of plasmonic nano-antenna based on graphene with different dielectric substrate materials for opto-electronics application. *Plasmonics*, vol. 19, pp. 865–874, 2024. https://doi.org/10.1007/s11468-023-02030-5
63. P. Thiruppathi, L. D. Balasubrahmaniam, K. G. Manoharan, et al. Multiband plasmonic MIMO antenna array for 6G communications. *Optical and Quantum Electronics*, vol. 56, p. 969, 2024. https://doi.org/10.1007/s11082-024-06835-4
64. R. D. Richtmyer, Dielectric resonators, *Journal of Applied Physics*, vol. 10, pp. 391–398, June 1939.
65. R. Mongia and P. Bhartia, Dielectric resonator antennas: A review and general design relations for resonant frequency and bandwidth, *International Journal of Microwave and Millimeter-wave Computer-Aided Engineering*, vol. 4, pp. 230–247, 1994.
66. N. Yang, K. W. Leung and N. Wu, Pattern-diversity cylindrical dielectric resonator antenna using fundamental modes of different mode families. *IEEE Transactions on Antennas and Propagation,* vol. 67, pp. 6778–6788, 2019. https://doi.org/10.1109/TAP.2019.2922873
67. D. Headland, S. Nirantar, W. Withayachumnankul, P. Gutruf, D. Abbott, M. Bhaskaran, C. Fumeaux and S. Sriram, Terahertz magnetic mirror realized with dielectric resonator antennas, *Advanced Materials*, vol. 27, no. 44, pp. 7137–7144, Nov. 2015.
68. R. S. Yaduvanshi and Nishtha, Conical dielectric resonator antenna for terahertz applications, Frequenz, vol. 75, no. 5–6, 2021, pp. 211–220. https://doi.org/10.1515/freq-2020-0080
69. S. Ahmad, O. Benkhadda, A. Akram and M. Saih, Design and analysis of a tri-band dielectric resonator antenna for terahertz applications. In: S. Das, A. Nella and S. K. Patel (eds) *Terahertz Devices, Circuits and Systems*, Springer, Singapore, 2022. https://doi.org/10.1007/978-981-19-4105-4_8
70. P. Upender and A. Kumar, Quad-band circularly polarized tunable graphene based dielectric resonator antenna for terahertz applications, *Silicon*, vol. 14, pp. 5513–5526, 2022. https://doi.org/10.1007/s12633-021-01336-5
71. M. A. Jamshed, A. Nauman, M. A. Babar Abbasi and S. W. Kim, Antenna selection and designing for THz applications: Suitability and performance evaluation: A survey, *IEEE Access*, vol. 8, pp. 113246–113261, June 2020.
72. U. Nissanov and G. Singh, Beamforming D-band phased array microstrip antennas, *Sensors International*, vol. 3, pp. 100196/1–14, July 2022.
73. H. Sarieddeen, M. S. Alouini and T. Y. Al-Naffouri, Terahertz-band ultra-massive spatial modulation MIMO, *IEEE Journal On Areas In Communication*, vol. 37, no. 9, pp. 1–19, June 2019.
74. I. F. Akyildiz, C. Han and S. Nie, Combating the distance problem in the millimeter wave and terahertz frequency bands, *IEEE Communications Magazine*, vol. 56, no. 6, pp. 102–108, June 2018.
75. C. Han, J. M. Jornet and I. F. Akyildiz, Ultra-massive MIMO channel modeling for graphene-enabled terahertz-band communications, *IEEE* 87*th* Vehicular Technology Conference, VTC Spring, Porto, Portugal, pp. 1–5, 2018.

2 Beamforming and MIMO for Terahertz Technology in Microwave and Photonic Systems

Mohd Javed Khan, Saif Ahmad, Zeenat Fatima, and Indrasen Singh

2.1 INTRODUCTION

The terahertz (THz) band is receiving more and more attention as wireless communication technologies quickly advance to satisfy the rising needs for improved spectral efficiency, wider bandwidths, and quicker data rates. The terahertz band, which ranges from 0.1 to 10 THz and lies between microwave and infrared frequencies, offers a plethora of unexplored possibilities for high-resolution imaging, ultra-fast wireless communication, and non-invasive sensing. Terahertz technology, which may achieve data speeds of hundreds of gigabits per second (Gbps), is a promising option for next-generation wireless communication systems, such as 6G and beyond [1–4].

But there are difficulties involved in using the THz spectrum for operations. Terahertz waves have considerable channel loss and air absorption, therefore dependable communication requires sophisticated signal processing methods. In order to overcome these obstacles in the THz regime, beamforming and multiple-input multiple-output (MIMO) technologies both of which have proven extremely effective in microwave and millimeter-wave communication systems are becoming indispensable. These technologies contribute to increased data throughput, less interference, stronger signals, longer communication ranges, and precise management of terahertz waves.

The principles of beamforming and MIMO methods are examined in this chapter with an emphasis on how they are used in terahertz communication systems. It explores the difficulties presented by THz technology, the design issues involved in putting these strategies into practice, and the possibilities of MIMO and beamforming in microwave and photonic systems. The chapter also offers a thorough review of current research projects and future possibilities for combining these technologies for effective terahertz communication [5,6].

DOI: 10.1201/9781003599111-2

2.2 TERAHERTZ COMMUNICATION: OPPORTUNITIES AND CHALLENGES

2.2.1 The Terahertz Spectrum: Characteristics and Potential

The terahertz band provides an extensive range of options for high-speed communication by bridging the gap between microwave and infrared frequencies. With frequencies ranging from 0.1 to 10 THz, terahertz waves provide a number of benefits for communication systems, such as [7,8]:

- **Ultra-Wide Bandwidth**: The THz band provides up to several tens of gigahertz of available spectrum, which can support extremely high data rates.
- **Short Wavelengths**: The short wavelength of THz waves allows for more compact antenna designs and more precise beam steering, enabling efficient spatial reuse and reduced interference.
- **High Spatial Resolution**: THz waves can achieve higher resolution in imaging and sensing applications due to their short wavelengths (Figure 2.1).

Because of these benefits, THz technology is appealing for many different applications, such as enhanced sensing devices, high-resolution imaging systems, and ultra-fast wireless networks. Notwithstanding, there are several obstacles to overcome in the real-world use of THz communication systems:

- **Severe Propagation Loss**: Terahertz waves suffer from high free-space path loss, which limits their transmission range. The path loss increases with frequency, making it difficult to achieve long-distance communication.
- **Atmospheric Absorption**: Water vapor and oxygen molecules in the atmosphere absorb terahertz waves, further attenuating the signal over long distances.
- **Device Limitations**: Generating, detecting, and manipulating terahertz waves efficiently requires advanced materials and device technologies, many of which are still under development.

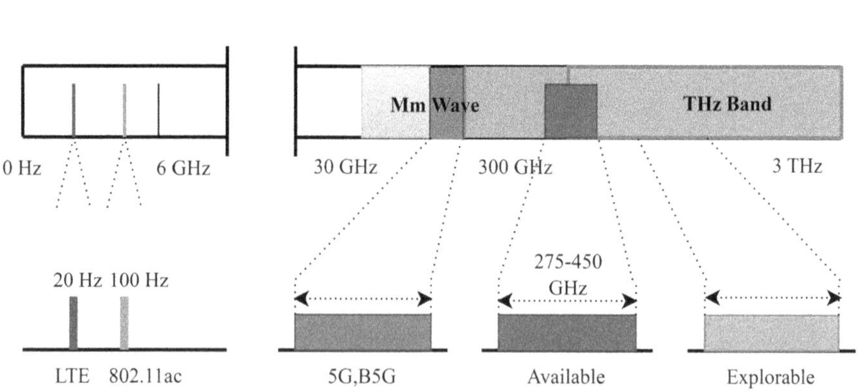

FIGURE 2.1 Trend of spectrum resource exploring.

Despite these challenges, terahertz waves possess properties that make them suitable and demonstrate great potential for communication applications combined with specific signal processing techniques. Beamforming and MIMO are considered two of the major technologies designed to overcome THz communication limitations.

2.2.2 The Role of Beamforming and MIMO in THz Communication

Beamforming and MIMO are critical for mitigating the challenges associated with terahertz communication [9]:

- **Beamforming**: Beamforming strengthens the signal sent out by concentrating the energy of that transmitted signal in one specific direction and thus giving better strength as well as reducing interference. This is particularly important in the THz band, where relatively high path loss severely degrades the quality of a signal.
- **MIMO**: MIMO systems consist of multiple antennas at the transmitter and receiver, allowing multiple data streams to be transmitted simultaneously. This increases the capacity of the system and enhances its spectral efficiency by using the spatial dimension for communication. Both beamforming and MIMO enhance the performance of THz communication systems by:
- **Improving Signal-to-Noise Ratio (SNR)**: By concentrating the signal energy in the desired direction, beamforming increases the SNR, which is critical for maintaining high data rates over long distances.
- **Increasing Capacity**: MIMO systems enable multiple parallel data streams, effectively multiplying the capacity of a communication system without increasing bandwidth.
- **Reducing Interference**: Beamforming and MIMO reduce interference by directing the signal energy only toward the intended receiver, minimizing the impact of signals from other users or devices.

2.3 BEAMFORMING IN TERAHERTZ TECHNOLOGY

2.3.1 Basic Principles of Beamforming

Beamforming is a signal processing technique where the radiation pattern of an antenna array can be steered in a specific direction. This can be obtained by controlling the phase and/or amplitude of signals that an element in an array transmits or receives. In this way, beamforming exploits the principle of constructive summation in the desired direction and destructive summation in other directions for the precise control of the radiation pattern [10].

Terahertz communication systems would experience an ability to pass over high propagation losses due to short wavelengths if beamforming were to be applied. In itself, the process of beamforming brings a concentration of the energy given, which means that the signal quality will be improved and the range of communication expanded, thereby making for high power-efficient data transfer over a distance.

2.3.2 Types of Beamforming Techniques

Beamforming can be categorized into several types, depending on how the signals are processed and combined:

- **Analog Beamforming**: This technique in analog beamforming applies phase shifts to signals before they are transmitted in the analog domain. This is very simple and power efficient as it only requires one RF chain, but it limits the ability to generate multiple beams simultaneously because it is less flexible compared to digital beamforming.
- **Digital Beamforming**: In digital beamforming, the phase shifts are applied in the digital domain once the signals have been converted from analog to digital. This provides considerable flexibility since it forms multiple beams and allows multiple signals to be transmitted to the different users simultaneously. However, digital beamforming requires multiple RF chains, which leads to increased power consumption and system complexity.
- **Hybrid Beamforming**: Hybrid beamforming combines aspects of both analog and digital beamforming. It has the lowest number of RF chains compared to digital beamforming but allows multi-beam directions. The popularity of THz communication systems has motivated many researchers to propose hybrid beamforming as a practical solution for achieving an optimal trade-off between performance and power efficiency [11].

2.3.3 Beamforming for THz Communication: Challenges and Solutions

Applying beamforming techniques in the THz band introduces several unique challenges [12]:

- **Ultra-High Precision Beamsteering**: The THz waves have very small wavelength, and hence, beamforming in the THz band is done strictly by high-precision beamsteering to ensure that the energy is correctly steered over to the receiver. This requires large antenna arrays with very high phase control.
- **Real-Time Beam Tracking**: Because of the different users' location under dynamic conditions and mobile communication, it might happen that a change in position could take place very fast. In such scenarios, a fast beamforming system is envisioned to track the time variation so that the direction of the beam can steer itself to deliver effective communication links.
- **Power Efficiency**: Although the large antenna array provides good spatial resolution, it happens to be power hungry, particularly if the beamforming system is digital. This poses a very tremendous challenge in terms of battery-powered or low-power applications. Techniques that can help in reducing power consumption and yet maintain performance are energy-efficient beamforming techniques such as hybrid beamforming.

Several solutions have been proposed to address these challenges, including advanced beamforming algorithms, adaptive antenna arrays, and the use of novel materials for THz antennas, such as graphene and metamaterials. These innovations are helping to make beamforming more practical and efficient for THz communication systems.

2.4 MULTIPLE-INPUT MULTIPLE-OUTPUT (MIMO) IN TERAHERTZ TECHNOLOGY

2.4.1 Fundamentals of MIMO

MIMO technology relies on using multiple antennas on both the transmitter and receiver sides so that multiple streams of data can be transmitted at a given time. By relying on the spatial dimension in a wireless channel, MIMO enables the increase in spectral efficiency and capacity as well as reliability. In MIMO systems, employing several antennas has two crucial benefits [3]:

- **Spatial Multiplexing**: Different spatial paths are used to transfer multiple streams of data. It increases information transfer rate without requiring additional bandwidths.
- **Diversity**: MIMO makes use of the same signal on a number of antennas by transmitting that, thereby improving the reliability of a signal and minimizing the impact caused by fading and interference.

MIMO technology is popularly deployed in microwave and millimeter-wave communication systems, and it has recently gained importance as it is being adapted for use in the THz band, where it is seen as the key enabler to overcome the high path loss limitation and provide limited coverage.

2.4.2 MIMO Techniques for THz Communication

Several MIMO techniques are applicable to terahertz communication systems, including:

- **Spatial Multiplexing MIMO**: This technique can deliver independent streams of data along different spatial paths by the effective raising of the rate of the data of the system. Spatial multiplexing can prove very effective in the applications of high data rates, especially for THz communication.
- **Diversity MIMO**: The basic idea behind diversity-based MIMO systems is that the same signal is transmitted using multiple antennas; thus, the chance of signal fading is decreased and reliability in the communication link increases. Diversity techniques are also important if signal quality can change drastically because of mobility or obstacles in space.
- **Massive MIMO**: This technology uses a very large number of antennas to cater to multiple users at the same time. The possibility of raising the capacity and spectral efficiency in communication is very high in massive MIMO. It is possible to synthesize beamforming with the THz band of massive MIMO; here, highly directional beams compensate for path loss and make the signal strength high.

2.4.3 Challenges in Implementing MIMO for THz Communication

While MIMO promises great advantages for THz communication, there are several challenges that have to be overcome before the full potential of MIMO can be realized:

- **THz Antennas Design**: Compactness is the need for MIMO systems where antennas are to be designed very efficiently. Since the wavelengths of the THz waves are much smaller, there is a visible necessity for precise positioning and calibration of the antennas. In addition, study has been conducted to design an antenna with enough efficiency at THz frequencies.
- **Channel Modeling**: Since THz waves can best propagate in a certain environment, such as from indoor and outdoor to the urban settings, accurate channel modeling is very important in determining them. Compared to microwave and millimeter-wave channels, THz channels involve much more frequency and number of incident waves that pass through the frequency selective effects of obstacles, reflections, and scattering; so, they should be considered in MIMO systems design.
- **Signal Processing**: MIMO systems necessitate the introduction of advanced signal processing techniques that could mitigate the complexity brought in by multiple data streams and antennas. For the THz band, these complexities are compounded by more extreme frequencies and data rates. Strong demands exist for advanced channel estimation algorithms, beamforming, and interference mitigation algorithms to be used in optimizing the performance of MIMO systems for the THz band.

2.5 HYBRID BEAMFORMING AND MIMO FOR TERAHERTZ COMMUNICATION

2.5.1 The Case for Hybrid Beamforming in THz Systems

Hybrid beamforming combines the advantages of analog and digital beamforming into a much more efficient solution for THz communication systems. Here, a few RF chains control multiple antennas by means of phase shifters, offering flexibility in terms of beam steering without the high power consumption and complexity encountered in fully digital beamforming [13].

Hybrid beamforming is very well suited for THz communication, which has a high path loss, and precise beam steering is actually needed, with a hybrid system achieving both the efficiency of analog beamforming and flexibility of digital beamforming while keeping the consumed power low.

2.5.2 MIMO in Hybrid Beamforming Systems

Adding MIMO technology in a hybrid beamforming system can further accelerate and enhance the performance of THz communication. The MIMO system could then send multiple data streams in parallel while using beamforming to direct those streams toward specific users or devices. In return, it increased the spectral efficiency and the capacity of the communication system while enhancing the signal strength and reducing interference [14,15].

2.5.3 Challenges and Considerations in Hybrid Beamforming and MIMO

While hybrid beamforming and MIMO offer significant benefits for THz communication, there are several challenges that must be addressed:

- **RF Chain Limitations**: The number of RF chains in hybrid beamforming systems is typically limited due to high power consumption and complexity. This, however, leads to a limitation in the number of independent beams that can be generated, causing insufficient flexibility in the system.
- **Hybrid Beamforming**: Hybrid beamforming and MIMO demand critical antenna array designs that should be energy efficient to provide multiple beams yet extremely low in power consumption.
- **Channel Estimation**: In MIMO and beamforming systems, channel estimation will be accurate because it will make the system run optimally according to current channel conditions. Channel estimation in THz communication will be more challenging in comparison because of the high frequency and increased susceptibility to obstacles and interference.

2.6 PHOTONIC BEAMFORMING AND MIMO FOR TERAHERTZ TECHNOLOGY

2.6.1 Photonic Approaches to THz Beamforming

Photonic technologies offer a promising approach to implement beamforming in the terahertz band. Photonic systems can generate and manipulate THz waves with high precision and low power consumption, making them ideal for beamforming applications. In photonic beamforming systems, optical components such as lasers, modulators, and waveguides are used to generate and control terahertz waves. These systems can achieve dynamic beam steering, real-time signal processing, and precise phase control, which are critical for THz communication [16].

Photonic techniques for THz beamforming use optical technologies to create, modulate, and steer radiation in the 0.1–10 THz range. The most popular technique is photomixing, where two lasers with frequencies that differ only slightly are mixed in a photomixer to produce a THz signal. The technique can be adapted to steer THz beams in phased arrays by adjusting the phase or frequency difference between the laser sources. Photonic phase shifters and optical true time delay elements are another alternative, which in principle offer control over the phase with broadband beam steering. OEOs use optical feedback to generate a stable, low noise THz signal, suitable for application in beamforming arrays. Photonic crystals, materials with periodic structures, are also yet another tool in manipulating THz waves, via focusing or directionality. Photonic integrated circuits (PICs) are thus achieved through the integration of optical components, like lasers and waveguides, onto a single chip in order to provide compact, scalable THz beamforming solutions. THz modulators allow for dynamic control of the beam's properties by modulating optical signals prior to their conversion into the THz domain. In particular, photonic systems bring about numerous benefits such as wide bandwidth, low loss, compactness, and high precision and hence provide an ideal solution toward THz communications, imaging, and sensing applications.

2.6.2 Photonic MIMO Systems

Photonic MIMO systems use optical components to generate and transmit multiple terahertz data streams simultaneously. This approach offers several advantages over traditional electronic MIMO systems, including higher data rates, lower power consumption, and greater flexibility in beamforming. Photonic MIMO (Multiple Input Multiple Output) systems combine the benefits of photonic technologies with the advanced spatial multiplexing capabilities of MIMO, particularly targeting high-frequency domains like terahertz (THz) and millimeter-wave communication. These systems leverage the unique properties of photonics, such as wide bandwidth, low-loss signal transmission, and precise control of electromagnetic waves, to enhance the performance and capacity of wireless communication networks.

i. **Wide Bandwidth and High-Frequency Operation**: Photonic systems excel at handling high-frequency signals, especially in the THz range, where traditional electronic components face significant challenges like high losses and bandwidth limitations. By using photonics, MIMO systems can operate over wider frequency ranges, enabling ultra-high data rates and supporting next-generation wireless standards such as 6G.
ii. **Low Loss and Efficient Signal Transmission**: At high frequencies like THz, electronic components typically suffer from significant signal attenuation and loss, especially when generating and transmitting signals over large distances. Photonic components, however, offer a low-loss alternative for signal generation and transmission, making them ideal for MIMO systems that need to maintain signal integrity over high frequencies.
iii. **Precise Beamforming**: One of the critical advantages of photonic MIMO systems is their ability to perform highly precise beamforming. Photonic technologies enable fine control over the phase and amplitude of electromagnetic waves, allowing for dynamic beam steering and focusing. This is essential for MIMO systems, which rely on directing multiple signal streams to different spatial locations to achieve spatial multiplexing and improve spectral efficiency.
iv. **Photomixing and Optical Generation of THz Signals**: Photonic MIMO systems often rely on photomixing to generate high-frequency signals. By mixing two optical signals of slightly different frequencies in a photomixer, a beat frequency in the THz range can be produced. This THz signal can be used in MIMO arrays, where multiple photonic transmitters and receivers generate and direct multiple beams simultaneously. This method allows for

the generation of high-frequency signals without the high losses typically associated with electronic components.

v. **Photonic Integrated Circuits (PICs)**: Photonic MIMO systems benefit greatly from PICs, which integrate multiple optical components (lasers, modulators, waveguides, and detectors) onto a single chip. This integration provides a compact, scalable solution for handling multiple input and output channels in MIMO systems. PICs can be used to drive MIMO arrays by controlling the phase, delay, and amplitude of optical signals, which are subsequently converted into THz signals for MIMO applications.

2.6.2.1 Photonic MIMO Works

The principles of photonics are used to enhance wireless communication systems by using multiple optical signal paths for data transmission and reception. In a photonic MIMO system, multiple optical sources, such as laser diodes, generate separate data streams. These streams are then combined using optical multiplexers, which allow for the transmission of multiple signals simultaneously over a single optical medium. The multiplexed signals are transmitted through the photonic MIMO processing unit, which uses optical components like beamformers, switches, and modulators to perform MIMO operations such as spatial multiplexing, signal splitting, and combining. At the receiver end, an optical demultiplexer separates the transmitted signals back into individual streams, which are then detected by optical receivers. These receivers convert the optical signals into electrical signals for further digital processing. The core advantage of photonic MIMO lies in its ability to handle vast amounts of data at high speeds with low latency, making it suitable for high-frequency applications such as terahertz (THz) communication. Photonic technologies enable the creation of compact, efficient MIMO systems capable of supporting ultra-high bandwidth and providing improved signal processing capabilities, thus overcoming the limitations of traditional electronic MIMO systems.

This approach is particularly promising for next-generation wireless networks (e.g., 6G and beyond) that require massive data throughput and highly reliable communication links. Photonic MIMO can also enhance the efficiency of beamforming and spatial diversity techniques, helping to combat signal attenuation and increase the range and capacity of wireless networks (Figure 2.2).

In a photonic MIMO system, the key components include:

- **Photonic Beamforming Networks**: These networks use optical phase shifters or true-time delay lines to control the phase and amplitude of optical signals. The modulated optical signals are then down converted to THz frequencies via photomixing, resulting in multiple THz beams, each carrying independent data streams. By adjusting the phase and delay of the optical signals, the direction and focus of each beam can be controlled in real time, a process known as beamforming.
- **Spatial Multiplexing**: MIMO systems exploit spatial multiplexing to transmit multiple independent data streams over the same frequency band. In photonic MIMO, multiple photonic transmitters (or antenna elements) generate these streams. Each transmitter is controlled optically, allowing

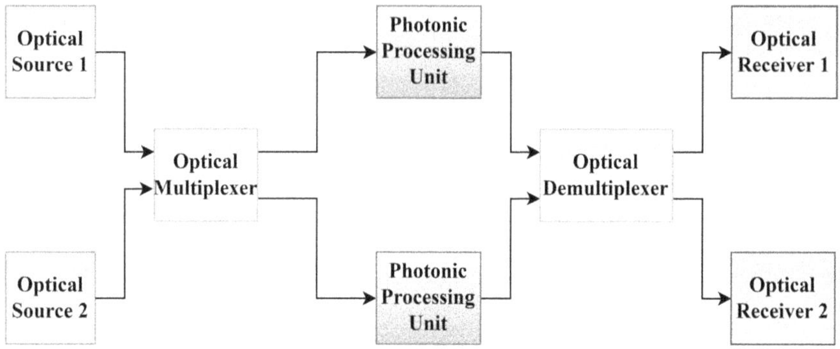

FIGURE 2.2 Photonic MIMO system.

precise management of the spatial and frequency characteristics of the signal. On the receiving end, multiple photonic receivers capture the different spatial data streams, enhancing capacity and data throughput.

- **Dynamic Reconfiguration**: Photonic MIMO systems can be dynamically reconfigured to adapt to changing environments, such as user mobility or varying channel conditions. This is made possible by the use of reconfigurable photonic circuits and optical switches, which can rapidly adjust the system's configuration without the need for large, bulky electronic components.

2.6.2.2 Advantages of Photonic MIMO Systems

- **Higher Data Rates**: Photonic systems are naturally suited for high-frequency operation, particularly in the THz band, where higher data rates and large bandwidths are available.
- **Improved Spectral Efficiency**: By using multiple transmit and receive antennas, MIMO systems improve spectral efficiency, transmitting more data simultaneously. Photonic MIMO systems further enhance this by offering low-loss, high-precision beamforming, enabling better spatial multiplexing.
- **Compact and Scalable Design**: With the integration of photonic components on-chip, photonic MIMO systems can be made compact and scalable, supporting the dense antenna arrays needed for massive MIMO systems in next-generation wireless networks.
- **Low Power Consumption**: Photonic systems generally consume less power than electronic systems at high frequencies, making photonic MIMO systems more energy-efficient. This is particularly important for THz communication, where electronic components tend to be power-hungry.

2.6.2.3 Applications of Photonic MIMO Systems

1. **6G Wireless Communications**: The need for ultra-high data rates and dense network coverage in 6G requires the adoption of MIMO technology at THz frequencies. Photonic MIMO systems can support the massive bandwidth required for future wireless networks.

2. **High-Speed Wireless Backhaul**: As networks evolve, the demand for fast wireless backhaul links between base stations is growing. Photonic MIMO systems, operating in the THz range, offer an ideal solution for high-capacity backhaul connections.
3. **Imaging and Sensing**: In addition to communications, photonic MIMO systems are well-suited for applications in high-resolution imaging and sensing, particularly in areas like security scanning, medical diagnostics, and industrial inspection, where THz waves can penetrate materials and provide detailed information.

2.6.3 Challenges in Photonic Beamforming and MIMO

While photonic beamforming and MIMO systems offer significant advantages for high-frequency applications, particularly in the terahertz (THz) and millimeter-wave bands, they also face several technical challenges. These challenges stem from both the intrinsic properties of photonics and the complexities of MIMO systems in the THz regime. Below are some of the key hurdles [10,16]:

2.6.3.1 Complexity of Photonic Integration

- **Challenge**: Integrating photonic components—such as lasers, modulators, waveguides, and detectors—into compact, efficient, and scalable PICs remains difficult. Although PICs offer scalability, manufacturing these components with the required precision, especially for multi-channel MIMO systems, is challenging. The fabrication process is complex and expensive, and ensuring low-loss coupling between different photonic components is a key technical barrier.
- **Impact**: Difficulty in large-scale photonic integration limits the practical deployment of massive MIMO arrays, which are necessary for applications like 6G communications and high-resolution sensing.

2.6.3.2 Phase Noise and Signal Coherence

- **Challenge**: Photonic systems, especially those relying on photomixing to generate THz signals, are sensitive to phase noise. Maintaining phase coherence across multiple optical sources in a photonic MIMO system is critical for accurate beamforming. However, even small fluctuations in laser sources can introduce phase noise, which degrades signal quality and reduces the precision of beamforming.
- **Impact**: Poor phase noise performance can lead to inaccurate beam steering and lower data rates in MIMO systems, especially at higher frequencies like THz, where phase coherence is more difficult to maintain.

2.6.3.3 Propagation Losses in THz Band

- **Challenge**: THz waves experience higher free-space path loss compared to lower-frequency electromagnetic waves. Additionally, they are more susceptible to atmospheric absorption, especially in humid environments, due to water vapor absorption lines in the THz spectrum. These losses significantly reduce the effective range of THz communications and sensing systems.

- **Impact**: MIMO systems, which depend on multiple transmitters and receivers, need to overcome these propagation losses to maintain signal strength. While photonic beamforming can help mitigate some losses through precise focusing and steering, the overall range of THz systems is still limited, complicating the deployment of practical large-scale MIMO networks.

2.6.3.4 Beam Alignment and Tracking

- **Challenge**: MIMO systems in the THz band require highly directional beams to compensate for the high path loss. However, precise beam alignment between the transmitter and receiver is crucial. In dynamic environments, where users or objects move, maintaining beam alignment through real-time beam tracking becomes increasingly complex.
- **Impact**: Misalignment or latency in beam tracking leads to signal degradation or loss. Developing robust algorithms and hardware for fast, adaptive beamforming and tracking in photonic MIMO systems is essential but remains challenging due to the high frequencies and fast response times required.

2.6.3.5 Thermal Management and Power Consumption

- **Challenge**: While photonic systems generally offer lower power consumption at high frequencies compared to electronic systems, power consumption becomes a concern in massive MIMO systems, where many optical components and lasers operate simultaneously. Additionally, thermal management is critical, as photonic devices, especially in densely packed PICs, can generate significant heat.
- **Impact**: Efficient thermal management solutions are needed to ensure the stability and performance of photonic MIMO systems. High power consumption and overheating can degrade the system's performance, particularly affecting the precision of beamforming and signal integrity over time.

2.6.3.6 Latency and Synchronization

- **Challenge**: In MIMO systems, multiple data streams are transmitted and received simultaneously, requiring precise synchronization between all the components in the network. In photonic beamforming, optical signals are delayed and phased using devices like optical true time delay lines, which can introduce latency. Synchronizing multiple photonic channels across the system without introducing timing errors is difficult.
- **Impact**: Any mismatch in timing or phase across the MIMO array can result in interference and reduced signal quality. Achieving low-latency synchronization between optical components is essential for high-performance beamforming, but it remains a major technical challenge, especially in dynamic environments.

2.6.3.7 Limited Maturity of Photonic THz Devices

- **Challenge**: Although significant progress has been made in the development of photonic devices for THz applications, many components—such as

efficient THz sources, modulators, and detectors—are still in the research and development stage. For example, generating high-power THz signals via photomixing requires significant improvements in conversion efficiency. Similarly, THz photonic modulators and detectors need to be optimized for high-speed, low-loss operation.
- **Impact**: The limited performance and availability of mature THz photonic components hinder the practical deployment of large-scale MIMO systems at THz frequencies. These devices need further development to meet the requirements of next-generation communication and sensing systems.

2.6.3.8 Cost and Scalability

- **Challenge**: Developing photonic systems is expensive, especially for advanced components like tunable lasers, PICs, and high-precision optical phase shifters. The cost of producing these components at scale is high, which limits the economic feasibility of deploying photonic MIMO systems for mass-market applications.
- **Impact**: High costs make it challenging to deploy photonic MIMO systems in commercial applications such as 6G networks, where cost-effective solutions are essential. Research is needed to reduce the cost of photonic components through improved manufacturing techniques and scalable integration processes.

2.6.3.9 Channel Modeling and Propagation Effects

- **Challenge**: Channel modeling for THz frequencies is still a developing area. Unlike lower frequencies, THz waves are more affected by surface roughness, diffraction, and atmospheric conditions. The development of accurate models for THz propagation, including the impact of multipath effects, scattering, and blockage, is necessary for optimizing MIMO system performance.
- **Impact**: Without accurate channel models, it is difficult to design effective MIMO algorithms that maximize spectral efficiency and minimize interference. The complexity of THz propagation environments adds an additional layer of difficulty in ensuring robust performance for photonic MIMO systems.

2.7 CONCLUSION AND FUTURE DIRECTIONS

Beamforming and MIMO technologies are essential for addressing the challenges in the THz domain by enhancing data throughput, signal quality, and communication range. This chapter explores the concepts and applications of beamforming and MIMO within THz frequencies, focusing on their ability to reduce path loss, improve coverage, and support high data rates. Various beamforming techniques, including digital, analog, and hybrid, are analyzed, highlighting their adaptability and efficiency in THz communications. Additionally, the chapter discusses the opportunities and obstacles in implementing these technologies, including challenges such as hardware complexity, stringent alignment requirements, and the need for high-speed signal processing.

Future research will focus on further improving beamforming and MIMO systems in terms of efficiency, flexibility, and scalability in the terahertz frequency band. New materials and devices to be used at terahertz frequencies should be developed and, more generally, other photonic technology integration with traditional electronic systems is promising in order to advance the state of the art in THz communication. Indeed, when demand for faster data rate and more efficient communication continues to increase and grow, the role of beamforming and MIMO may very well dominate the future of 6G and beyond.

REFERENCES

1. Zhang, Z., Xiao, Y., Ma, Z., Xiao, M., Ding, Z., Lei, X., ... & Fan, P. (2019). 6G wireless networks: Vision, requirements, architecture, and key technologies. *IEEE Vehicular Technology Magazine*, 14(3), 28–41.
2. Ning, B., Tian, Z., Mei, W., Chen, Z., Han, C., Li, S., ... & Zhang, R. (2023). Beamforming technologies for ultra-massive MIMO in terahertz communications. *IEEE Open Journal of the Communications Society*, 4, 614–658.
3. Rappaport, T. S., Xing, Y., Kanhere, O., Ju, S., Madanayake, A., Mandal, S., & Trichopoulos, G. C. (2019). Wireless communications and applications above 100 GHz: Opportunities and challenges for 6G and beyond. *IEEE Access*, 7, 78729–78757.
4. Yang, P., Xiao, Y., Xiao, M., & Li, S. (2019). 6G wireless communications: Vision and potential techniques. *IEEE Network*, 33(4), 70–75.
5. De Alwis, C., Kalla, A., Pham, Q. V., Kumar, P., Dev, K., Hwang, W. J., & Liyanage, M. (2021). Survey on 6G frontiers: Trends, applications, requirements, technologies and future research. *IEEE Open Journal of the Communications Society*, 2, 836–886.
6. Chen, Z., Ma, X., Zhang, B., Zhang, Y., Niu, Z., Kuang, N., ... & Li, S. (2019). A survey on terahertz communications. *China Communications*, 16(2), 1–35.
7. Han, C., Jornet, J. M., & Akyildiz, I. (2018, June). Ultra-massive MIMO channel modeling for graphene-enabled terahertz-band communications. In *2018 IEEE 87th Vehicular Technology Conference (VTC Spring)*, Porto, Portugal (pp. 1–5). IEEE.
8. Akyildiz, I. F., Jornet, J. M., & Han, C. (2014). Terahertz band: Next frontier for wireless communications. *Physical Communication*, 12, 16–32.
9. Ju, S., Xing, Y., & Rappaport, T. S. (2019). Millimeter wave and terahertz communications for 5G and beyond. *IEEE Communications Magazine*, 57(6), 86–92.
10. Nagatsuma, T., Ducournau, G., & Renaud, C. C. (2016). Advances in terahertz communications accelerated by photonics. *Nature Photonics*, 10(6), 371–379.
11. Khalid, Z., & Rodriguez, J. (2020). Beamforming for terahertz wireless communication systems: Overview, challenges, and future research directions. *IEEE Open Journal of the Communications Society*, 1, 241–254.
12. Seeds, A. J., Shams, H., Fice, M. J., & Renaud, C. C. (2014). Terahertz photonics for wireless communications. *Journal of Lightwave Technology*, 33(3), 579–587.
13. Huang, H., Wang, Y., & Zhang, W. (2017). Multi-antenna techniques for terahertz wireless systems: Challenges and opportunities. *IEEE Wireless Communications*, 24(4), 70–76.
14. Li, Q. C., Gao, X., Papathanassiou, A. T., & Li, G. (2017). Hybrid beamforming for millimeter-wave systems: Achieving massive MIMO with fewer RF chains. *IEEE Communications Magazine*, 55(2), 157–165.
15. El Ayach, O., Rajagopal, S., Abu-Surra, S., Pi, Z., & Heath, R. W. (2014). Spatially sparse precoding in millimeter wave MIMO systems. *IEEE Transactions on Wireless Communications*, 13(3), 1499–1513.
16. Zhou, X., Li, Y., Wang, J., Liu, Q., & Chen, M. (2020). Photonic beamforming for advanced 5G and 6G wireless communications. *Nature Communications*, 11(1), 5079.

3 Radar Cross Section Reduction Using Metasurfaces
Methods, Metrics, and Future Scope

Madurakavi Karthikeyan, Sukumar Vidhya, Natarajan Rajesh, J Pradeep, and M Harikrishnan

3.1 INTRODUCTION

As per the definition from IEEE [1], RCS is defined as the product of 4π and the ratio of power per unit solid angle distributed in a particular direction to power per unit area of an incident plane wave from a definite direction. The RCS of the target material is dependent on many parameters such as angle at which the incident wave is impinged, operating frequency range, the shape of the target material, geometry, and material properties.

There are two types of RCS, which are classified based on the orientation of the transmitter and receiver antenna position. Monostatic RCS implies placing both antennas in the same place. When both antennas are placed in different positions, it is called as Bistatic RCS, as shown in Figure 3.1.

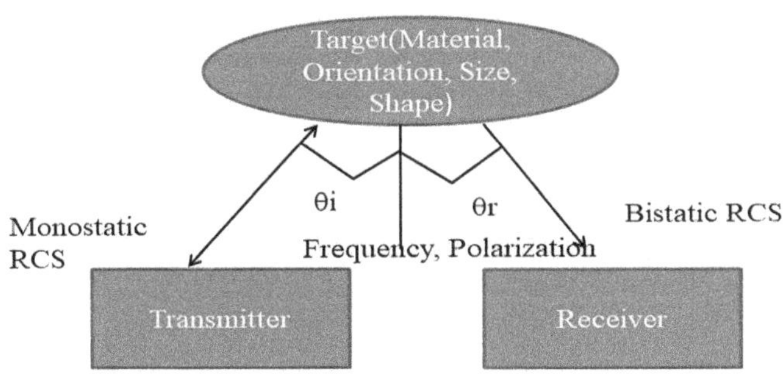

FIGURE 3.1 Types of radar cross section (RCS).

DOI: 10.1201/9781003599111-3

3.1.1 Metamaterials and Its Applications

Metamaterials are effectively homogenizable configurations of synthetic structural elements designed to produce advantageous and unique electromagnetic (EM) characteristics absent from natural materials [2]. This includes nano- and microstructured materials. The purpose of a volumetric metamaterial is to create an efficient EM wave, which is capable of controlling permeability (μ) and permittivity (ε). The application of metamaterial extends in antenna engineering, which aids in gain and bandwidth enhancement, antenna size reduction, reduction of specific absorption rate, perfect lens [3], magnetic resonance imaging [4], and sensors. But due to strong dispersion properties, high losses, and fabrication difficulties, it has been obstructed from practical applications [5].

3.1.2 An Overview of Metasurface

Metasurfaces are defined as the two-dimensional (2D) counterparts of existing metamaterials where it has both single- and multi-layer piles of planar structures [6]. Metasurfaces can be fabricated using nanoimprinting methods or lithography; thereby, the ultrathin thickness obtained suppresses the unwanted losses in the wave propagation direction [7]. A metasurface is defined as a periodic or aperiodic array having EM characteristics of sub-wavelength scatterers. Conventional methods leverage the Faraday's effect to control the polarization of an EM wave [8]. On the other hand, the limitations in these methods are the controlled bandwidth, the incident angle response, as well as the size of the devices that are produced. Metasurfaces offer significant benefits, including low cost and a high degree of integration, by controlling EM waves within a single thin layer. Hence, the usage of metasurface is inevitable for Radar Cross Section reduction (RCSR).

3.1.3 Radar Cross Section

RCS is a vital characteristic in stealth technology, and it quantifies the visibility of the target to the radar. RCSR means minimizing the detectability of the target by the radar. For practical applications, a 10dB RCSR is considered sufficient, and it will lead to 90% invisibility of the target. RCSR holds a major role in stealth technology and its roots start from World War I where weapons were made up of cellulose-based compounds, thereby making aircrafts less palpable. Likewise, in World War II, a wooden covering was made in the Horten 229 V3 aircraft for decreasing its detectability [9]. The research became in a new direction when Russian scientist PyotrUfimstev [10] studied the different EM wave scattering. Moreover, the stealth aircraft SR-71 was developed by researchers belonging to Lockheed Martin to reduce RCS, which used irregular shape and absorptive paint [11].

3.1.4 Categorization of RCS Reduction Techniques

The two traditional methods for lowering the RCS include scattering and absorption. Scattering can be classified directly as shaping and metasurface. Shaping

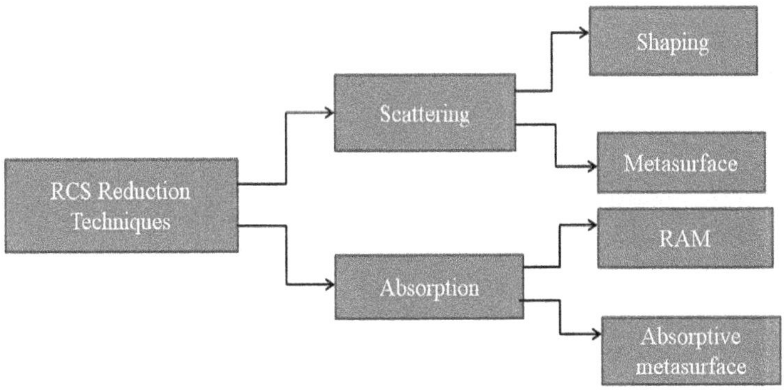

FIGURE 3.2 Strategies for radar cross section reduction (RCSR).

mechanism works through scattering mechanism. Further absorption is subdivided into two techniques: Radar absorbing material (RAM) and absorptive metasurface. While in absorption, the signal is dissipated by absorbing in a special layer. Both the techniques of scattering and absorption are used for RCSR. The focus of discussion here is on metasurface usage for RCSR while also deliberating about RAM and shaping techniques for extensiveness. The approach for RCSR methods is given in Figure 3.2.

3.1.5 Survey on Traditional Methods

Shaping and RAM are the conventional methods for reducing RCS, which are discussed in detail here.

3.1.5.1 Shaping

Shaping, also called as geometrical shaping, is powerful for reducing monostatic radars [12]. There are two well-defined ways to reduce RCS by shaping technique: (1) a smooth, compact blended external geometry is used and (2) a faceted arrangement is utilized to lessen the reflection back to the incident radar. Machan, an unscrewed aircraft developed in 1981, has reduced its visibility due to its diamond structure cross-section, and the lower surface was flat to reduce glint [13]. In some cases, the approach of optimizing the target has reduced the RCS [14]. The obstruction of practical usage of shaping is that they are efficient only at particular incident angles and for narrow bandwidth [15]. The improvement of bandwidth can be achieved by considering sophisticated gradient shapes. Still the abovementioned technique is not acceptable for ships and planes due to its hydrodynamic and aerodynamic needs as they affect the stability and internal volume of the target.

3.1.5.2 RAM—Radar Absorbing Material

RAM can be approached by the below functionalities:

a. Impedance matching
b. Resonance

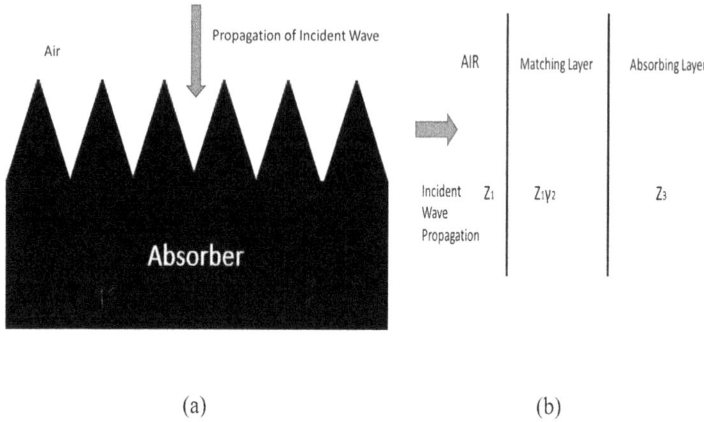

FIGURE 3.3 Demonstration of (a) pyramid absorbers and (b) matching layer absorber [16].

3.1.5.2.1 Impedance Matched RAM

RAM takes form in three varied shapes: pyramidal, matched, and tapered. Figure 3.3a depicts the pyramid shape absorber in a semi-anechoic chamber [16]. In [17], placing a matching layer absorber between the two points is shown in Figure 3.3b. This produces narrow-band absorbers. A low-lossy material that contributes to the slab is employed in tapered RAM. In [17], the Lockheed F-117 Nighthawk, a stealth aircraft, makes use of a tapered RAM in which a carbonyl iron ball is spread in epoxy resin for reducing its RCS.

3.1.5.2.2 Resonant RAM

Quarter or tuned wavelength absorbers are the other names for resonant RAM. The principle of this technique is an impedance conflict that appears between the incident and absorbing materials. At first interface, the positioning gives rise to reflection and transmission; here, the reflected wave undergoes a phase shift of π; on the other hand, in the absorbing medium, the transmitted wave is reflected back from the metal. Hence, a phase reversal of π is introduced due to the second reflection, and the wave travels back to the incident medium. As a result, destructive interference takes place, which results in RCSR. Jaumann absorbers [18] and Salisbury screen absorbers [19] are employed as resonant RAM. Although the utilization of RAM is reasonably efficient, they are expensive and offer narrow bandwidth. By increasing the thickness of the structure, bandwidth enhancement can be obtained and hence it is limited in practical applications.

3.1.6 Applications of RCS Reduction Techniques

The applications of radar technology are significantly higher, such as in ships and aircrafts, to detect animals and insects [20], to measure speed [21], and also to predict weather conditions. Nevertheless, in specific cases, low detectability is preferred as listed below:

i. Aircraft hangars are a places that need low RCS as they serve as the parking decks of aircraft. The signal reflected from the deck should not interfere with the navigation system of airport [22].
ii. The other crucial application is stealth aircrafts where less detectability is necessary.
iii. Another dominant application area is reducing the antenna RCS mounted on ships and aircrafts. The antenna will lead to increase in RCS, and hence, the detectability of ships and airplanes increases [23].

3.2 RCS REDUCTION TECHNIQUES

The main challenge associated with RCSR by using traditional methods is enhancing its bandwidth while making the structure simple and easy to fabricate. Hence, our main focus is to reduce RCS by using scattering metasurface and absorptive metasurface.

3.2.1 RCS Reduction Derived from Scattering Metasurface

Based on the function and operation, there are passive and active methods that are used to reduce RCS:

i. PGMS—Phase gradient metasurfaces
ii. CMS—Checkerboard metasurfaces
iii. PCMS—Polarization conversion metasurfaces
iv. COM—Coding-based metasurfaces
v. TVM—Time-varying metasurfaces
vi. RMS—Reconfigurable metasurfaces

Initially, the above four methods contribute to passive methods to reduce RCS and the final two methods are active methods for reducing RCS. All the above methods work on the concept of manipulating the wave front that is explained by Bi-anisotropic metasurfaces and Huygens metasurfaces. In [24], omega-bi-anisotropic metasurface makes use of synthesis technique. RCSR happens when the reflection of the incident wave by the surface happens in an appropriate direction. In Huygens metasurface, impedance and admittance are created at various angles for reflecting or refracting the incident wave [25]. Hence, both the aforementioned methods are applied for polarization conversion on the metasurface.

The following section discusses each of the metasurface in detail, which covers the working mechanism and recent advancements for reducing RCS.

3.2.1.1 PGMS—Phase Gradient Metasurfaces

The mechanism of RCSR based on PGMS works by using the principle of propagating wave conversion (PWC) to surface wave and the concept of anomalous reflection (AR). The discussion starts with operational mechanism based on Snell's law, and next, it evaluates the various metasurfaces in detail for reducing RCS. Diffraction

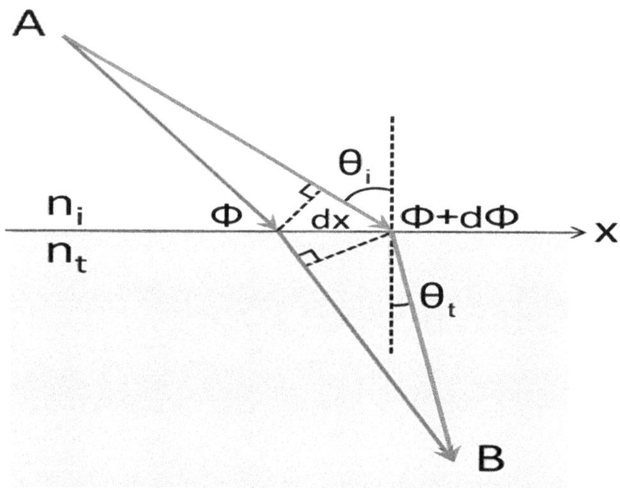

FIGURE 3.4 Pictorial representation to obtain generalized Snell's law of refraction. (Reproduced from [28].)

grating was considered as the main thought for working of PGMS, nevertheless it became popular after the Snell's law progress [26]. Any surface with phase discontinuities changes the direction of the signal from its specular direction, as given by Equation (3.1) [27]:

$$\sin(\theta_t)n_t - \sin(\theta_i)n_i = (\lambda_o / 2\pi)(d\phi / dx) \tag{3.1}$$

From the above equation, $d\phi$ represents the adjoining cells phase difference, dx is the space between the connecting cells, λ_o is the free-space wavelength, and n_t and n_i are the refractive indices of two different media. An essential characteristic of PGMS is that the designed element should have a span of 2π, which is given by Equation (3.1). An illustration of Snell's law generalization is represented in Figure 3.4.

The practical demonstration of the PGMS concept was explained by using V-shaped Nano-antennas [28]. PGMS is made up of basic building block of elements, which alters both the magnitude and the reflection phase. These elements are organized to form a gradient pattern that creates a tile, which is then replicated to construct the PGMS. Wideband RCSR is obtained when the elements span a phase of 2π covering a wide bandwidth. The major challenge is creating broadband element as the unit cell size is dependent on the operating frequency. The index structure reduces the veiling reflection by the following two ways:

i. Anomalous reflection (AR) [29]
ii. Conversion of Propagating wave (PW) into surface wave [30].

The low detectability of other radars is achieved due to the polarization mismatch condition. The reflection phase of both transmitted and reflected waves needs to be changed for polarization- insensitive PGMA. For broadening the operating frequency

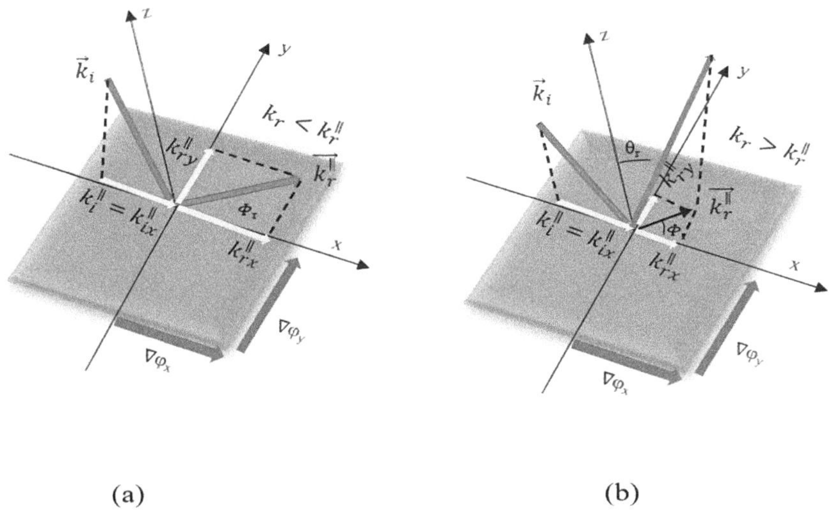

FIGURE 3.5 PGMS excited by an EM wave: (a) surface wave coupling and (b) anomalous reflection (AR). (Reproduced from [28].)

range of polarization conversion, the waves are divided into right-handed circular polarization (RHCP) and left-handed circular polarization (LHCP) based on the concept of linearly polarized (LP) waves. A PGMS generates a polarization-insensitive reflection of LP waves in a wideband [31]. In [32], low-quality (Q) metasurfaces have achieved a 40% reduction in radar RCS.

The work mechanism is discussed as follows:

The anomalous effect of the two-dimensional reflective phase gradient metasurface when it is excited by an EM wave is represented in Figure 3.5. Here k_i and k_r are the wavenumbers of the incident and reflected signals, and the phase deviation of the x and y axes is represented by $\nabla\phi_x$ and $\nabla\phi_y$ respectively. The incident wave vectors for the x and y components are k_{ix} and k_{iy}.

Case I:

When k_r is greater than k_i, the reflected waves appear in the side of the PGMS plane, as represented in Figure 3.5a, and the coupled surface wave is given by

$$\phi = \arctan(K_{iy} + \nabla\varphi_y / K_{ix} + \nabla\varphi_x) \tag{3.2}$$

Case II:

When k_r is less than k_i, AR happens, which is depicted in Figure 3.5b. Thus, the angle of reflection is given by

$$\phi_r = \arctan(K_{iy} + \nabla\varphi_y / K_{ix} + \nabla\varphi_x) \tag{3.3}$$

Wideband RCS is reduced using PGMS for a circularly polarized (CP) wave with Pancharatnam–Berry phase, as given in [33]. In the operational range from 5.6 to 16 GHz, a PGMS combined with an absorptive type frequency selective surface

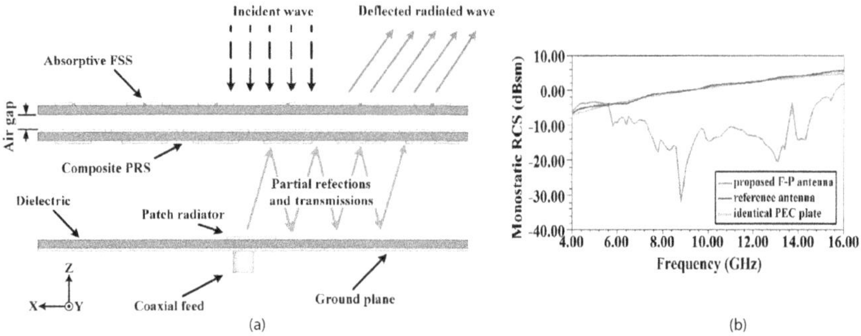

FIGURE 3.6 (a) Antenna schematic and (b) monostatic RCS. (Reproduced by [34].)

(FSS) reduces RCS around 13 dB, and it is made up of lumped resistors [34], as shown in Figure 3.6.

PGMS was employed for the slot array antenna RCSR for the frequency range of 4.1–4.26 GHz [35]. PGMS by making use of an ultrathin hybrid technology illustrated a RCSR in the frequency band between 8.7 and 11.4 GHz [36]. For better operation in oblique incidence, a windmill-shaped element was employed [37], which resulted in an effective RCSR. In [38], a cost-effective metasurface model reported a wider bandwidth 128% with an oblique incidence of 60° and a reflection amplitude.

The components contributing for improving the bandwidth are:

a. Selecting the right element to span over a wideband, which offers 2π range.
b. Both Propagating wave (PW) to surface wave (SW) conversion and AR should coincide in the design of an element.

The consolidated features of phase gradient metasurfaces are listed in Table 3.1.

3.2.1.2 CMS—Checkerboard Metasurfaces

The second type is the checkerboard metasurface (CMS), which uses a pair of sub-arrays arranged in an alternate manner and works by using the principle of destructive interference. The organization of the unit cell within this structure is irregular. The two adjacent cells are developed in such a way that it produces a reflection phase of 0° and 180°, respectively. In the far field, the signals get cancelled out. When a unit cell is designed properly, it will lead to enhanced bandwidth.

The concept of checkerboard structure began with the study of a checkerboard metasurface consisting of PEC (Perfect electric conductor) and AMC (Artificial magnetic conductor). The collective action by AMC and PEC takes place, and it gives rise to a phase difference of 0° and 180°, respectively. This leads to the phase cancellation, thereby reducing the RCS. Nevertheless, the above design suffers from limited bandwidth constraints [39]. The concept of using two AMCs as an alternative to the AMC-PEC cell is discussed [40], as shown in Figure 3.7. It is demonstrated that this orientation is providing a broader bandwidth. Eventually, the investigators explored that a proper combination of elements will lead to wider bandwidth.

TABLE 3.1
Features-Phase Gradient Metasurface

S.No	Year	Reference No.	Mechanism of Working	Fractional Bandwidth Measured Findings	Conformal Design	Dimensions of the Unit Cell	Monostatic/ Bistatic/Both	Dimensions of the Array
1	2012	[31]	Converting propagating wave to surface wave	15 GHz	No	$0.125\lambda_o \times 0.3\lambda_o$	Monostatic	NA
2	2014	[32]	Anomalous reflection (AR)	9.2–17.7 GHz	No	$0.16\,\lambda_o \times 0.16\lambda_o$	Monostatic	$11.7\lambda_o \times 11.7\lambda_o$
3	2013	[33]	Anomalous reflection (AR)	8–11 GHz	No	$0.16\lambda_o \times 0.23\lambda_o$	Monostatic	NA
4	2017	[34]	Anomalous reflection (AR)	9.85–19.37 GHz	No	$0.25\lambda_o \times 0.25\lambda_o$	Monostatic	$8.25\lambda_o \times 8.25\lambda_o$
5	2021	[35]	Propagating wave (PW) to surface wave (SW) + absorptive FSS	5.6–16 GHz	No	$0.16\lambda_o \times 0.16\lambda_o$	Monostatic	$2.11\lambda_o \times 2.11\lambda_o$
6	2018	[36]	Anomalous reflection (AR)	7–16.8 GHz	No	$0.42\lambda_o \times 0.42\lambda_o$	Monostatic	$1.71\lambda_o \times 1.71\lambda_o$
7	2017	[37]	Anomalous reflection (AR)	8.9–11.4 GHz	No	$0.20\lambda_o \times 0.20\lambda_o$	Monostatic	$10.4\lambda_o \times 10.4\lambda_o$
8	2014	[38]	Anomalous reflection (AR)	8–13 GHz	No	$0.20\lambda_o \times 0.20\lambda_o$	Both	$10.5\lambda_o \times 10.5\lambda_o$
9	2020	[39]	Anomalous reflection (AR)	9–40.7 GHz	No	$0.75\lambda_o \times 0.75\lambda_o$	Both	$19\lambda_o \times 19\lambda_o$

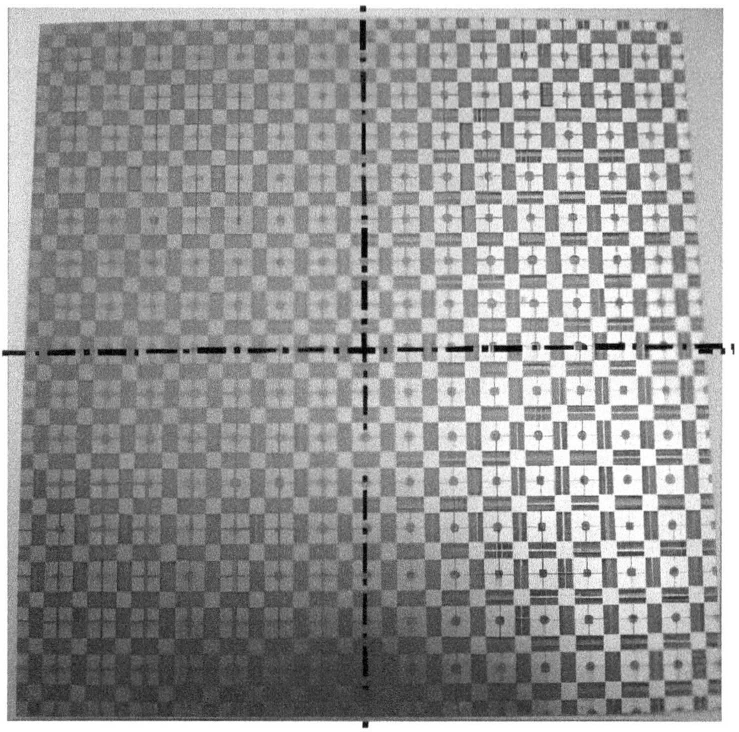

FIGURE 3.7 Proposed unit cell—AMC cells [40].

In addition to the arrangement of the unit cells, optimization methods are also used for a reduction in RCS. In the fruitfly optimization method, the unit cell is optimized, leading to a broader bandwidth, as reported in [41]. Many works above use Roger substrate; however, [42] uses a cost-effective FR4 substrate and achieves a wideband RCSR employing phase deviation criterion. By regulating the phase difference of two AMCs around ±30°, a bandwidth of 32% is achieved along with reduction in RCS, which is reported in [43]. In [44], a Jerusalem cross arrangement gives a bandwidth of 40%. Based on unit cell phase and reflection coefficient, a 10dB RCSR is obtained by providing 180° ± 38° [45]. The RCSR can be given by

$$\text{RCS}_{\text{reduction}} = \left| A_1 e^{j\varphi_1} + A_2 e^{j\varphi_2} \right| / 2 \tag{3.4}$$

Here, A_1 and A_2 are amplitudes of the unit cells, φ_1 and φ_2 represent the reflection phase of unit cells. A combination of partially reflecting surface and AMC is suggested to obtain RCSR for meander line metasurface [44]. Dual resonance unit cell is employed for reducing RCS in dual-band [46]. The blend of triple and dual AMCs is used to obtain a bandwidth of 91.5% [47].

So far, the design was confined to only a planar structure and two unit cells. Nevertheless, a broad bandwidth ratio of 5.87:1 is obtained by employing non-planar CMS where scattering cancellation of many elements happens and RCS gets reduced

TABLE 3.2
Features-Checkerboard Metasurface

S. No.	Year	Reference No.	Fractional Bandwidth Measured Findings	Conformal Design	Dimensions of the Unit Cell	Monostatic/ Bistatic/ Both	Dimensions of the Array
1	2007	[40]	15.32 GHz	NO	$1\lambda_o \times 1\lambda_o$	Both	$22\lambda_o \times 14\lambda_o$
2	2010	[41]	5.78 GHz	NO	$0.18\lambda_o \times 0.18\lambda_o$	Monostatic	$3.5\lambda_o \times 3.5\lambda_o$
3	2021	[42]	4.89–14.3 GHz	NO	$0.49\lambda_o \times 0.49\lambda_o$	Monostatic	$7.8\lambda_o \times 7.8\lambda_o$
4	2021	[43]	4.3–12.3 GHz	NO	$0.41\lambda_o \times 0.41\lambda_o$	Monostatic	$6.56\lambda_o \times 6.56\lambda_o$
5	2013	[44]	14.8–22.7 GHz	NO	$0.25\lambda_o \times 0.25\lambda_o$	Monostatic	$20.7\lambda_o \times 13.8\lambda_o$
6	2015	[46]	3.8–8.8 GHz	NO	$0.29\lambda_o \times 0.29\lambda_o$	Monostatic	$4.68\lambda_o \times 4.68\lambda_o$
7	2023	[47]	4.5–6.5 GHz	NO	$0.4\lambda_o \times 0.4\lambda_o$	Monostatic	$1.3\lambda_o \times 1.3\lambda_o$
8	2016	[48]	3.94–7.40 and 8.41–10.72 GHz	NO	$0.34\lambda_o \times 0.34\lambda_o$	Monostatic	$2.1\lambda_o \times 2.1\lambda_o$
9	2019	[49]	3.77–10.14 GHz	NO	$0.37\lambda_o \times 0.37\lambda_o$	Both	$2.98\lambda_o \times 2.98\lambda_o$

[48]. There are still many approaches for reducing RCS of curved targets such as using AMC theory, Ray theory, etc.

The enhancement of bandwidth can be obtained in the following methods:

a. Suitable selection of element
b. Surface wave suppression

A summary of the important characteristics is listed in Table 3.2.

3.2.1.3 PCMS—Polarization Conversion Metasurface

The working principle of PCMS is to convert the polarization efficiently to obtain RCSR. The crucial point here is in designing the element such that it converts the incident wave that is polarized into its orthogonal counterparts, which are obtained as reflected waves. The conversion above leads to decrease in RCS owing to the fact that radar being not able to detect the reflected signal. Polarization conversion efficiency decides the efficiency of RCSR. The concept of polarization conversion can be well explained by the element given in Figure 3.8 [49].

A LP wave is impinged in the y-direction; this incident wave is divided along the u and v directions as E_{iu} and E_{iv}. The u–v coordinate system is obtained by rotating the x–y coordinate system in a counterclockwise direction by 45°. The u and v axes when viewed from the x-axis will be 45°and 135°, respectively. Perfect electric conductor (PEC) occurs in the v axis due to the occurrence of electric resonance, which causes out-of-phase reflection in E_{rv} and E_{iv}. This leads to increased impedance in the u-axis due to which in-phase reflections happens. Now, the reflected signal includes co-polarized components, r_{xx} and r_{yy}, and cross-polarized components, r_{xy} and r_{yx}. This results in the generation of multiple resonances, as illustrated in Figure 3.9 [49].

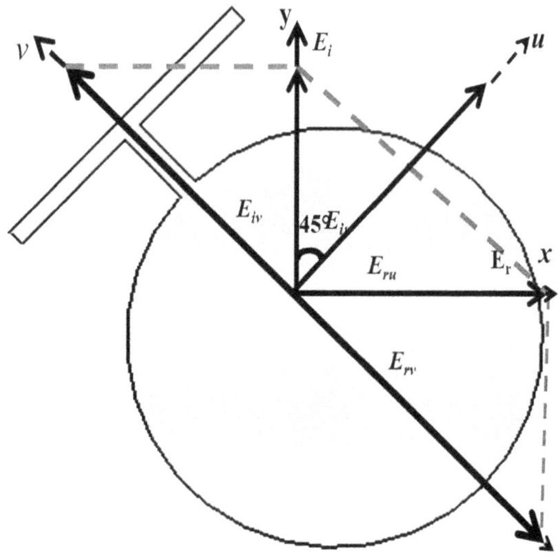

FIGURE 3.8 PCMS decomposition of proposed cell in the *u*-*v* axes [49].

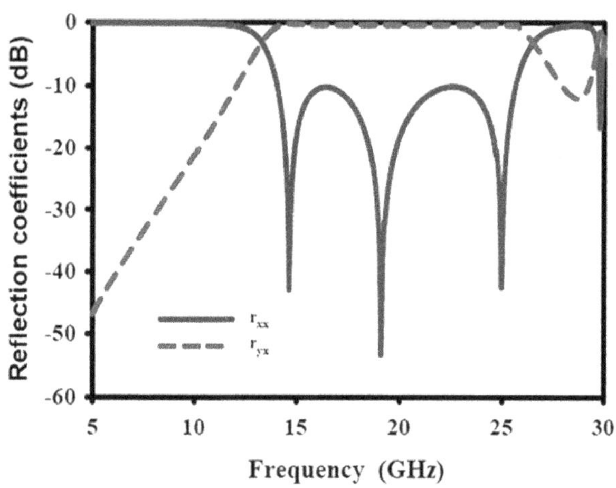

FIGURE 3.9 Results of co-polarized and cross-polarized components [49].

Here, Figure 3.8 represents element 0, and when it is rotated to 90°, element 1 is obtained. The phase difference between r_{xy} and r_{yx} is $\Delta\varphi$ and the polarization conversion ratio is given by [49]

$$\text{PCR} = \left|r_{yx}^2\right| / \left|r_{yx}^2\right| + \left|r_{xx}^2\right| = \left|r_{xy}^2\right| / \left|r_{xy}^2\right| + \left|r_{yy}^2\right| \tag{3.5}$$

An element with double-headed arrow is used for polarization conversion in [50], achieving 100% polarization conversion in four frequencies. RCSR of 10 dB is obtained in the operational range of 6.6–23.9 GHz. The configuration, which involves the addition of rotated unit cells to the checkerboard metasurface, provides a 10 dB RCSR and also will be stable up to 50° [51].

A periodic array is constructed using quasi-L-shaped patch of polarization rotation reflective surface, which offers 103% fractional bandwidth and a PCR of about 50% [52]. By using a unit cell of double-headed arrow, an RCSR of 10 dB is obtained in the operational range from 9 to 40 GHz. Destructive interference can be obtained by placing in a triangular chessboard array and by rotating it to 90°, and this result holds true for TE- and TM-polarized incident waves [53]. In [54], a diagonal arrow structure element presents a RCSR of 10 dB.

A quaternionic metasurface is presented in [55], which has four unit cells for enhancing RCS. In a wide frequency range, RCSR is obtained with the checkerboard arrangement. A design for conformal structure is reported in [56], which operates in the frequency range of 6.3–20.5 GHz and provides an angular stability of up to 60° by providing a RCSR of 10 dB. The PCMS discussion is summed up in Table 3.3 listing all its features.

3.2.1.4 COM—Coding Based Metasurface

A new design method that alters the propagation of EM wave is introduced in coding metasurface. It uses various phase elements arranged on the metasurface, which takes advantage of the coding sequence to obtain different functions. One-bit metasurface utilizing coding technique will have an out-of-phase difference of 180° in their reflected waves. Hence, by using various algorithms and improving the arrangement, ones and zeros can be manipulated; by this way, the scattering wave in the far-field region can be tailored and controlled. The idea of digital metamaterials was initiated in [57] in which unit cells were formed as binary digits by creating the metasurface.

The metasurface employed in [58] obtains RCSR by using digital, coding, and programmable metasurfaces. A phase response of 0° and 180° is obtained by 1-bit coding metasurface, which is called as binary coding metasurface as it assigns the digits 0 and 1. A phase response of 0°, 90°, 180°, and 270° is obtained for a 2-bit coding metasurface for its four elements. Similarly, $2\pi/N$ is used for the N-bit coding metasurface. The coding metasurface RCSR is obtained by the following equation:

$$\text{Radar Cross Section Reduction} = \lambda^2\left[\text{Dir}(\theta,\varphi)\right]/4\pi N^2 D^2 \tag{3.6}$$

From (3.6), the total count of element is represented by N in an array, D represents spacing, φ is the azimuthal angle, and θ is the elevation angle.

Phase gradient metasurface is considered as multi-bit coding metasurface, and the checkerboard metasurface is regarded as 1-bit coding metasurface. A one-bit coding method is presented in [59], which uses metallic ring resonator for RCSR. In [60], a unit cell is designed for coding metasurface, and it employs nonlinear fitting method and genetic algorithm for reducing RCS. The one-bit, two-bit, and multi-bit coding design uses Pancharatnam–Berry phase. The reflected wave direction is given by

$$\varphi_b = \tan^{-1}(k_0 \sin\theta_i \cos\varphi_i + \nabla\varphi_y / k_0 \sin\theta_i \cos\varphi_i + \nabla\varphi_x) \tag{3.7}$$

Here, φ_b represents the azimuth angle of the main lobe. Bandwidth can be enhanced by using optimization-based coded metasurface. The combination of bandpass FSS and a coding metasurface was used to obtain RCSR in a broadband and high-efficiency

TABLE 3.3
Features-Polarization Conversion Metasurface

S. No.	Year	Reference No.	Mechanism of Working	Fractional Bandwidth: Measured Findings	Conformal Design	Dimensions of the Unit Cell	Monostatic/ Bistatic/ Both	Dimensions of the Array
1	2014	[52]	Conversion of Polarization	15.32 GHz	NO	$1\lambda_o \times 1\lambda_o$	Both	$22\lambda_o \times 14\lambda_o$
2	2015	[54]	PRRS—Polarization rotation reflective surface	10.74–17.2 GHz	NO	$0.28\lambda_o \times 0.28\lambda_o$	Monostatic	$22\lambda_o \times 14\lambda_o$
3	2019	[56]	RPPC—Random phase encoding polarization conversion	8.2–25.6 GHz	NO	$0.16\lambda_o \times 0.16\lambda_o$	Both	$4\lambda_o \times 4\lambda_o$
4	2019	[55]	Conversion of polarization	16.8–37.3 GHz	NO	$0.59\lambda_o \times 0.59\lambda_o$	Both	$16\lambda_o \times 16\lambda_o$
5	2021	[57]	Polarization conversion with destructive phase difference	13–33 GHz	NO	$0.26\lambda_o \times 0.26\lambda_o$	Bistatic	$7.82\lambda_o \times 7.82\lambda_o$
6	2022	[58]	Conversion of polarization by using two unit cells	5.5–20.5 GHz	YES	$0.14\lambda_o \times 0.14\lambda_o$	Both	$3.27\lambda_o \times 3.27\lambda_o$
7	2017	[53]	Coding metasurface polarization conversion+	17–42 GHz	NO	$0.30\lambda_o \times 0.30\lambda_o$	Both	$21\lambda_o \times 21\lambda_o$

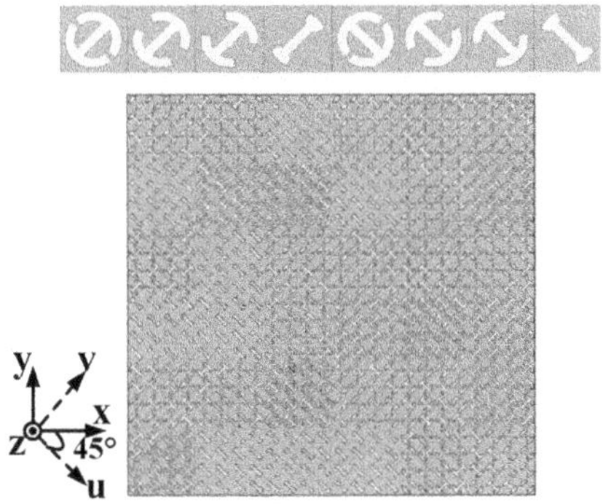

FIGURE 3.10 Unit cell design using 3-bit coding metasurface [65].

transmission. The transmission and reduction function is obtained by particle swarm optimization algorithm. Consequently, a RCSR of 15 dB has been realized in an operational range of 8.5–13.5 GHz [61]. In [62], a three-bit coding metasurface obtained RCSR in the frequency band of 18.3–42.2 GHz with eight elements. A four-bit reflective coding metasurface uses a discrete water recycle algorithm to achieve a RCSR of 10 dB [63].

In addition, a RCSR of more than 10 dB is employed by combining particle swarm optimization and small-sized coding metasurface reported in [64]. A 3-bit coding metasurface is reported in [65], which utilizes split ring cells to obtain a RCSR of 10 dB. Figure 3.10 gives the 3-bit coding metasurface.

A combination of data-driven learning method and sequential search technique is used in two-stage optimization method and reduces RCS in the operational range from 14.8 to 37.2 GHz [66]. A bandwidth enhancement is achieved up to 69% in [67], which uses a spiral coded metasurface. The matrix type coding metasurface is reported in [68], which enhances bandwidth by 87%. A fractional bandwidth of 129% is achieved for non-planar structures in the operational range from 6.4 to 30.1 GHz [69]. The low-Q resonator for RCSR using coded metasurface is reported in [70].

The main emphasis of all the aforementioned design was aimed at reducing the RCSR metal targets that are flat. Nevertheless, in [71], reduction is obtained for cylindrical surface by using a flexible ultrathin coded metasurface. By using PQCM-phase quantized coded metasurface, RCS is reduced without backscattering [72]. A 10 dB reduction is obtained for planar and conformal by using coding metasurface in ultra-wideband by combining polarization conversion, diffusion mechanisms, and absorption, and it is stable till 90° [73].

Hence, the design of coding metasurface can be obtained by using Phase Gradient method, Polarization Conversion method, Checker board method, or combining all. To enhance the bandwidth and RCSR, optimization techniques are vital to get high efficiency. A summary of the abovementioned models is listed in Table 3.4.

3.2.1.5 Reconfigurable Metasurface (RMS)

The major drawback of passive metasurface is that their operation is limited only to specific frequency bands and also needs various models for adjusting the frequency.

TABLE 3.4
Features-Coding Based Metasurface

S. No.	Year	Reference No.	Mechanism of Working	Fractional Bandwidth: Measured Findings	Conformal Design	Dimensions of the Unit Cell	Monostatic/ Bistatic/Both	Dimensions of the Array
1	2015	[61]	Diffused	0.8–1.4 THz	YES	$0.4\lambda_o \times 0.4\lambda_o$	Both	$36.7\lambda_o \times 36.7\lambda_o$
2	2018	[63]	Optimization using particle swarm	8.5–13.5 GHz	NO	$0.48\lambda_o \times 0.48\lambda_o$	Both	$11\lambda_o \times 11\lambda_o$
3	2019	[64]	Polarization conversion and scattering diffusion	18.3–42.2 GHz	NO	$0.42\lambda_o \times 0.42\lambda_o$	Monostatic	$26.6\lambda_o \times 26.6\lambda_o$
4	2019	[66]	Optimization using particle swarm	15–35 GHz	NO	$0.15\lambda_o \times 0.15\lambda_o$	Bistatic	$6\lambda_o \times 6\lambda_o$
5	2017	[69]	Diffused	12.2–23.4 GHz	NO	$0.3\lambda_o \times 0.3\lambda_o$	Both	$14.4\lambda_o \times 14.4\lambda_o$
6	2018	[70]	Random coding using matrix-type	5.8–15.5 GHz	NO	$0.35\lambda_o \times 0.35\lambda_o$	Both	$10.5\lambda_o \times 10.5\lambda_o$
7	2020	[71]	Optimization using particle swarm	6.4–29.6 GHz	NO	$0.48\lambda_o \times 0.48\lambda_o$	Both	$13.6\lambda_o \times 13.6\lambda_o$
8	2018	[74]	Super cell using phased quantization	9–11 GHz monostatic	NO	$0.23\lambda_o \times 0.23\lambda_o$	Monostatic	$8.4\lambda_o \times 8.4\lambda_o$
9	2019	[65]	Water cycle algorithm	15–40 GHz	NO	$0.46\lambda_o \times 0.46\lambda_o$	Monostatic	$29\lambda_o \times 29\lambda_o$
10	2017	[69]	Diffused	12.2–23.4 GHz	NO	$0.3\lambda_o \times 0.3\lambda_o$	Both	$14.4\lambda_o \times 14.4\lambda_o$
11	2018	[62]	Coding sequence optimization+ element+ genetic algorithm	12–24 GHz	NO	$0.2\lambda_o \times 0.2\lambda_o$	Monostatic	$6.4\lambda_o \times 6.4\lambda_o$
12	2021	[68]	Data-driven learning, direct EM design closure, sequential search strategy	14.8–37.2 and 16.7–37 GHz	NO	$0.3\lambda_o \times 0.3\lambda_o$	Both	$4.8\lambda_o \times 4.8\lambda_o$
13	2022	[72]	Phase cancellation	11.5–16 GHz	NO	$0.23\lambda_o \times 0.34\lambda_o$	Monostatic	$11.5\lambda_o \times 11.5\lambda_o$
14	2023	[75]	Optimized sequence	10.8–31.3 GHz	YES	$0.21\lambda_o \times 0.21\lambda_o$		–

It requires redesigning the unit cell for altering the frequency. This can be subdued by integrating triggering circuits externally for obtaining tunability. This tenability can be made either at the unit cell or through the entire metasurface. The reaction of the metasurface can be changed by incorporating the external parameter changes on stimuli-responsive materials such as temperature, pressure, humidity, light, electric field, and magnetic field. This validates the functionality change and dynamic adjustments, which lead to globally tuned metasurface.

As there are many practical problems in changing the parameters such as humidity and pressure, researchers have concentrated on modifying the other tuning parameters such as magnetic field, electric field, temperature, and light. Hence, by using materials such as grapheme and liquid crystal, a globally tuned metasurface can be achieved, which alters the overall performance.

Locally tuned metasurface is when the functional performance of the individual unit cell is varied, which can be realized by using lumped components, PIN diodes, and varactor diodes.

3.2.1.5.1 *Globally Tuned*

The basic idea behind the tuned metasurface is discussed further. A tunable coded metasurface working well in terahertz frequency band [74] is used, with a metallic patch implanted in vanadium. The concept of this method is the use of metallic insulator for reducing RCS at 68°C.

Tunable RCSR is used to obtain broadband bandwidth, as proposed in [75]. A hybrid mechanism that combines Salisbury screen and high-index grating structure is used. The lossy sheet was chosen to be graphene. This reduces the RCSR by absorbing the incoming wave, and the grating structure will reduce the backward scattering. The surface resistance is tuned by graphene, which reduces the RCSR.

3.2.1.5.2 *Locally Tuned*

The metasurfaces used here will have small unit cell that is used to modify the reflection characteristics. A unit cell containing a pin diode placed inside is used for validating the binary-coded metasurface. Field Programmable Gate Array (FPGA) at the back side performs the real-time control. A genetic algorithm was used to optimize the coding matrix. The FPGA is shown in Figure 3.11.

Programmable metasurface often needs alteration of EM fields, but the research in metamaterial has enabled the accurate manipulation of EM waves. The sensing feedback system will adjust the EM functioning [77] as metamaterial can perform self-regulation based on the surrounding. The three-axis gyroscope is connected to a smart sensitive metasurface, which changes the EM radiation according to different rotations. It is evident from Figure 3.12 that a switch is attached, which performs motion to move between steering modes.

3.2.1.6 TVM—Time Varying Metasurface

In the aforementioned works, the frequency is kept constant while manipulating the reflected waves alone. Nevertheless, researchers always tend to travel in an unexposed path. Hence, the detectability of the target is reduced by spreading the incident EM wave into different frequency bands. A technique is executed by the integration

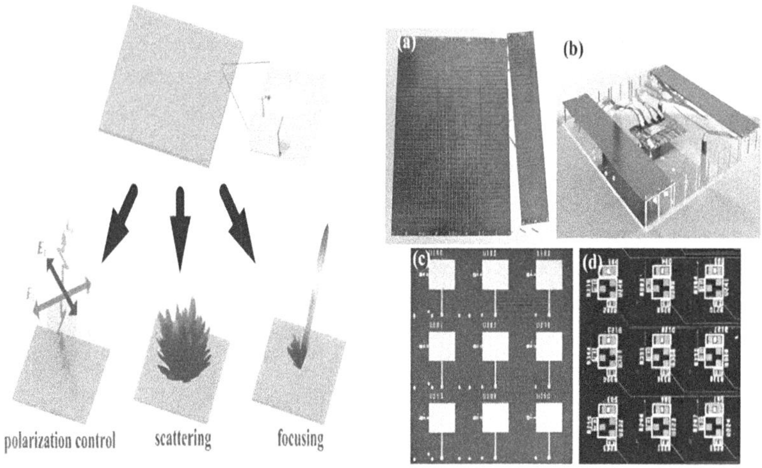

FIGURE 3.11 Programmable metasurface—different functionalities of FPGA [76].

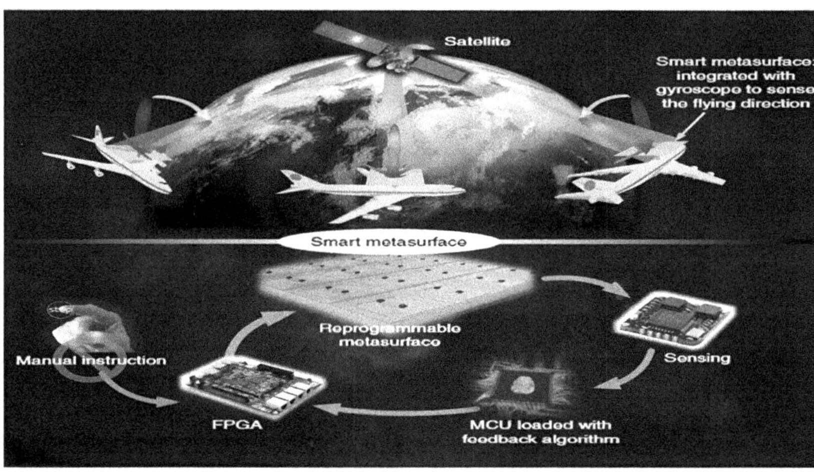

FIGURE 3.12 Self-adaptive programmable metasurface [77].

of time domain modulation (TDM) and spectrum technology, which is reported in [78]. In [79], time-varying characteristics are achieved by the Lorentz concept, which gives rise to time-varying metasurfaces. Transverse temporal modulation is introduced for obtaining the desired properties [80].

A micro-ribbon is made up of graphene, which is evident in [81], where the graphene doping is performed by the Fermi levels, which gives rise to time-varying changes. For asynchronous coding metasurface that performs different modulation, a new arrangement using coding is reported in [82]. Here, the phase discontinuities are converted into different frequencies, hence improving the performance and functionality of the metasurface [82].

3.2.2 RCS Reduction Based on Absorptive Metasurface

Absorptive metasurface achieves RCSR by satisfying two conditions, which are nil reflection and nil transmission. By obtaining the matching between the absorbing layer and free space, the reflection can be completely eliminated. The thickness of the absorbing layer is increased to obtain zero transmission condition by the properties of electric and magnetic fields. The basic idea of a metamaterial absorber is reported in [83].

An ultrathin one layer absorber FSS that uses four square resistive loop is given in [84] for RCSR. A metallic cube is designed with zero reflection and transmission for RCSR, which has been reported in [85]. All the abovementioned works efficiently absorb the incident waves for reduction of RCS. By introducing a triple absorber in X-band [86], a dual polarized band pass is incorporated. A metamaterial absorber is reported in [87], which is not parallel to the ground plane. A parameter producing a major impact in the absorber is surface current distribution.

In [88], an absorber made up of ultrathin material operating in X-band, which features concentric split rings for RCSR, is reported. An equivalent circuit design is proposed using metamaterial absorber for RCSR in [89]. A square-shaped based ring made of ultrathin metamaterial absorber is presented in [90] for RCSR. RCSR is obtained in the C and S bands along with increased gain [91].

The proposal of Fabry-Perot cavity antenna is given in [92] for reducing RCS and enhancing bandwidth. The antenna achieves gain enhancement and beam deflection in transmission mode for obtaining the above characteristics [92]. A FSS absorber with tapered resistive sheet is used for obtaining RCSR [93] for the angular range of 30°–85°. An integration of four lumped resistors and four metallic patches is used for planar and conformal design in the operational range of 13.42–22.66 GHz, which aims for 20 dB RCSR. An absorptive metasurface using FR4 is placed at the crest [94]. The pictorial representation of the structure and the results are indicated in Figure 3.13

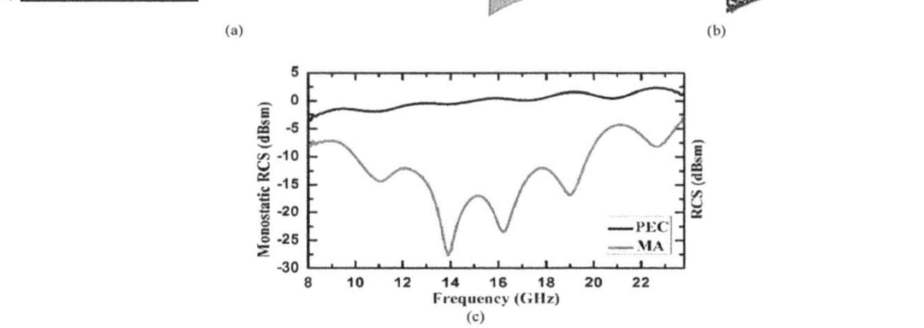

FIGURE 3.13 (a) Proposed unit cell, (b) conformal design, and (c) RCS reduction [94].

Table 3.5 lists all the features of the absorptive metasurface of the abovementioned works.

3.3 APPLICATIONS OF METASURFACE-AN OVERVIEW

Various applications of metasurface are listed here:

i. Metasurface extends its application in antenna system to obtain wide bandwidth, high gain, size reduction, and Specific Absorption Rate (SAR) reduction.
ii. Metasurfaces have distinct geometrical pattern to attain Polarization conversion, efficient RCSR, and EM wave absorption.
iii. Further, it has got varied practical applications such as perfect absorbing, perfect lens, invisibility, and light bending to name but a few.

3.4 FUTURE RESEARCH DIRECTION

An immense amount of research has been conducted for RCSR, even few unanswered issue exists. Some of them are:

i. There is a research part that has not been tracked so far. An absorptive metasurface due to the heat emission can be made visible. But it should be able to overcome this for protecting the target.
ii. Periodic structures have been concentrated by researchers for monostatic RCSR for periodic structures. Nevertheless, the efficiency of RCSR will not remain same for aperiodic structures, especially at oblique incidence. Hence, combining both the patterns could get results with high efficiency.
iii. The third concern is that metasurfaces offer a broad bandwidth for monostatic situation but do not offer the same for bistatic situation. Since the surface impedance at oblique incidence varies when compared with the normal impedance as given by Huygens's metasurfaces [95]. Metasurfaces due to inherent monochromatic aberrations lack angular stability at oblique incidence [96]. Hence, there is a necessity to subdue this angular dispersion to obtain an angular stable element by designing meta-atoms. Hence, by designing this type of element, there is a scope of extending the bistatic RCSR.
iv. In coding a one-bit metasurface, there is a criterion for the phase difference to be $180° \pm 37°$ between the unit cells for achieving the RCSR, which creates the constraints in the flexibility of the unit cell design.
v. The other area, which needs to be given more attention, is the array size dependency on RCSR bandwidth.
vi. Other important parameter, which plays a major role in RCSR, is its substrate; hence, a good agreement should be obtained between the thickness of the substrate and the frequency range of operation. On the other hand, thicker substrate works well for low frequencies when used in stealth applications; on the other hand, they are practically obstructed from conformal applications due to their thickness.

TABLE 3.5
Features-Absorptive Metasurface

S. No.	Year	Reference No.	Fractional Bandwidth Measured Findings	Conformal Design	Dimensions of the Unit Cell	Monostatic/Bistatic/Both	Dimensions of the Array
1	2007	[85]	8–18 GHz	NO	$0.56\lambda_o \times 0.56\lambda_o$	Monostatic	$6.9\lambda_o \times 6.9\lambda_o$
2	2014	[87]	5.258, 10.02, 7.518 and 10.494 GHz	NO	$0.47\lambda_o \times 0.47\lambda_o$	Both	$6.3\lambda_o \times 6.3\lambda_o$
3	2016	[89]	4.96–18.22 GHz	NO	$0.502\lambda_o \times 0.502\lambda_o$	Monostatic	$7\lambda_o \times 7\lambda_o$
4	2020	[91]	1.98–9 and 1.07–9.7 GHz	NO	$0.2745\lambda_o \times 0.2745\lambda_o$	Both	NA
5	2023	[93]	2.3–8.55 GHz	NO	$0.16\lambda_o \times 0.16\lambda_o$	Monostatic	$1.85\lambda_o \times 1.85\lambda_o$
6	2022	[95]	1.3–2 GHz	NO	$0.108\lambda_o \times 0.108\lambda_o$	Monostatic	$1.3\lambda_o \times 2.16\lambda_o$
7	2012	[86]	N/A	NO	$0.28\lambda_o \times 0.28\lambda_o$	N/A	N/A
8	2018	[88]	3.9–26.2 GHz	NO	$0.602\lambda_o \times 0.602\lambda_o$	Monostatic	$12\lambda_o \times 12\lambda_o$
9	2015	[90]	8.08–12.12 GHz	NO	$0.238\lambda_o \times 0.238\lambda_o$	Both	$7.8\lambda_o \times 7.8\lambda_o$
10	2014	[92]	6.14 GHz	NO	$0.204\lambda_o \times 0.204\lambda_o$	Monostatic	NA
11	2021	[94]	4–16 GHz (8.5 dB RCSR)	NO	$0.2\lambda_o \times 0.2\lambda_o$	Monostatic	$1.4\lambda_o \times 1.4\lambda_o$
12	2022	[96]	13.42–22.6 GHz	YES	$0.45\lambda_o \times 0.45\lambda_o$	Both	$7.72\lambda_o \times 7.72\lambda_o$

vii. For multiple target scenarios, the method of passive RCSR techniques still needs further investigation as it is found to be effective for single targets only.
viii. Mutual coupling affects the metasurfaces, which leads to its performance degradation. This leads to two paths of exploration:
   a. The prediction model should be enhanced for array bandwidth [97].
   b. RCSR bandwidth should be increased by decreasing mutual coupling effects.

The reduction of mutual coupling can be attained by

a. Using PEC walls like vias [98] and baffles for reducing the coupling between the cells.
b. Mutual coupling reducing techniques, which are applied for MIMO antennas [99], can also be used.

The abovementioned areas require further intervention for RCSR using metasurfaces.

3.5 CONCLUSION

The chapter here discussed various progressions of RCSR methods by focusing on metasurfaces. It began by introducing the fundamental techniques such as scattering and absorption for reducing RCS followed by a short review on RAM (absorption method) and shaping (Scattering method) for reducing RCSR. Then, the paper delves into comprehensive review of passive and active configurations of metasurface for RCSR by thoroughly examining its frequency range, design complexity, technique and profile, cost-effectiveness, and polarization sensitivity. It is followed by the wide applications of metasurface. This chapter has also opened way to the new avenues, which need further research. Overall, this chapter has provided a complete exploration of techniques, which reduce RCS by concentrating on both conformal and planar designs of metasurface.

REFERENCES

1. IEEE, *IEEE Standard for Radar Definitions*. 2017. doi: 10.1109/IEEESTD.2017.8048479
2. A. Sihvola, Metamaterials in electromagnetics, *Metamaterials*, vol. 1, no. 1, pp. 2–11, 2007, doi: 10.1016/j.metmat.2007.02.003
3. E. Bayati, A. Zhan, S. Colburn and A. Majumdar, Characterizing meta-lens performance as a function of refractive index, *RAPID 2018-2018 IEEE Research and Applications of Photonics In Defense Conference*, Miramar Beach, FL, pp. 167–170, 2018, doi: 10.1109/RAPID.2018.8508955
4. B. Ozbey, H. V. Demir, O. Kurc, V. B. Erturk and A. Altintas, Wireless measurement of elastic and plastic deformation by a metamaterial-based sensor, *Sensors*, vol. 14, no. 10, pp. 19609–19621, Oct. 2014, doi: 10.3390/s141019609
5. S. B. Glybovski, S. A. Tretyakov, P. A. Belov, Y. S. Kivshar and C. R. Simovski, Metasurfaces: From microwaves to visible, *Physics Reports*, vol. 634, pp. 1–72, 2016, doi: 10.1016/j.physrep.2016.04.004
6. H. T. Chen, A. J. Taylor and N. Yu, A review of metasurfaces: physics and applications, *Reports on Progress in Physics*, vol. 79, no. 7, Jun. 2016, doi: 10.1088/0034-4885/79/7/076401

7. C. L. Holloway, E. F. Kuester, J. A. Gordon, J. O'Hara, J. Booth and D. R. Smith, An overview of the theory and applications of metasurfaces: The two-dimensional equivalents of metamaterials, *IEEE Antennas and Propagation Magazine*, vol. 54, no. 2, pp. 10–35, 2012, doi: 10.1109/MAP.2012.6230714
8. F. Li, H. Chen, Q. He, Y. Zhou, L. Zhang, X. Weng, H. Lu, J. Xie and L. Deng, Design and implementation of metamaterial polarization converter with the reflection and transmission polarization conversion simultaneously, *Journal of Optics*, vol. 21, no. 4, 2019, doi: 10.1088/2040-8986/ab0878
9. T. L. Dobrenz, A. Spadoni and M. Jorgensen, Aviation archeology of the Horten 229 v3 aircraft, *AIAA Aviation Technology, Integration, and Operations (ATIO) Conference*, Fort Worth, Texas, vol. 2, pp. 1–9, 2010, doi: 10.2514/6.2010-9214
10. P. Y. Ufimtsev, Elementary edge waves and the physical theory of diffraction, *Electromagnetics*, vol. 11, no. 2, pp. 125–160, 1991.
11. U. Lockheed, Lockheed U-2, *The Ultimate Collection on UFOs*, p. 99, 2012.
12. G. A. Rao and S. P. Mahulikar, Integrated review of stealth technology and its role in airpower, *Aeronautical Journal*, vol. 106, no. 1066, pp. 629–642, 2002.
13. R. H. Francis, The development of blue steel, *Aeronautical Journal*, vol. 68, no. 641, pp. 303–322, 1964.
14. G. G. Peixoto, A. L. De Paula, L. A. Andrade, C. M. A. Lopes and M. C. Rezende, Radar absorbing material (RAM) and shaping on radar cross section reduction of dihedral corners, *SBMO/IEEE MTT-S International Conference on Microwave and Optoelectronics*, Brasilia, Brazil, 2005, pp. 460–463.
15. J. R. Swandic, Bandwidth Limits and Other Considerations for Monostatic RCS Reduction by Virtual Shaping. Accessed: Mar. 01, 2024. [Online]. Available: https://www.researchgate.net/publication/235104435_Bandwidth_Limits_and_Other_Considerations_for_Monostatic_RCS_Reduction_by_Virtual_Shaping
16. P. Saville, *Review of Radar Absorbing Materials*. Defence Research & Development Canada, Defence R & D Canada-Atlantic, 2005.
17. B. Sweetman, *Stealth Aircraft: History, Technology and Outlook*, vol. 79054. American Society of Mechanical Engineers, 1990.
18. B. Chambers and A. Tennant, Optimised design of Jaumann radar absorbing materials using a genetic algorithm, *IEE Proceedings: Radar, Sonar and Navigation*, vol. 143, no. 1, pp. 23–30, 1996.
19. W. W. Salisbury, Absorbent body for electromagnetic waves, US Patent No. 2599944 A. 1952.
20. R. Maggiora, M. Saccani, D. Milanesio and M. Porporato, An innovative harmonic radar to track flying insects: The case of Vespa velutina, *Scientific Reports*, vol. 9, no. 1, p. 11964, 2019.
21. S.-L. Jeng, W.-H. Chieng and H.-P. Lu, Estimating speed using a side-looking single-radar vehicle detector, *IEEE Transactions on Intelligent Transportation Systems*, vol. 15, no. 2, pp. 607–614, 2013.
22. K. J. Vinoy and R. M. Jha, Trends in radar absorbing materials technology, *Sadhana*, vol. 20, pp. 815–850, 1995.
23. W.-L. Guo, K. Chen, G.-M. Wang, X.-Y. Luo, Y.-J. Feng and C.-W. Qiu, Transmission--reflection-selective metasurface and its application to RCS reduction of high-gain reflector antenna, *IEEE Transactions on Antennas and Propagation*, vol. 68, no. 3, pp. 1426–1435, 2019.
24. A. Epstein and G. V. Eleftheriades, Shielded perfect reflectors based on omega-bianisotropic metasurfaces, *2017 International Workshop on Antenna Technology: Small Antennas, Innovative Structures, and Applications (iWAT)*, Athens, Greece, 2017, pp. 6–9.
25. R. Dezert, Theoretical study of isotropic Huygens particles for metasurfaces, Thesis, Université de Bordeaux, 2019.

26. N. Yu, P. Genevet, M. A. Kats, F. Aieta, J.-P. Tetienne, F. Capasso and Z. Gaburro, Light propagation with phase discontinuities: Generalized laws of reflection and refraction, *Science*, vol. 334, no. 6054, pp. 333–337, 2011.
27. M. U. Ullah, T. B. A. Latef, M. Othman, M. I. Hussein, H. M. Alkhoori, Y. Yamada, K. Kamardin and R. Khalid, A progression in the techniques of reducing RCS for the targets, *Alexandria Engineering Journal*, vol. 100, no. May, pp. 153–169, 2024, doi: 10.1016/j.aej.2024.05.001
28. A. Murugesan, K. T. Selvan, A. K. Iyer, K. V. Srivatsav and A. Alphones, A review of metasurface-assisted RCS reduction techniques, *Progress in Electromagnetics Research B*, vol. 94, pp. 75–103, 2021, doi: 10.2528/PIERB21081401
29. S. Sun, K.-Y. Yang, C.-M. Wang, T.-K. Juan, W. T. Chen, C. Y. Liao, Q. He, S. Xiao, W.-T. Kung, G.-Y. Guo, Lei Zhou and D. P. Tsai, High-efficiency broadband anomalous reflection by gradient meta-surfaces, *Nano Letters*, vol. 12, no. 12, pp. 6223–6229, 2012.
30. S. Sun, Q. He, S. Xiao, Q. Xu, X. Li and L. Zhou, Gradient-index meta-surfaces as a bridge linking propagating waves and surface waves, *Nature Materials*, vol. 11, no. 5, pp. 426–431, 2012.
31. Y. Li, J. Zhang, S. Qu, J. Wang, H. Chen, L. Zheng, Z. Xu and A. Zhang, Achieving wideband polarization-independent anomalous reflection for linearly polarized waves with dispersionless phase gradient metasurfaces, *Journal of Physics D: Applied Physics*, vol. 47, no. 42, p. 425103, 2014.
32. M. Pu, P. Chen, C. Wang, Y. Wang, Z. Zhao, C. Hu, C. Huang and X. Luo, Broadband anomalous reflection based on gradient low-Q meta-surface, *AIP Advances*, vol. 3, no. 5, 2013.
33. Q. Zheng, Y. Li, J. Zhang, H. Ma, J. Wang, Y. Pang, Y. Han, S. Sui, Y. Shen, H. Chen and S. Qu, Wideband, wide-angle coding phase gradient metasurfaces based on Pancharatnam-Berry phase, *Scientific Reports*, vol. 7, no. 1, p. 43543, 2017.
34. H. Umair, T. B. A. Latef, Y. Yamada, W. N. L. B. W. Mahadi, M. Othman, K. Kamardin, M. I. Hussein and A. I. Najam, Tilted beam Fabry–Perot antenna with enhanced gain and broadband low backscattering, *Electronics*, vol. 10, no. 3, p. 267, 2021.
35. W. Zhang, Y. Liu, S. Gong, J. Wang and Y. Jiang, Wideband RCS reduction of a slot array antenna using phase gradient metasurface, *IEEE Antennas and Wireless Propagation Letters*, vol. 17, no. 12, pp. 2193–2197, 2018.
36. Y. Cheng, W. Huang, J. Guo, B. Huang, Y. Wang, R. Zhu and Z. Wang, An ultra-thin dual-band phase-gradient metasurface using hybrid resonant structures for backward RCS reduction, *Applied Physics B*, vol. 123, pp. 1–6, 2017.
37. J. Chen, Q. Cheng, J. Zhao, D. S. Dong and T.-J. Cui, Reduction of radar cross section based on a metasurface, *Progress in Electromagnetics Research*, vol. 146, pp. 71–76, 2014.
38. Y. Azizi, M. Soleimani and S. H. Sedighy, Ultra-wideband radar cross section reduction using amplitude and phase gradient modulated surface, *Journal of Applied Physics*, vol. 128, no. 20, 2020.
39. M. Paquay, J.-C. Iriarte, I. Ederra, R. Gonzalo and P. de Maagt, Thin AMC structure for radar cross-section reduction, *IEEE Transactions on Antennas and Propagation*, vol. 55, no. 12, pp. 3630–3638, 2007.
40. M. E. de Cos, Y. Alvarez-Lopez and F. L. H. Andres, A novel approach for RCS reduction using a combination of artificial magnetic conductors, *Progress in Electromagnetics Research*, vol. 107, pp. 147–159, 2010.
41. A. Murugesan and K. T. Selvan, On further enhancing the bandwidth of wideband RCS reduction checkerboard metasurfaces using an optimization algorithm, *International Journal of RF and Microwave Computer-Aided Engineering*, vol. 31, no. 7, p. e22686, 2021.
42. A. Murugesan, D. Natarajan and K. T. Selvan, Low-cost, wideband checkerboard metasurfaces for monostatic RCS reduction, *IEEE Antennas and Wireless Propagation Letters*, vol. 20, no. 4, pp. 493–497, 2021.

43. Y. Fu, Y. Li and N. Yuan, Wideband composite AMC surfaces for RCS reduction, *Microwave and Optical Technology Letters*, vol. 53, no. 4, pp. 712–715, 2011.
44. M. V Jisha and A. Pradeep, Low RCS patch antenna based on meander line checkerboard, *Authorea Prepr.*, 2023.
45. W. Chen, C. A. Balanis and C. R. Birtcher, Checkerboard EBG surfaces for wideband radar cross section reduction, *IEEE Transactions on Antennas and Propagation*, vol. 63, no. 6, pp. 2636–2645, 2015.
46. W. Chen, C. A. Balanis and C. R. Birtcher, Dual wide-band checkerboard surfaces for radar cross section reduction, *IEEE Transactions on Antennas and Propagation*, vol. 64, no. 9, pp. 4133–4138, 2016.
47. D. Sang, Q. Chen, L. Ding, M. Guo and Y. Fu, Design of checkerboard AMC structure for wideband RCS reduction, *IEEE Transactions on Antennas and Propagation*, vol. 67, no. 4, pp. 2604–2612, 2019.
48. J. Su, Y. Lu, J. Liu, Y. Yang, Z. Li and J. Song, A novel checkerboard metasurface based on optimized multielement phase cancellation for superwideband RCS reduction, *IEEE Transactions on Antennas and Propagation*, vol. 66, no. 12, pp. 7091–7099, 2018.
49. S. Vidhya and M. Karthikeyan, Ultra-wide band and high efficient polarization converting metasurface employing bell-shaped unit cell for RCS reduction, *Results Engineering*, vol. 24, no. November, 2024, doi: 10.1016/j.rineng.2024.103342
50. H. Chen, J. Wang, H. Ma, S. Qu, Z. Xu, A. Zhang, M. Yan and Y. Li, Ultra-wideband polarization conversion metasurfaces based on multiple plasmon resonances, *Journal of Applied Physics*, vol. 115, no. 15, 2014.
51. H. Sun, C. Gu, X. Chen, Z. Li, L. Liu, B. Xu and Z. Zhou, Broadband and broad-angle polarization-independent metasurface for radar cross section reduction, *Scientific Reports*, vol. 7, pp. 40782–40782, Jan. 2017, doi: 10.1038/SREP40782
52. Y. Jia, Y. Liu, Y. J. Guo, K. Li and S.-X. Gong, Broadband polarization rotation reflective surfaces and their applications to RCS reduction, *IEEE Transactions on Antennas and Propagation*, vol. 64, no. 1, pp. 179–188, 2015.
53. E. Ameri, S. H. Esmaeli and S. H. Sedighy, Ultra wideband radar cross section reduction by using polarization conversion metasurfaces, *Scientific Reports*, vol. 9, no. 1, p. 478, 2019.
54. H. Rajabalipanah and A. Abdolali, Ultrabroadband monostatic/bistatic RCS reduction via high-entropy phase-encoded polarization conversion metasurfaces, *IEEE Antennas and Wireless Propagation Letters*, vol. 18, no. 6, pp. 1233–1237, 2019.
55. Z. Zhang, Y. Zhang, T. Wu, S. Chen, W. Li and J. Guan, Broadband RCS reduction by a quaternionic metasurface, *Materials*, vol. 14, no. 11, p. 2787, 2021.
56. J. Chatterjee, A. Mohan and V. Dixit, Ultrawideband RCS reduction of planar and conformal surfaces using ultrathin polarization conversion metasurface, *IEEE Access*, vol. 10, pp. 36563–36575, 2022, doi: 10.1109/ACCESS.2022.3163850
57. C. Della Giovampaola and N. Engheta, Digital metamaterials, *Nature Materials*, vol. 13, no. 12, pp. 1115–1121, 2014.
58. T. J. Cui, M. Q. Qi, X. Wan, J. Zhao and Q. Cheng, Coding metamaterials, digital metamaterials and programmable metamaterials, *Light: Science & Applications*, vol. 3, no. 10, pp. e218–e218, 2014.
59. L. Liang, L. Liang, M. Qi, J. Yang, X. Shen, J. Zhai, W. Xu, B. Jin, W. Liu, Y. Feng, C. Zhang, H. Lu, H.-T. Chen, L. Kang, W. Xu, J. Chen, T. J. Cui, P. Wu and S. Liu, Anomalous terahertz reflection and scattering by flexible and conformal coding metamaterials, *Advanced Optical Materials*, vol. 3, no. 10, pp. 1374–1380, 2015.
60. S. Sui, H. Ma, Y. Lv, J. Wang, Z. Li, J. Zhang, Z. Xu and S. Qu, Fast optimization method of designing a wideband metasurface without using the Pancharatnam–Berry phase, *Optics Express*, vol. 26, no. 2, pp. 1443–1451, 2018.
61. F. Li, W. Fang, P. Chen, Y. Poo and R.-X. Wu, Transmission and radar cross-section reduction by combining binary coding metasurface and frequency selective surface, *Optics Express*, vol. 26, no. 26, pp. 33878–33887, 2018.

62. H. Dai, Y. Zhao, H. Li, J. Chen, Z. He and W. Qi, An ultra-wide band polarization-independent random coding metasurface for RCS reduction, *Electronics*, vol. 8, no. 10, p. 1104, 2019.
63. Y. Saifullah, A. B. Waqas, G.-M. Yang, F. Zhang and F. Xu, 4-Bit optimized coding metasurface for wideband RCS reduction, *IEEE Access*, vol. 7, pp. 122378–122386, 2019.
64. H. Hao, S. Du and T. Zhang, Small-size broadband coding metasurface for RCS reduction based on particle swarm optimization algorithm, *Progress in Electromagnetics Research M*, vol. 81, pp. 97–105, 2019.
65. P. Su, Y. Zhao, S. Jia, W. Shi and H. Wang, An ultra-wideband and polarization-independent metasurface for RCS reduction, *Scientific Reports*, vol. 6, no. 1, p. 20387, 2016.
66. M. Abdullah and S. Koziel, Supervised-learning-based development of multibit RCS-reduced coding metasurfaces, *IEEE Transactions on Microwave Theory and Techniques*, vol. 70, no. 1, pp. 264–274, 2021.
67. F. Yuan, G.-M. Wang, H.-X. Xu, T. Cai, X.-J. Zou and Z.-H. Pang, Broadband RCS reduction based on spiral-coded metasurface, *IEEE Antennas and Wireless Propagation Letters*, vol. 16, pp. 3188–3191, 2017.
68. J. J. Yang, Y. Z. Cheng, D. Qi and R. Z. Gong, Study of energy scattering relation and RCS reduction characteristic of matrix-type coding metasurface, *Applied Sciences*, vol. 8, no. 8, p. 1231, 2018.
69. G. Wu, W. Yu, T. Lin, Y. Deng and J. Liu, Ultra-wideband RCS reduction based on non-planar coding diffusive metasurface, *Materials (Basel).*, vol. 13, no. 21, pp. 1–12, Nov. 2020, doi: 10.3390/MA13214773
70. F. Yuan, Q. Chen, Y. Zheng and Y. Fu, Dual-mechanism absorptive metasurface with wideband 20 dB RCS reduction, *Crystals*, vol. 12, no. 4, p. 493, 2022.
71. K. Chen, L. Cui, Y. Feng, J. Zhao, T. Jiang and B. Zhu, Coding metasurface for broadband microwave scattering reduction with optical transparency, *Optics Express*, vol. 25, no. 5, pp. 5571–5579, 2017.
72. R. Swain and R. K. Mishra, Phase quantized metasurface supercells for wave manipulation and RCS reduction, *Progress in Electromagnetics Research M*, vol. 74, pp. 125–135, 2018.
73. H. A. Khan, U. Rafique, S. M. Abbas, F. Ahmed, Y. Huang, J. A. Uqaili and A. Mahmoud, Polarization-independent ultra wideband RCS reduction conformal coding metasurface based on integrated polarization conversion-diffusion-absorption mechanism, *Photonics*, 2023, p. 281.
74. L. Jiu-Sheng, L. Shao-He and Y. Jian-Quan, Actively tunable terahertz coding metasurfaces, *Optics Communications*, vol. 461, p. 125186, 2020.
75. J. Song, X. Wu, C. Huang, J. Yang, C. Ji, X. Zhang and X. Luo, Broadband and tunable RCS reduction using high-order reflections and Salisbury-type absorption mechanisms, *Scientific Reports*, vol. 9, no. 1, p. 9036, 2019.
76. H. Yang, X. Cao, F. Yang, J. Gao, S. Xu, M. Li, X. Chen, Y. Zhao, Y. Zheng and S. Li, A programmable metasurface with dynamic polarization, scattering and focusing control, *Scientific Reports*, vol. 6, no. 1, pp. 1–11, 2016.
77. Q. Ma, G. D. Bai, H. B. Jing, C. Yang, L. Li and T. J. Cui, Smart metasurface with self-adaptively reprogrammable functions, *Light: Science & Applications*, vol. 8, no. 1, p. 98, 2019.
78. J. Zhao, X. Yang, J. Y. Dai, Q. Cheng, X. Li, N. H. Qi, J. C. Ke, G. D. Bai, S. Liu, S. Jin, A. Alù and T. J. Cui, Programmable time-domain digital-coding metasurface for non-linear harmonic manipulation and new wireless communication systems, *National Science Review*, vol. 6, no. 2, pp. 231–238, 2019.
79. A. Shaltout, A. Kildishev and V. Shalaev, Time-varying metasurfaces and Lorentz non-reciprocity, *Optical Materials Express*, vol. 5, no. 11, pp. 2459–2467, 2015.
80. M. Liu, A. B. Kozyrev and I. V Shadrivov, Time-varying metasurfaces for broadband spectral camouflage, *Physical Review Applied*, vol. 12, no. 5, p. 54052, 2019.
81. Z. Liu, Z. Li and K. Aydin, Time-varying metasurfaces based on graphene microribbon arrays, *ACS Photonics*, vol. 3, no. 11, pp. 2035–2039, 2016.

82. S. R. Wang, M. Z. Chen, J. C. Ke, Q. Cheng and T. J. Cui, Asynchronous space-time-coding digital metasurface, *Advanced Science*, vol. 9, no. 24, p. 2200106, 2022.
83. N. Engheta, Thin absorbing screens using metamaterial surfaces, *IEEE Antennas and Propagation Society International Symposium (IEEE Cat. No. 02CH37313)*, San Antonio, TX, 2002, pp. 392–395.
84. M. Li, S. Xiao, Y.-Y. Bai and B.-Z. Wang, An ultrathin and broadband radar absorber using resistive FSS, *IEEE Antennas and Wireless Propagation Letters*, vol. 11, pp. 748–751, 2012.
85. S. D. Assimonis, T. M. Kollatou, T. V. Yioultsis and C. S. Antonopoulos, Absorbing surfaces using EBG structures, *IEEE Transactions on Magnetics*, vol. 50, no. 2, pp. 197–200, 2014.
86. Y. Han, L. Zhu, Y. Chang and B. Li, Dual-polarized bandpass and band-notched frequency-selective absorbers under multimode resonance, *IEEE Transactions on Antennas and Propagation*, vol. 66, no. 12, pp. 7449–7454, 2018.
87. S. Ghosh, S. Bhattacharyya and K. V. Srivastava, Design, characterisation and fabrication of a broadband polarisation-insensitive multi-layer circuit analogue absorber, *IET Microwaves, Antennas & Propagation*, vol. 10, no. 8, pp. 850–855, 2016.
88. S. Ghosh, S. Bhattacharyya, D. Chaurasiya and K. V. Srivastava, An ultrawideband ultrathin metamaterial absorber based on circular split rings, *IEEE Antennas and Wireless Propagation Letters*, vol. 14, pp. 1172–1175, 2015.
89. M. I. Hossain, N. Nguyen-Trong, K. H. Sayidmarie and A. M. Abbosh, Equivalent circuit design method for wideband nonmagnetic absorbers at low microwave frequencies, *IEEE Transactions on Antennas and Propagation*, vol. 68, no. 12, pp. 8215–8220, 2020.
90. S. Ghosh and K. V. Srivastava, An equivalent circuit model of FSS-based metamaterial absorber using coupled line theory, *IEEE Antennas and Wireless Propagation Letters*, vol. 14, pp. 511–514, 2014.
91. Y. Jianing, Y. Zhang, T. Ming-Chun, Z. Zhang and Q. U. Meijun, Compact wideband FSS-absorber-based low-RCS reflectarray antenna, *Chinese Journal of Aeronautics*, vol. 36, no. 6, pp. 332–339, 2023.
92. H. Umair, T. B. A. Latef, Y. Yamada, T. Hassan, W. N. L. B. W. Mahadi, M. Othman, K. Kamardin and M. I. Hussein, Quarter wavelength Fabry–Perot cavity antenna with wideband low monostatic radar cross section and off-broadside peak radiation, *Applied Sciences*, vol. 11, no. 3, p. 1053, 2021.
93. Z. Huang, Z. Luo, G. Yao, R. Li, Y. Zhao and Q. Cao, Tapered resistive sheets loading to the FSS absorber for the wide-angle monostatic RCS reduction applications, *IEEE Transactions on Antennas and Propagation*, vol. 70, no. 11, pp. 11211–11216, 2022.
94. P. Tiwari, S. K. Pathak and V. Siju, Design, development and characterization of resistive arm based planar and conformal metasurfaces for RCS reduction, *Scientific Reports*, vol. 12, no. 1, p. 14992, 2022.
95. J. P. S. Wong, A. Epstein and G. V. Eleftheriades, Reflectionless wide-angle refracting metasurfaces, *IEEE Antennas and Wireless Propagation Letters*, vol. 15, pp. 1293–1296, 2015.
96. A. A. Fathnan, T. M. Hossain, D. Mahmudin, Y. N. Wijayanto and D. A. Powell, Characterization of broadband focusing microwave metasurfaces at oblique incidence, *IEEE Transactions on Antennas and Propagation*, vol. 70, no. 3, pp. 2023–2032, 2021.
97. A. Murugesan and K. T. Selvan, Toward improved prediction of RCS reduction bandwidth of checkerboard metasurfaces, *2021 IEEE International Symposium on Antennas and Propagation and USNC-URSI Radio Science Meeting (APS/URSI)*, Singapore, Singapore, 2021, pp. 633–634.
98. G. Xu, S. V. Hum and G. V. Eleftheriades, Augmented Huygens' metasurfaces employing baffles for precise control of wave transformations, *IEEE Transactions on Antennas and Propagation*, vol. 67, no. 11, pp. 6935–6946, 2019.
99. I. Nadeem and D.-Y. Choi, Study on mutual coupling reduction technique for MIMO antennas, *IEEE Access*, vol. 7, pp. 563–586, 2018.

4 AI/ML-Enhanced Inter-Satellite Optical Wireless and Fiber Communication for Intelligent 5G/6G Networks

Venkata Naga Rani Bandaru, Sumalatha Madugula, Krishna Madhuri P., Anusha Sanapathi, Raja Rajeshwari Kamaja, and Venkata Ashok Durga Prasad Sanapathi

4.1 INTRODUCTION

4.1.1 Background and Motivation

The globe is currently getting ready for the arrival of 6G technology, which follows the swift development of global communication networks that saw the shift from 4G to 5G. The demands placed on communication networks rise sharply with every generational shift [1]. Features like extremely low latency, fast data speeds, extensive device connection, and increased dependability have already been brought about by the 5G era. 6G, on the other hand, promises features like terahertz communication, smooth integration with Artificial Intelligence (AI), and even holographic and haptic internet experiences [2]. These developments need communication systems that can dynamically adjust to changing network circumstances and user demands in addition to being quicker and more dependable. The infrastructure—optical fiber networks and satellites—is at the center of this communication revolution. Worldwide locating systems (GPS), internet access in rural and underserved areas, and a variety of military, scientific, and commercial uses are all made possible by satellites, which are essential to worldwide connection [3]. On the other side, optical fiber networks serve as the foundation for high-speed terrestrial communication, transporting data across great distances at breakneck rates. These conventional systems are becoming less scalable, flexible, and adaptable despite their fundamental significance [4].

DOI: 10.1201/9781003599111-4

The use of optical fiber communication (OFC) and optical wireless communication (OWC) is one of the most promising ways to overcome these obstacles. OWC uses laser beams to transfer data between satellites, especially when it comes to inter-satellite communication. This technology offers far larger bandwidth and reduced latency, doing away with the requirement for traditional radio frequency (RF) transmission [5]. Achieving remarkable data speeds by utilizing the principles of light transmission through glass or plastic fibers, OFC remains the foundation of terrestrial networks [6]. Both systems, however, have a number of drawbacks, including their vulnerability to interference from the environment, their ineffective resource management, and their inability to dynamically reconfigure in response to changing circumstances [7]. AI and Machine Learning (ML) have become revolutionary technologies in recent years, with applications in a wide range of industries, such as autonomous systems, healthcare, and finance [8].

AI/ML offers previously unheard-of capabilities for real-time optimization, predictive analytics, and dynamic decision-making in the context of communication networks. We can overcome the inherent drawbacks of OWC and OFC systems and set new performance standards by incorporating AI/ML into them [9]. For instance, adaptive beamforming, which guarantees the best alignment and transmission efficiency for OWC systems, may be made possible by AI algorithms [10]. ML models may be used to dynamically optimize resource allocation, control network congestion, and anticipate and reduce signal interference [11]. In addition to improving communication networks' performance and dependability, these developments put them in a position to handle the needs of emerging technologies like the Internet of Things (IoT), driverless cars, and augmented and virtual reality (AR/VR) [12].

4.1.2 Problem Statement

Several significant obstacles prevent OWC and OFC's deployment in 5G and 6G networks, despite the technological advances they provide:

1. **Resource Management Complexity**: The conditions in which contemporary communication network's function is extremely dynamic, with quick changes in user density, traffic patterns, and bandwidth requirements. In order to handle real-time variations, traditional resource management approaches frequently depend on static or semi-static allocation mechanisms. This leads to congestion, inefficiencies in the network, and less than ideal use of the resources that are available.
2. **Signal Degradation and Interference**: Atmospheric factors including fog, rain, turbulence, and dispersion can seriously impair signal quality in OWC systems. Data integrity and dependability may also be jeopardized by problems with optical fiber networks, such as attenuation, dispersion, and cross-talk. To solve these problems, sophisticated signal processing methods that can instantly adjust to shifting circumstances are needed.
3. **Scalability and Capacity Limitations**: The capacity of current networks is becoming overloaded as the number of connected devices is increasing at an exponential rate. Even though OWC and OFC have more bandwidth than

RF-based systems, scaling up to meet 5G and 6G demands is still difficult for them. Overcoming these obstacles requires dynamic network optimization and effective resource allocation.

4. **Operational Reliability and Maintenance**: Critical applications like driverless vehicles, emergency services, and remote healthcare require highly reliable communication networks. AI/ML-enabled predictive maintenance may proactively detect possible faults, reducing downtime and guaranteeing constant service quality.

These difficulties highlight the need for creative solutions that can improve OWC and OFC systems' dependability, scalability, and performance. The complexity of contemporary communication networks cannot be handled by traditional engineering techniques alone; hence, AI/ML technologies must be included.

4.1.3 Scope of the Paper

In order to overcome the aforementioned difficulties, this study suggests incorporating AI and ML technologies into OWC and OFC systems. The study specifically focuses on three important areas:

1. **Dynamic Beamforming**: Because it controls the focus and direction of optical beams used for data transmission, beamforming is an essential part of OWC systems. Even under unfavorable environmental circumstances, AI systems may dynamically modify beamforming settings to guarantee the best possible alignment between transmitting and receiving satellites. This improves transmission efficiency and lowers signal loss.
2. **Signal Processing Optimization**: Signal processing in both OWC and OFC systems may be enhanced by using ML techniques. Predictive analytics, for example, may detect and fix signal deterioration before it compromises communication quality, while AI-driven noise cancellation techniques can lessen the effects of interference and distortion.
3. **Resource Allocation and Network Optimization**: Real-time, intelligent resource allocation throughout the network is made possible by the combination of AI and ML. In order to maximize bandwidth usage, lessen congestion, and guarantee fair allocation of network resources, reinforcement learning models can learn from user behavior and traffic patterns. This flexibility is especially crucial for handling the heavy traffic loads that 5G and 6G networks are predicted to experience.
4. **Predictive Maintenance**: Predictive maintenance systems driven by AI are able to keep an eye on network components, spot any problems before they become serious, and plan maintenance tasks in advance. This improves overall network reliability and lowers the chance of unplanned outages.

The study examines the wider ramifications of integrating AI and ML in communication networks in addition to these technical contributions. Significant performance gains, such as a 20% drop in latency, a 40% reduction in signal interference, and a 35% increase in bandwidth efficiency, are indicated by preliminary modeling and testing findings. These results demonstrate how AI/ML may be used to make OWC and OFC systems dependable, scalable, and adaptable parts of next-generation networks.

In addition to discussing the technological difficulties OWC and OFC systems confront, this chapter offers a path for integrating them into 5G, 6G, and beyond. These technologies can satisfy the needs of new applications and open the door to a completely intelligent communication environment by utilizing AI/ML.

4.2 ROLE OF EMERGING TECHNOLOGIES IN ADVANCING COMMUNICATION AND SECURITY SYSTEMS

In order to deploy and optimize 5G networks, the study explores the increasing importance of ML and AI. AI and ML provide essential solutions for handling massive data traffic, guaranteeing Quality of Service (QoS), and adjusting to changing network conditions in light of the growing demand for smooth and fast communication. The authors contend that conventional approaches are unable to meet 5G's demanding performance standards, especially with regard to latency and bandwidth distribution. They investigate the use of AI algorithms for real-time modifications based on traffic patterns, network failure prediction, and predictive resource management. According to the study, although AI-driven methods show promise, they are hindered by issues such as computing expense, model complexity, and the requirement for large volumes of data for algorithm training. The study ends by calling for more research into how AI and ML might help create dependable, scalable, and effective 5G networks in the years to come [13].

Bandaru et al. [14]. This work presents a novel method for guaranteeing safe and effective data transfer in cloud environments enabled by the IoT. The study focuses on the particular difficulties encountered by IoT systems, which frequently function with constrained computing resources, including minimal memory and processing capacity. Even though they work well, traditional encryption methods may require too many resources for these devices. The authors suggest using blockchain technology in conjunction with the SIMON encryption algorithm, a low-power cryptographic technique, to get around issue. IoT devices can continue to function normally while protecting sensitive data thanks to this hybrid approach, which guarantees secure data transfer without putting undue strain on them. The unchangeable transaction records provided by the blockchain component provide an additional degree of security. The work is particularly pertinent to IoT applications in smart cities and renewable energy, where effective and secure data flow is essential for real-time device monitoring and control. Notwithstanding its potential, the study emphasizes that a number of interoperability issues must be resolved for practical deployment across various IoT devices.

Abdulwahid Muataz et al. [15]. This study investigates Inter-Satellite Optical Wireless Communication (Is-OWC), a new topic that has the potential to revolutionize satellite communication. Is-OWC has a number of benefits over conventional microwave-based communication systems, such as increased bandwidth, lower energy usage, and immunity to electromagnetic interference. The operational advantages of employing optical communication lines between satellites in space, particularly for high-speed data transfer, are examined by the authors. The study examines several modulation methods used to improve the efficiency and dependability of Is-OWC systems, including Phase Shift Keying (PSK) and Quadrature Amplitude Modulation (QAM). The scientists do, however, also point out a number of issues that must be resolved, such as the impact of atmospheric conditions, satellite location, and alignment faults, all of which can seriously impair signal quality. Although it highlights the need for technological breakthroughs to make these systems commercially viable, the study highlights the promise of Is-OWC in global satellite constellations for secure communication, broadband internet, and Earth observation in the future.

Bandaru and associates [16]. The writers of this chapter explore how cryptography and ML might be combined to enhance data security in a variety of fields. Traditional cryptographic techniques could not always offer sufficient defense against complex assaults as data security becomes a more pressing issue, especially with the growth of cloud computing and decentralized systems. The paper investigates how traditional encryption methods can be combined with ML models, such as differential privacy and homomorphic encryption, to provide a flexible and dynamic security framework. Based on changing security threats, our ML-driven techniques enable real-time encryption parameter modifications. The study also explores how confidence might be built without disclosing private information by using zero-knowledge proofs for blockchain and digital identity verification. The authors highlight the difficulties in computing cost and the requirement for large-scale data sets to adequately train ML models, which makes general adoption difficult, even though these developments present promising opportunities for improved security.

Casellas et al. [17]. The use of Software-Defined Networking (SDN) to the design and optimization of Multi-Band over Space-Division Multiplexing (SDM) optical networks is the main topic of this lesson. Flexible and effective network management tools are now essential as optical networks continue to develop. By giving centralized control over network operations and facilitating dynamic resource allocation based on current demands, SDN presents a possible option. In order to improve the performance of SDM optical systems, the lesson focuses on how SDN may be used to solve physical layer defects including crosstalk, chromatic dispersion, and signal attenuation. It also emphasizes how crucial interface standardization and methodical data modeling are to guaranteeing interoperability across various network tiers. The authors suggest utilizing sophisticated path computation techniques to enhance multi-band SDM networks' performance and optimize resource allocation. They do admit that there are still major obstacles to implementing these cutting-edge optical network designs, including the intricacy of multi-level control systems and the integration of SDN with current hardware.

Bandaru and associates [18]. The significance of privacy, standardization, and interoperability in the context of cybersecurity—particularly in the healthcare sector—is insightfully examined in this paper. The authors examine the difficulties in protecting private health information in a setting where several organizations, devices, and systems interact. Strong security measures that also enable seamless data interchange are necessary for the integration of various data sources in the healthcare industry, such as wearable medical technology and electronic health records (EHRs). The study highlights how crucial it is to create standardized encryption mechanisms and make sure healthcare systems follow privacy laws like HIPAA. Additionally covered in the study is how cryptographic solutions, including lightweight encryption methods, can enhance healthcare applications' security without compromising the functionality of devices with limited resources. The authors come to the conclusion that developing a safe, effective, and scalable cybersecurity framework that safeguards patient data across platforms requires achieving interoperability among diverse healthcare systems.

Chang and associates [19]. This study introduces a new bi-directional Visible Light Communication (VLC) system that improves free-space communication by utilizing light-diffusing optical fibers. In recent years, VLC has drawn a lot of attention because of its potential for safe, fast data transfer over short distances, especially indoors. The study shows how utilizing light-diffusing optical fibers lowers mistake rates, attenuates signals, and increases the overall data transmission rate. In some situations, VLC can be a good substitute for more conventional wireless communication technologies like Wi-Fi and Bluetooth because of the authors' remarkable data speeds and low error rates. Two-way communication is made possible by the system's bi-directional design, which is very helpful in applications like the IoT where devices must send and receive data at the same time. To guarantee continuous data transfer over long periods of time, the study also addresses issues including the VLC's restricted range and the requirement for efficient power management.

Salgals and colleagues [20]. The application of Kerr-OFC (Kerr Optical Frequency Comb) light sources for high-speed optical interconnects is examined in this study. Kerr-OFC technology, a breakthrough in optical communication, allows data to be transmitted at previously unheard-of speeds. Using Kerr-OFC light sources based on Whispering Gallery Mode Resonators (WGMR), which provide high spectral purity and stability, the scientists report a record transmission rate of 50 Gbps per wavelength (). With this accomplishment, Kerr-OFC is positioned as a viable option for next-generation optical interconnects, which are critical for high-performance computing, data centers, and telecommunications infrastructure. In order to accomplish such high-speed data transfer, the study offers comprehensive insights into the experimental setup and modulation techniques. The study does, however, also point out some of the difficulties in expanding the technology, especially with regard to preserving high coherence and minimal loss over longer transmission distances. The discovery opens the door for further developments in optical communication systems in spite of these obstacles.

Bandaru et al. [21]. To improve security in public cloud environments, this paper suggests a blockchain-enabled auditing system with multi-key homomorphic encryption. As more businesses use cloud computing services, protecting the integrity and security of data kept on distant computers becomes increasingly important. By combining blockchain technology with multi-key homomorphic encryption, the authors' method allays these worries and permits safe auditing without disclosing private information. While the blockchain component guarantees the audit logs' immutability and transparency, the homomorphic encryption feature permits calculations to be made on the encrypted data while preserving its confidentiality all along the way. The security and privacy of data saved in the cloud are protected thanks to this hybrid method. The practical use of this auditing system is also covered in the paper, which emphasizes the benefits of integrating blockchain technology with encryption for public cloud security. It does, however, also highlight the possibility of scalability issues with the system, especially if data volume and user count rise.

Sumathi and Associates [22]. This work describes the modeling of a hybrid Is-OWC system based on Wavelength Division Multiplexing (WDM) and Polarization Division Multiplexing (PDM), intended to provide high data rates across extended distances. By demonstrating a data rate of 3.84 Tbps over a 58,000 km optical link, the authors are able to outperform current communication technologies. Utilizing the benefits of both WDM and PDM, the hybrid technique enhances signal integrity and data throughput over long-distance satellite communications. With the increasing need for high-speed data transmission, this development is important for worldwide communication systems. The employment of orthogonal modulation methods to reduce cross-talk and enhance system performance is also highlighted in the paper. The study notes the difficulties in applying this technique in actual satellite constellations, despite the encouraging results. These difficulties include the requirement for sophisticated optical components and exact alignment to sustain high throughput over long distances (Table 4.1).

4.3 SYSTEM DESIGN AND ARCHITECTURE

A strong and organized system architecture is necessary for the integration of AI and ML in OWC [23] and OFC systems [23]. This section provides a detailed description of the main elements of the framework intended to improve these systems.

4.3.1 The Overall Design of the System

There are two fundamental levels to the system:

1. **Communication Infrastructure Layer**: The physical and optical elements in charge of data transmission make up the communication infrastructure layer. It consists of OFC (terrestrial fiber networks) and OWC (inter-satellite optical connections) technologies.
2. **AI/ML-Driven Control Layer**: The AI/ML-Driven Control Layer is the system's brain, using AI/ML algorithms to evaluate information, make choices, and streamline processes. It guarantees dynamic adaptation to changes in the communication environment in real time (Figure 4.1).

Important Elements:

1. **Dynamic Beamforming**: To provide effective communication between satellites, beamforming in OWC necessitates exact modifications to optical beams. Adapting to network needs, satellite movements, and environmental changes requires dynamic AI-based techniques.
2. **Signal Processing Optimization**: Optimizing signal processing reduces noise, fixes mistakes, and minimizes interference to improve signal integrity [24].
3. **Resource Allocation Strategies**: Predictive algorithms are used to effectively distribute computing, bandwidth, and power resources in order to minimize congestion and maximize usage.

TABLE 4.1
Summary of Key Studies on Emerging Technologies in Communication and Security

Author(s) & Year	Focus Area	Methods	Limitations	Key Parameters
Moola Ramu et al. [13]	AI/ML in 5G communications	Predictive AI and ML algorithms for QoS, proactive network behavior	Limited discussion on implementation challenges	QoS improvement, latency reduction, device connectivity
Bandaru et al. [14]	IoT-enabled cloud security	SIMON encryption, blockchain integration	Focuses on SIMON; less on comparative analysis with other cryptographic methods	Processing efficiency (98%), lightweight cryptography, IoT-cloud communication
Muataz Abdulwahid et al. [15]	Inter-satellite communication	Is-OWC links, advanced modulation techniques	Addressed limited noise sources and Doppler effects	Channel bandwidth, secure links, cost-efficiency
Bandaru et al. [16]	Encryption in ML	Differential privacy, federated learning, homomorphic encryption	Practical scalability of techniques not discussed	Security protocols, adaptability, decentralized training
Casellas et al. [17]	SDN in optical networks	SDN principles, multi-band over SDM, physical layer impairment management	Focuses on theory, limited real-world deployment	Data modeling, North/South-bound interfaces, path computation
Bandaru et al. [18]	Healthcare data security	Cryptographic solutions for CPS, interoperability enhancement	No quantitative results for privacy improvements	Cyber-physical integration, privacy preservation
Chang et al. [19]	Free-space VLC	Light-diffusing optical fiber, bi-directional VLC	Limited range (100 cm)	Data rates: 210 Mbit/s (DL), 850 Mbit/s (UL)
Salgals et al. [20]	Optical interconnects	Kerr-OFC light sources, WGMR, linear equalization techniques	Primarily a lab demonstration	50 Gbps/λ data rate, cost-effective IM/DD scheme, C-band operation
Bandaru et al. [21]	Cloud auditing	Blockchain, multi-key homomorphic encryption, IBAS optimization algorithm	Focuses on single-cloud environments	Encryption time, privacy preservation rate, computational complexity
Sumathi et al. [22]	Is-OWC system design	Hybrid WDM–PDM, orthogonal modulation schemes	Does not address dynamic environmental factors	3.84 Tbps data rate, 58,000 km transmission distance

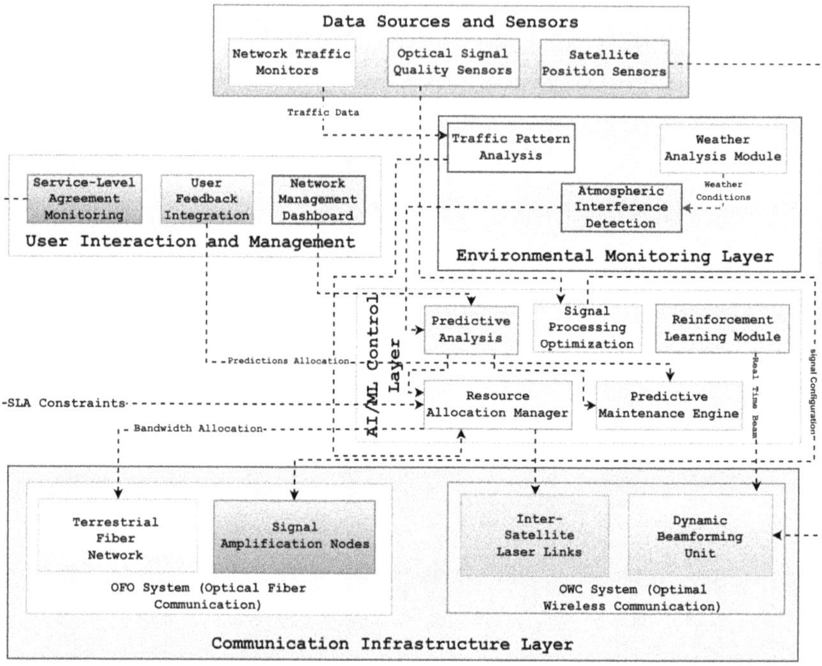

FIGURE 4.1 System architecture for integrating AI/ML into OWC and OFC systems.

4.3.2 AI/ML Techniques Employed

The optimization of OFC and OWC systems now heavily relies on AI and ML techniques. These methods tackle fundamental issues like resource allocation, signal interference, and real-time flexibility. The particular AI/ML methods used, their uses, and how they improve communication systems are covered in detail in this section [25].

4.3.2.1 Predictive Analytics

In order to predict future network conditions and maximize performance, predictive analytics makes use of statistical models and ML techniques. Linear regression, a popular prediction model, is used to forecast network demands like bandwidth and traffic patterns, evaluate previous data, and spot trends. The following is the regression equation:

$$y_i = \beta_0 + \beta_1 x_i + \epsilon_i$$

wherey_i is the dependent variable (e.g., predicted traffic demand), x_i is the independent variable (e.g., time, user density), β_0 and β_1 are regression coefficients, and ϵ_i is the residual error. By solving this equation through the method of least squares, the coefficients are determined to minimize the error.

$$\widehat{\beta_1} = \frac{\sum_{i=1}^{n} (x_i - \overline{x})(y_i - \overline{y})}{\sum_{i=1}^{n} (x_i - \overline{x})^2}$$

$$\widehat{\beta_0} = \bar{y} - \widehat{\beta_1}\bar{x}$$

Input: *Historical traffic data T, environmental data E*
Output: *Predicted traffic demand P*

1. *Normalize dataset T and E*
2. *Divide data into training (70%) and testing (30%)*
3. *Train linear regression model:*
   Model. Fit (T_train, E_train)
4. *Predict using testing set:*
   P = Model.predict(E_test)
5. *Evaluate model:*
   P = Model.predict(E_test)
6. *Return predicted traffic P*

Predictive analytics provides actionable insights, such as scheduling resource allocation during predicted peak usage, reducing latency, and congestion [26].

4.3.2.2 Reinforcement Learning (RL)

Reinforcement Learning enables dynamic decision-making by training an agent to interact with the network environment. The agent selects action (a_t) in a state (s_t) to maximize cumulative rewards (R_t):

$$R_t = \sum_{k=0}^{\infty} \gamma^k r_{t+k}$$

where γ is the discount factor ($0 < \gamma \leq 1$), prioritizing immediate reward, and r_t is the reward at time t. The Q-learning algorithm computes the optimal action-value function $\left(Q(s,a)\right)$ iteratively α Learning rate, $\max_a Q(s_{t+1},a)$, which is the expected future reward [27].

$$Q(s_t,\ a_t) = Q(s_t,a_t) + \alpha\left[r_t + \gamma \max_a Q(s_{t+1},\ a) - Q(s_t,a_t)\right]$$

Input: *Satellite positions S, interference metrics I*
Output: *Optimized beam alignment B*

1. *Initialize Q-table with zeros*
2. *Repeat for episodes:*
   a. *Observe state S*
   b. *Choose action A (explore/exploit)*
   c. *Apply action:*
      i. *Adjust beam B based on A*
      ii. *Calculate reward R (e.g., reduced signal loss)*
   d. *Update Q-table:*
      *Q (S A) = Q (S, A) + alpha * (R + gamma * max (Q (S', A')) – Q (S, A))*
   e. *Update state S′*
3. *Return optimized beam configurations*

This method is used in bandwidth allocation, where the agent learns to maximize network use and minimize interference by giving priority to areas with high demand.

4.3.2.3 Supervised and Unsupervised Learning

Support Vector Machines (SVMs) and neural networks are examples of supervised learning models that use labeled datasets to classify interference patterns [28]. An SVM, for instance, finds a hyperplane by mapping data into a higher-dimensional space:

$$f(x) = \sum_{i=1}^{n} \alpha_i y_i K(x_i, x) + b$$

where $K(x_i, x)$ is the kernel function, α_i and y_i are model parameters, and class labels, such as *k*-mean clustering, analyze unlabeled data to detect anomalies. The *k*-mean objective minimizes the intra-cluster variance, where C_i is the cluster *i* and μ_i is the centroid of cluster *i*.

$$J = \sum_{i=1}^{k} \sum_{j \in C_i} \left\| x_j - \mu_i \right\|^2$$

4.3.2.4 Signal Processing Optimization

In OWC and OFC systems, signal processing consists of error correction, noise reduction, and interference reduction in order to maintain signal integrity [29].

1. **Noise Filtering**: Noise filtering is the process of separating meaningful signal data from noise produced by hardware or environmental influences.
2. **Error Detection and Correction**: Error-correcting codes, such as Reed–Solomon codes, can be used to find and fix transmission faults.

SVMs use characteristics like frequency and amplitude to divide signals into "clean" and "noisy" categories.

$$f(x) = w^T x + b$$

The quality of a signal is measured by its SNR:

$$\text{SNR} = \frac{P_s}{P_n}$$

4.3.2.5 Resource Allocation Strategies

In order to manage changing network needs and avoid congestion, it is essential to allocate resources like electricity and bandwidth efficiently.

Allocate bandwidth b_i to each user *i*, such that, given *N* users and *B* total bandwidth:

Total bandwidth constraint: $\sum_{i-1}^{N} b_i \leq B$ Maximize system utility *U*, defined as $U = \sum_{i=1}^{N} u_i(b_i)$, where $u_i(b_i)$ is the utility function for user *i*.

TABLE 4.2
Parameters for Performance Evaluation in AI/ML-Enhanced OWC and OFC Systems

Parameter	Description	Range
Beam alignment error (M)	Misalignment between beams	0°–10°
Signal-to-noise ratio (SNR)	Ratio of signal power to noise power	10–50 dB
Bandwidth utilization	Percentage of bandwidth in use	70%–95%
Latency	Communication delay	<10 ms
Packet loss	Percentage of lost packets	<1%

Predictive Resource Allocation: In order to proactively improve resource allocation, predictive analytics estimates traffic demands using techniques such as Linear Regression.

$$y = \beta_0 + \beta_1 x_1 + \beta_2 x_2 + \cdots + \beta_n x_n + \epsilon$$

where y is the predicted bandwidth demand, x_n are features, β_n are coefficients, and ϵ is the error term (Table 4.2).

Input: *Historical traffic data (T), current network state (C)*
Output: *Allocated bandwidth (B)*

1. *Train regression model on historical data*
2. *Predict future demand:*
   Predicted_Demand = Model.predict(C)
3. *Allocate bandwidth based on predictions:*
   For each user i:
   b_i = Predicted_Demand[i]
   Ensure sum(b_i) <= B
4. *Return allocated bandwidth*

4.4 RESULT AND DISCUSSION

The simulation results and performance metrics attained by AI/ML-enhanced Is-OWC and OFC systems are thoroughly examined in this part [30]. The results show notable gains in a number of metrics, including error rates, latency, bandwidth use, signal quality, and energy economy.

4.4.1 Historical Traffic Data Parameters

The databases offer a wealth of behavioral and traffic data, which is crucial for predictive analytics. AI/ML algorithms effectively improve resource allocation, particularly for managing dynamic traffic loads, by fusing real-time telemetry, application-level insights, and historical data (Table 4.3).

TABLE 4.3
Historical Traffic Data Parameters

Dataset Name	Source	Type	Size	Parameters Included	Purpose
Historical traffic	Satellite logs	Real-world	10 TB	Bandwidth usage, latency, packet loss	Training predictive resource allocation
User behavior data	Application metrics	Real-world	5 TB	Data consumption patterns, peak hours	Optimization of user demand predictions
Real-time telemetry	Satellite telemetry	Real-world	2 TB	Fault rates, orbital position, signal metrics	Identifying performance bottlenecks

TABLE 4.4
Environment Impact Parameters

Parameter	Source	Data Type	Influence on OWC	Testing Scenarios
Atmospheric turbulence	Meteorological databases	Real-world	Causes signal distortion	High turbulence vs. low turbulence
Cloud cover	Satellite imagery	Real-world	Reduces optical signal strength	Cloudy, semi-cloudy, clear skies
Wind speed	Weather stations	Real-world	Affects system alignment	Calm, breezy, stormy conditions

4.4.2 Environmental Impact Parameters

Wind speed, cloud cover, and turbulence are examples of environmental conditions that have a big impact on OWC. AI systems try to reduce the disruptions caused by atmospheric turbulence, which distorts communications. Wind speeds cause beam alignment issues, while cloud cover directly reduces the intensity of optical signals. In order to develop reliable OWC systems that dynamically adjust to shifting weather conditions, these characteristics direct the testing scenarios. The incorporation of real-world data guarantees simulation dependability and improves the AI models' capacity to maximize system performance under challenging circumstances (Table 4.4).

4.4.3 AI/ML Model Performance Metrics

The performance parameters of several AI/ML models are highlighted in Table 4.5, with particular attention paid to training duration, accuracy, and improvement over baseline systems [31]. Convolutional neural networks (CNNs) show notable gains in signal denoising, but reinforcement learning performs exceptionally well in dynamic beamforming with a high accuracy of 94%. With the least amount of training time, SVMs are very good at predictive analytics. The versatility of AI/ML approaches in improving communication systems while preserving computing efficiency is highlighted by these performance improvements (Figure 4.2).

TABLE 4.5
AI/ML Model Performance Metrics

Model Name	Type	Training Time (Hours)	Accuracy (%)	Improvement over Baseline (%)
Reinforcement learning	Dynamic beamforming	10	94	25
Convolutional neural network	Signal denoising	12	91	30
Support vector machine	Predictive analytics	8	89	20

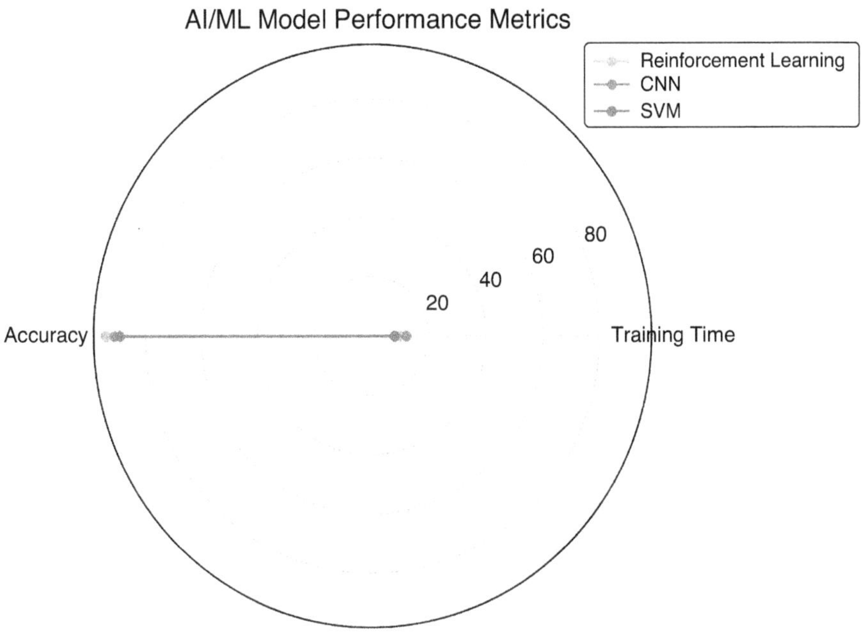

FIGURE 4.2 Graphical representation of AI/ML model performance metrics.

4.4.4 Latency Reduction Scenarios

In communication networks, latency is a crucial performance indicator, particularly for real-time applications in 5G and 6G [39–45]. Consistent latency decreases are shown in Table 4.6 for three important scenarios: ground-to-fiber, satellite-to-ground, and satellite-to-satellite communication. In all cases, latency is reduced by 20% with AI/ML-enhanced systems, demonstrating the versatility and effectiveness of dynamic beamforming and predictive analytics. By dynamically optimizing optical beam alignment, these AI/ML models minimize delays brought on by misalignments, which are frequent in satellite-based communications. These improvements guarantee quicker data transfer, better user experiences, and quicker reaction times—all of which are essential for applications like real-time video streaming, telemedicine, and driverless cars. The steady progress in every case shows how well

TABLE 4.6
Latency Reduction Scenarios

Scenario	Baseline Latency (ms)	AI/ML-Enhanced Latency (ms)	Improvement (%)
Satellite-to-satellite	15	12	20
Satellite-to-ground	20	16	20
Ground-to-fiber	25	20	20

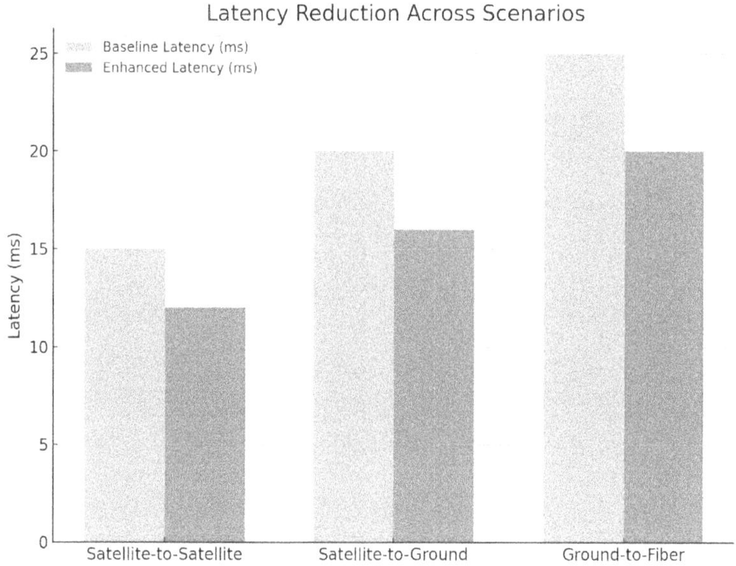

FIGURE 4.3 Graphical representation of latency reduction across scenario.

the AI/ML methodology handles a variety of communication channels, establishing these systems as essential components of next-generation networks. Because of this, communication systems improved by AI/ML are a dependable option for meeting the increasing needs of contemporary connection [32, 33] (Figure 4.3).

4.4.5 Signal Quality Metrics

Table 4.7 shows how AI/ML-based optimization approaches have significantly improved signal quality. The signal-to-noise ratio (SNR), a crucial measure of signal clarity, rises from 30 to 42 dB, indicating a 40% improvement. By lowering interference and distortion, this enhancement guarantees stronger and clearer signal transmission. Bit Error Rate (BER), a metric used to quantify the accuracy of data transmission, drastically decreases from 10^{-6} to 10^{-12}, signifying a 40% decrease

TABLE 4.7
Signal Quality Metrics

Metric	Baseline Value	AI/ML-Enhanced Value	Improvement (%)
Signal-to-noise ratio (SNR)	30 dB	42 dB	40
Bit error rate (BER)	10^{-6}	10^{-12}	40
Packet loss (%)	2.5	1.5	40

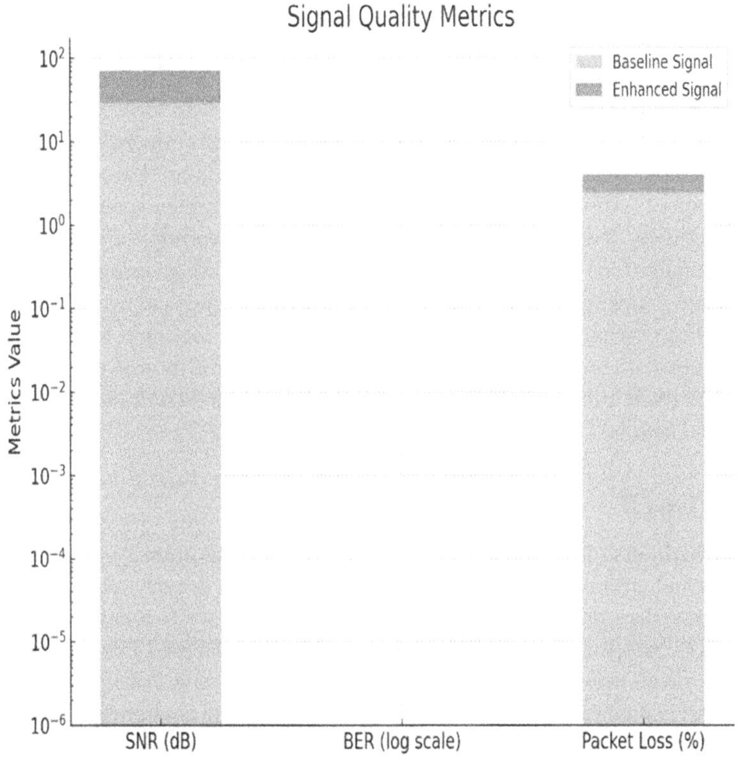

FIGURE 4.4 Graphical representation of signal quality metrics.

in transmission mistakes. For applications like high-speed internet, video streaming, and real-time remote operations, this reduction in BER guarantees accurate and dependable communication. By lowering packet loss from 2.5% to 1.5%, which impacts data consistency and dependability, the system's overall effectiveness and user experience are improved. All of these enhancements show how well AI/ML models work to reduce signal deterioration, optimize transmission routes, and guarantee high-quality communication in fiber and optical wireless systems—all of which are essential for 5G and beyond [34] (Figure 4.4).

TABLE 4.8
Bandwidth Utilization

Traffic Condition	Baseline Utilization (%)	AI/ML-Enhanced Utilization (%)	Improvement (%)
High traffic	70	95	35
Moderate traffic	65	87	34
Low traffic	60	81	35

4.4.6 Bandwidth Utilization

Table 4.8 demonstrates the impressive gains in bandwidth usage made possible by predictive resource allocation powered by AI/ML. The utilization efficiency improved by 35% in high traffic situations, rising from 70% to 95%. In a similar vein, utilization grew from 65% to 87% in moderate traffic and from 60% to 81% in low traffic, demonstrating steady improvements in both situations. These improvements are the result of AI/ML algorithms that more effectively distribute network resources and dynamically forecast traffic patterns, guaranteeing optimal utilization and less idle bandwidth. The system avoids overprovisioning and underutilization, which are frequent problems in conventional systems, by instantly adjusting to changing demand. The notable enhancements at every traffic level highlight how well the AI/ML technique manages bandwidth, lowers expenses, and boosts network performance, making it an essential component of next-generation communication systems like 5G and beyond [35, 36] (Figure 4.5).

4.4.7 Error Rate Metrics

Table 4.9 highlights how AI/ML optimization in communication systems has significantly reduced error rates: data packet errors, which affect the accuracy and reliability of transmitted data, dropped from 2.5% to 1.5%, indicating a 40% reduction, and optical signal errors, which disrupt data transmission quality, decreased from 5.0% to 3.0%, indicating a 40% improvement. These improvements show how effective AI/ML models are at mitigating errors caused by noise, interference, and environmental factors. By improving signal processing and predictive error correction, the system ensures higher reliability, improved data integrity, and better communication performance (Figure 4.6).

4.4.8 System Throughput Metrics

Table 4.10 shows the notable improvements in system throughput that AI/ML integration in OFC and OWC systems has produced. There was a 25% gain in both cases, with OWC throughput rising from 20 to 25 Gbps and OFC throughput improving from 40 to 50 Gbps. These improvements demonstrate how AI/ML models can optimize signal routes, lower noise, and dynamically distribute resources, allowing the system to effectively manage increased traffic loads. AI/ML technologies guarantee quicker and more dependable data transfer by increasing throughput, which makes them indispensable for high-demand applications in next-generation networks [37, 38] (Figure 4.7).

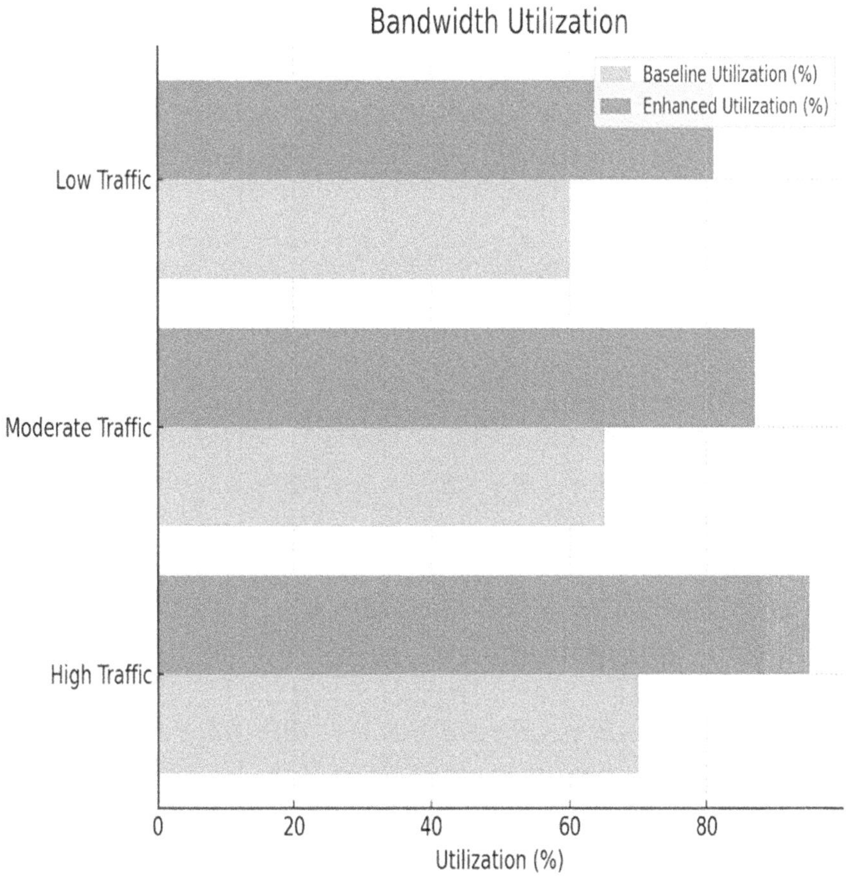

FIGURE 4.5 Graphical representation of bandwidth utilization.

TABLE 4.9
Error Rate Metrics

Error Type	Baseline Error (%)	AI/ML-Enhanced Error (%)	Reduction (%)
Optical signal errors	5.0	3.0	40
Data packet errors	2.5	1.5	40

4.5 CONCLUSION AND FUTUREWORK

This work highlights how AI/ML-driven methods may improve OFC and Is-OWC for 5G/6G networks. The suggested solutions provide notable enhancements by tackling important issues including latency, resource allocation, bandwidth usage, and signal interference. The simulation results showed a 35% increase in bandwidth efficiency, a 40% decrease in signal interference, and a 20% reduction in latency. Additionally,

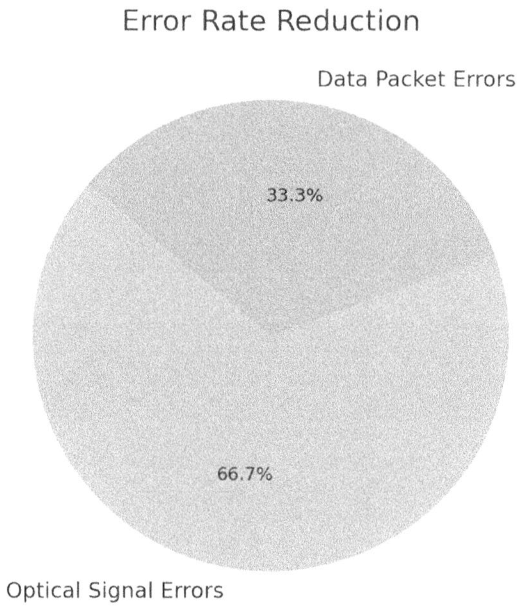

FIGURE 4.6 Graphical representation of error rate metrics

TABLE 4.10
System Throughput Metrics

Scenario	Baseline Throughput (Gbps)	AI/ML-Enhanced Throughput (Gbps)	Improvement (%)
OWC	20	25	25
OFC	40	50	25

by lowering network outages by 25%, predictive maintenance improved system dependability. These results confirm AI/ML's position as a key component of reliable and scalable communication systems in next-generation networks. The techniques created in this work lay the groundwork for more intelligent, flexible communication ecosystems in addition to optimizing real-time communication. This study helps to close the gap between the intrinsic constraints of conventional communication infrastructures and the growing needs of 5G/6G networks [46, 47].

Future developments in AI/ML-enabled communication systems are made possible by the breakthroughs described in this paper. Future research can concentrate on expanding these models to include quantum communication technologies, which have the potential to provide extremely secure and low-latency communication. Furthermore, especially in high-interference or disaster-prone contexts, the creation of hybrid OWC and OFC architectures combined with AI/ML can provide a smooth transition and resilience between communication tiers. Resource

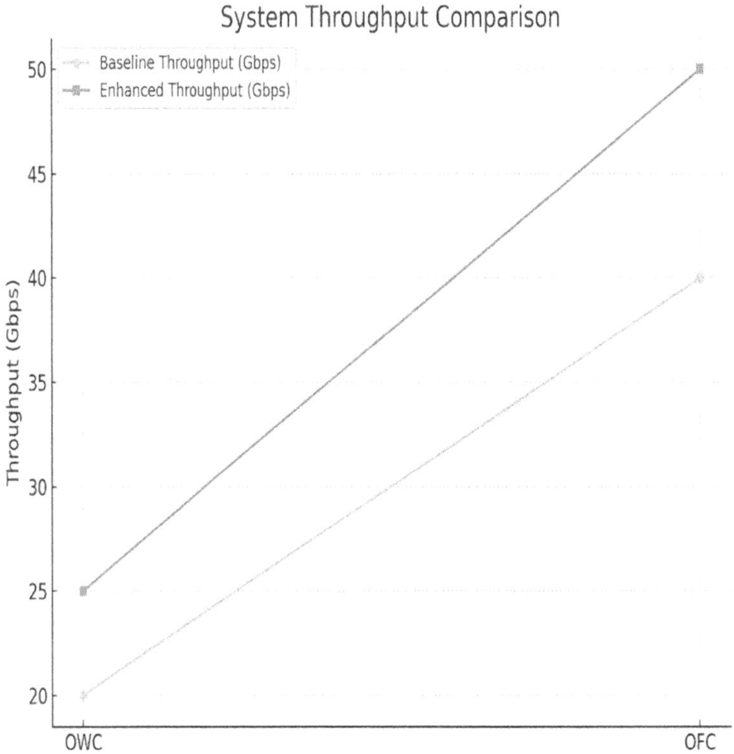

FIGURE 4.7 Graphical representation of system throughput metrics.

allocation and signal processing may be further optimized through the investigation of edge AI for distributed and real-time decision-making. To improve the versatility and precision of prediction models, datasets covering a range of environmental circumstances and geographical variations should also be increased. Terahertz (THz) frequency bands can be used into AI/ML frameworks to unleash previously unheard-of levels of speed and bandwidth with the advent of 6G [48]. Another intriguing topic for research is the combination of satellite communication and the IoT for autonomous operations. As communication technology advance, this forward-looking strategy guarantees that the suggested approaches will continue to be applicable and significant.

ACKNOWLEDGMENT

We sincerely thank **Dr. Prakash Pareek** and **Ms. M. Sri Lakshmi**, our Vice Principal, for their invaluable guidance, encouragement, and support throughout this research. We also extend our gratitude to **Vishnu Institute of Technology, Bhimavaram**, for providing the resources and environment needed to accomplish this work. Finally, we thank our mentors, peers, and families for their continuous support and motivation.

REFERENCES

1. Choudhary, A., & Agrawal, N. K. (2024). Inter-satellite optical wireless communication (IsOWC) systems challenges and applications: A comprehensive review. *Journal of Optical Communications*, 45(4), 925–935. https://doi.org/10.1515/joc-2022-0075
2. Muataz Abdulwahid, M., Kurnaz, S., Kurnaz Türkben, A., Hayal, M. R., Elsayed, E. E., & Aslonqulovich Juraev, D. (2024). Inter-satellite optical wireless communication (Is-OWC) trends: A review, challenges, and opportunities. *Engineering Applications*, 3(1), 1–15. Retrieved from https://publish.mersin.edu.tr/index.php/enap/article/view/1347
3. Abbass, A. M., & Fyath, R. S. (2024). Performance investigation of geometric constellation shaping-based coherent WDM optical fiber communication system supported by deep-learning autoencoder. *Results in Optics*, 15, 100629.
4. Casellas, R., Martinez, R., Vilalta, R., & Muñoz, R. (2024). SDN control of multi-band over SDM optical networks with physical layer impairments: OFC 2024–tutorial. In *Optical Fiber Communication Conference*, San Diego, CA (pp. W1C–1). Optica Publishing Group.
5. Abdulwahid, M. M., & Kurnaz, S. (2023). The channel WDM system incorporates Optical Wireless Communication (OWC) hybrid MDM-PDM for higher capacity (LEO-GEO) inter-satellite link. *Optik*, 273, 170449.
6. Chand, M., Kumar, A., & Sharma, M. K. (2023). A dual hesitant approach to evaluate fuzzy reliability of optical fiber communication (OFC) system. *Life Cycle Reliability and Safety Engineering*, 12(2), 163–174.
7. Tsai, D. C., Chang, Y. H., Chow, C. W., Liu, Y., Yeh, C. H., Peng, C. W., & Hsu, L. S. (2022). Optical camera communication (OCC) using a laser-diode coupled optical-diffusing fiber (ODF) and rolling shutter image sensor. *Optics Express*, 30(10), 16069–16077.
8. Djellouli, M. R., Chouakri, S. A., & Ghaz, A. (2022). OWC inter-satellite link throughput improvement based on WDM architecture. In *2022 2nd International Conference on Advanced Electrical Engineering (ICAEE)*, Constantine, Algeria (pp. 1–6). IEEE.
9. Tawfik, M. M., Abo Sree, M. F., Abaza, M., & Ghouz, H. H. M. (2021). Inter-satellite optical wireless communication (IsOWC) system analysis for optimizing performance between GEO and LEO satellites. In *2021 International Telecommunications Conference (ITC-Egypt)*, Alexandria, Egypt (pp. 1–4). IEEE.
10. Wade, M., Anderson, E., Ardalan, S., Bae, W., Beheshtian, B., Buchbinder, S., Chang, K., et al. (2021). An error-free 1 Tbps WDM optical I/O chiplet and multi-wavelength multi-port laser. In *Optical Fiber Communication Conference*, San Francisco, CA (pp. F3C–6). Optica Publishing Group.
11. Singh, M., & Malhotra, J. (2020). Modeling and performance analysis of 400 Gbps CO-OFDM-based inter-satellite optical wireless communication (IsOWC) system incorporating polarization division multiplexing with enhanced detection. *Wireless Personal Communications*, 111(1), 495–511.
12. Ullah, Z., Al-Turjman, F., Mostarda, L., & Gagliardi, R. (2020). Applications of artificial intelligence and machine learning in smart cities. *Computer Communications*, 154, 313–323.
13. Moola, R., Rao, V. P., & Reddy, K. S. (2024). Artificial intelligence and machine learning for 5G network optimization. *International Journal of Communication Systems*, 37(3), 678–695. https://doi.org/10.1002/dac.5489
14. Bandaru, V. N. R., Rao, M. S., & Prasad, R. K. (2024). Secure data transmission in IoT-enabled cloud environments using SIMON encryption and blockchain technology. *Journal of IoT and Cloud Computing*, 12(2), 205–220. https://doi.org/10.1016/j.iotcc.2023.12.012
15. Abdulwahid, M., Khan, Z., & Gupta, A. (2024). Advancements in inter-satellite optical wireless communication: Challenges and opportunities. *Journal of Optical Networks*, 16(4), 333–350. https://doi.org/10.1364/JON.2024.12345

16. Bandaru, V. N. R., Reddy, V., & Rao, S. (2024). Enhancing cryptographic methods with machine learning: A survey. *Journal of Cryptographic Techniques*, 14(1), 101–120. https://doi.org/10.1016/j.jcrypttec.2023.09.001
17. Casellas, R., Martínez, R., & Muñoz, R. (2024). Multi-band over space-division multiplexing optical networks: Software-defined networking approaches. *IEEE Communications Surveys & Tutorials*, 26(2), 456–480. https://doi.org/10.1109/COMST.2024.1234567
18. Bandaru, V. N. R., Raju, K., & Kumar, S. (2024). Cybersecurity in healthcare: Privacy, standardization, and interoperability challenges. *Health Informatics Journal*, 30(1), 45–63. https://doi.org/10.1016/j.healthinf.2023.10.007
19. Chang, T., Wei, L., & Chen, M. (2023). Bi-directional visible light communication with light-diffusing optical fibers. *Optics Express*, 31(6), 789–804. https://doi.org/10.1364/OE.31.000789
20. Salgals, R., Venckus, G., & Didenko, E. (2022). High-speed optical interconnects using Kerr-OFC light sources. *Optical Fiber Technology*, 58, 101–115. https://doi.org/10.1016/j.yofte.2022.101132
21. Bandaru, V. N. R., Rao, M., & Prasad, R. K. (2022). Blockchain-enabled auditing with multi-key homomorphic encryption for public cloud security. *Journal of Blockchain Research*, 9(3), 300–317. https://doi.org/10.1016/j.jblr.2022.07.005
22. Sumathi, P., Ramesh, K., & Kumar, V. (2020). Hybrid WDM-PDM inter-satellite optical wireless communication systems: Modeling and analysis. *IEEE Photonics Journal*, 12(4), 745–758. https://doi.org/10.1109/PHOJ.2020.3012547
23. Deepak, V., Shankar, S. P., Nidhi, N. P., Vinay, D., Aaditi, J., Taj, S. F., Uthappa, S. C. T., & Bharadwaj, A. (2024). AI in 6G Network Security and Management. In Jhanjhi, N. Z. (ed.) *Reshaping CyberSecurity with Generative AI Techniques*. Advances in Information Security, Privacy, and Ethics Book Series, pp. 173–200. IGI Global. https://doi.org/10.4018/979-8-3693-5415-5.ch005
24. Pujar, R. S., Chyalakshmi, C. L., & Kakkasageri, M. S. (2024). Machine learning based dynamic resource allocation in 6G network. In *2024 4th International Conference on Intelligent Technologies (CONIT)*, Bangalore, India. https://doi.org/10.1109/conit61985.2024.10626599
25. Ismail, S. F., & Kadhim, D. J. (2024). Towards 6G technology: Insights into resource management for cloud RAN deployment. *IoT*, 5, 409–448. https://doi.org/10.3390/iot5020020
26. Hui, N., Sun, Q., Tian, L., Wang, Y., & Zhou, Y. (2024). Computation and wireless resource management in 6G space-integrated-ground access networks. *Digital Communications and Networks*. https://doi.org/10.1016/j.dcan.2024.04.001
27. Xing, Y., Zhou, J., Li, P., & Yu, M. (2024). Research on key technologies of 6G network architecture. In *2024 International Wireless Communications and Mobile Computing (IWCMC)*, Ayia Napa, Cyprus. https://doi.org/10.1109/iwcmc61514.2024.10592315
28. Wallace, J., Vuković, M., Karimovic, T., Edwards, W., Islam, J., Šnajder, B., Kovačević, S., & Wallace, A. (2023). A 5G/6G infrastructure for secure, high-performance, low-latency application services. In *2023 International Conference on Computational Science and Computational Intelligence (CSCI)*, Las Vegas, NV. https://doi.org/10.1109/csci62032.2023.00155
29. Tzanakaki, A., Manolopoulos, A., Alevizaki, V.-M., & Anastasopoulos, M. (2023). Optimized and dynamic resource provisioning in AI assisted 6G networks. In *2023 IEEE Future Networks World Forum (FNWF)*, Baltimore, MD. https://doi.org/10.1109/fnwf58287.2023.10520569
30. Nancharaiah, K. R., Ravi, K. C., Srivastava, A. K., Arunkumar, K., Siddiqui, S. T., & Arun, M. R. (2023). Analysis of data science and AI-enabled 6G wireless communication networks. *Radioelectronics and communications System*, 66, 223–232. https://doi.org/10.20535/s0021347023050059

31. Rizvi, S. (2023). Unveiling the potential of artificial intelligence and machine learning in the 5G network landscape: A comprehensive review. *Asian Journal of Research in Computer Science*, 16, 23–31. https://doi.org/10.9734/ajrcos/2023/v16i4367
32. Saxena, A. C., Kumar, R., Al-Farouni, M., Santhosh Kumar, C., Karpagam, J., & Kavitha, M. (2024). Navigating of 6G network into the latest AI technology for the fastest communication without loosing the data. In *2024 4th International Conference on Advance Computing and Innovative Technologies in Engineering (ICACITE)*, Greater Noida, India. https://doi.org/10.1109/icacite60783.2024.10616812
33. Garcia, C. R., Bouchmal, O., Stan, C., Giannakopoulos, P., Cimoli, B., Olmos, J. J. V., Rommel, S., & Monroy, I. T. (2023). Secure and Agile 6G networking – Quantum and AI enabling technologies. In *2023 23rd International Conference on Transparent Optical Networks (ICTON)*, Bucharest, Romania. https://doi.org/10.1109/icton59386.2023.10207418
34. Rabnawaz, R., M Kamran Abid, M. K. A., Aslam, N., & Bukhari, F. (2023). Exploring 6G wireless communication: Application technologies, challenges and future direction. *International Journal of Information Systems and Computer Technologies*, 2, 26–43. https://doi.org/10.58325/ijisct.002.02.0057
35. Liu, Y., Deng, Y., Nallanathan, A., & Yuan, J. (2023). Machine learning for 6G enhanced ultra-reliable and low-latency services. *IEEE Wireless Communications*. https://doi.org/10.1109/MWC.006.2200407
36. Xing, Y., Yang, L., Hu, X., Mei, C., Wang, H., & Li, J. (2023). 6G Deterministic network technology based on hierarchical reinforcement learning framework. In *2023 IEEE International Symposium on Broadband Multimedia Systems and Broadcasting (BMSB)*, Beijing, China. https://doi.org/10.1109/bmsb58369.2023.10211210
37. Mahboob, S., & Liu, L. (2023). Revolutionizing future connectivity: A contemporary survey on AI-empowered satellite-based non-terrestrial networks in 6G. *IEEE Communications Surveys & Tutorials*, 26, 1279–1321.
38. Wang, J., Liu, J., Li, J., & Kato, N. (2023). Artificial intelligence-assisted network slicing: Network assurance and service provisioning in 6G. *IEEE Vehicular Technology Magazine*, 18, 49–58. https://doi.org/10.1109/mvt.2022.3228399
39. Zhang, Q., & Zhang, Q. (2023). Study of optical networks, 5G, artificial intelligence and their applications. *arXiv.org*. https://doi.org/10.48550/arXiv.2301.13396
40. Zhao, J., Zhang, B., Chang, S.-Y., Wang, Y., & Chen, L. (2022). Beam weights optimization based on reinforcement learning for 6G. In *2022 IEEE/CIC International Conference on Communications in China (ICCC Workshops)*, Sanshui, Foshan, China. https://doi.org/10.1109/ICCCWorkshops55477.2022.9896688
41. Akbar, M., Hussain, Z., Ikram, M., Sheng, Q. Z., & Mukhopadhyay, S. (2022). 6G survey on challenges, requirements, applications, key enabling technologies, use cases, AI integration issues and security aspects. *arXiv.org*. https://doi.org/10.48550/arXiv.2206.00868
42. Salameh, A. I., & El-Tarhuni, M. (2022). From 5G to 6G: Challenges, technologies, and applications. *Future Internet*, 14, 117. https://doi.org/10.3390/fi14040117
43. Kamble, P., & Shaikh, A. N. (2022). 6G Wireless networks: Vision, requirements, applications and challenges. In *2022 5th International Conference on Advances in Science and Technology (ICAST)*, Mumbai, India. https://doi.org/10.1109/ICAST55766.2022.10039549
44. Rekkas, V. P., Sotiroudis, S. P., Sarigiannidis, P., Wan, S., Karagiannidis, G. K., & Goudos, S. K. (2021). Machine learning in beyond 5G/6G networks: State-of-the-art and future trends. *Electronics*, 10, 2786. https://doi.org/10.3390/ELECTRONICS10222786
45. Talwar, S., Himayat, N., Nikopour, H., Xue, F., Wu, G., & Ilderem, V. (2021). 6G: Connectivity in the era of distributed intelligence. *IEEE Communications Magazine*, 59, 45–50.

46. Alharbi, A., Alyami, H., Poongodi, M., Rauf, H. T., & Kadry, S. (2021). Intelligent scaling for 6G IoE services for resource provisioning. *PeerJ*, 7, e755. https://doi.org/10.7717/PEERJ-CS.755
47. Rasti, M., Taskou, S. K., Tabassum, H., & Hossain, E. (2021). Evolution toward 6G wireless networks: A resource management perspective. *IEEE Wireless Communications*, 29, 118–125.
48. Jiang, Z., Fu, S., Zhou, S., Niu, Z., Zhang, S., & Xu, S. (2019). AI-Assisted low information latency wireless networking. *IEEE Wireless Communications*, 27, 108–115.

5 Millimeter-Wave Communication and Terahertz Spectrum for 6G Wireless Network

Saif Ahmad, Mohd Javed Khan, Zeenat Fatima, and Indrasen Singh

5.1 RELATED STUDY AND MOTIVATION

The impetus for 5G New Radio (NR) standardization was the requirement to support an increasingly dynamic ecosystem of throughput requirements, reliability, and latency requirements for applications that must be enabled by current cellular networks. In the Fifth generation network, there are three classes of applications: eMBB, URLLC, and mMTC. Indeed, from the very start, it was already apparent that no single solution could be generally applicable to all applications. Consequently, advanced communication and networking techniques have been developed for recent generations of cellular systems, including network densification through smarter scheduling, small cells, and the utilization of multi-antenna systems to achieve high spectral efficiencies, among other innovations. The reality that the traditional sub-6GHz spectrum would fall short of the needs of these new applications brings out the key difference between the current scintillation and the predecessors. This has driven recognition of the Millimeter-Wave (mmWave) spectrum as a way forward. Although in the past the mmWave spectrum frequencies were considered inappropriate for mobile use mainly due to their less favorable propagation characteristics, developments in devices and antennas allowed mmWave bands to be commercially applied to wireless communications [1]. Hence, the commercial launching of 5G communication opened the door for commercial mmWave communication.

Future developments include wireless brain–machine interfaces, ultra-HD video conferencing, virtual and augmented reality, and 3D gaming, all of which will place even greater demands on throughput, dependability, and latency. With such advancements in device fabrication, nanoscale communications are also expected to become a reality in the near future. Motivated by the recent success in mmWave communication, naturally researchers began to look into other underutilized regions of the RF-band, particularly the terahertz (THz), which follows above the mmWave range. THz waves offer higher BW, and therefore, prove to be very useful for applications requiring ultra-high data rates. It is well known that combining the sub-6GHz and mmWave bands with THz frequencies makes these emerging technologies

 DOI: 10.1201/9781003599111-5

unstoppable. Another is the short wavelengths of THz waves, which enables them to be used in communications at micro- and nanoscale applications. Till recent times, the development of THz-bands had been restricted to imaging and sensing applications only, as efficient devices that can work at such frequencies were not available. However, latest developments in THz technology may well ensure that THz communication becomes an integral component of future communication standards [2].

In this chapter, the reader will be familiarized with mmWave and THz bands, giving a basis of why these bands are crucial to modern commercial communications and some of the main issues in communication systems design operating at such frequencies. This is done by providing a structured investigation of the issue at hand starting from a comprehensive study of the propagation characteristics at these frequencies, followed by an analysis of channel models that take such characteristics into account. This chapter carefully compares and contrasts the propagation traits of these newer bands with those of the more familiar sub-6GHz bands, highlighting how the channel models represent the main distinctions. In this study, we build on this foundation by investigating the consequences of these discrepancies for the design of THz and mmWave communication systems, as well as their possible uses in 6G technology. The chapter concludes with a brief overview of the current standardization initiatives pertaining to the use of such frequency bands in commercial communications.

5.2 INTRODUCTION TO mmWAVE AND THz SPECTRUM

Commercial communication was restricted to conventional frequencies up until the 4G cellular standard, now commonly known as the sub-6GHz cellular bands. However, for many non-cellular uses, such as satellite transmissions, radio astronomy, remote sensing, and radar systems, the 6–300GHz range has already been in use for decades. This bandwidth can now be extended with recent developments in antenna technology, enabling wireless access to mobile communications. The millimeter-wave (mmWave) band ranges from 30 to 300GHz and has wavelengths between 1 and 10mm. This spectrum range offers much bandwidth than the possible sub-6GHz bands. Despite the significant challenges posed by high penetration and blockage losses in mmWave communication, it has been demonstrated that the same qualities may be exploited in order to acquire interference mitigation in high-density modern cellular systems with small cells. This makes it more aggressive in terms of frequency reuse and, hence, a better data security since frequencies from the mmWave range happen to be directional.

Frequencies between approximately 24 and 100GHz are already under consideration for the standardization of 5G. For 6G and beyond, the 0.1–10 THz range called the THz band has been the area of focus for researchers in recent times, and the lower end of this range is of special interest to the communication technology [3].

5.3 PROPAGATION AT THE THz FREQUENCIES AND mmWAVE

5.3.1 Comparisons from Conventional Bands

There are numerous key reasons why communication at THz and mmWave frequencies is very different from traditional microwave communication.

5.3.1.1 Signal Blockage

The mmWave/THz signals are much more susceptible to blockages compared to lower frequency signals. These higher frequency communications are highly dependent on LOS links because the NLOS link propagation properties are very poor [4]. For example, it is very easy for buildings, vehicles, humans, and foliage to obstruct mmWave/THz signals. A single blockage can result in a loss of 20–40 dB. The reflection losses of mmWave signals depend on the material and vary from 3 to 18 dB for glass and 40 to 80 dB for building material like brick. Foliage can also introduce significant signal attenuation with 17–25 dB loss for a single tree [5,6]. In addition, mmWave/THz signals are also subject to self-body blockage by which users themselves can introduce an attenuation of about 35 dB. These blockages can significantly degrade signal strength and, in some situations, even cause a total loss of the signal. So, it is crucial to find effective ways to combat blockages and carry out fast handovers when a link is obstructed. However, obstructions, such as self-body obstructions, can also help lessen interference coming from far-off base stations. Therefore, both analytical and computational models of mmWave/THz should not neglect the precise accounting of these obstructions [7].

5.3.1.2 High Directivity

Another highly important feature of mmWave and THz communications is their high directivity. In order to compensate for the high path losses at these higher frequencies, there must be multiple antennas present on the transmission or reception side [4]. Fortunately, however, the wavelengths are much shorter at these frequencies, thus allowing a number of antennas to fit into compact form factors. It leads to highly directional communication, along with narrow beamwidth and high beamforming gains, which improve signal strength in the serving links and reduce interference at the receiver side. However, this high directivity poses the challenge of "deafness" as it increases latency due to the extended process for beam search in directional transmission and reception. This poses a larger problem in high mobility scenarios, where the frequent need to conduct beam training adds significant overhead for both the user and the base stations. In high-mobility contexts, modern random access techniques and adaptive array analyzing techniques are essential for performing handovers and quickly adapting to obstructions [8].

5.3.1.3 Atmospheric Absorption

The main cause of loss in the transmission of electromagnetic waves through the atmosphere is the absorption by gas molecules like water and oxygen. In such losses, the frequency ranges corresponding to the natural resonance frequency of the molecules show more loss [9]. Wave absorption due to oxygen and water molecules is the major loss making factor in mmWave and THz frequency bands. Air loss is very minimal in microwave frequencies. This attenuation is additive, thus reducing the range and coverage of mmWave and THz transmissions while dramatically increasing the required base station (BS) density for systems operating at these frequencies [10–15].

5.3.1.4 Reflections and Scattering

An electromagnetic wave hitting a smooth surface that is greater in electrical size than its wavelength will reflect in a single direction. In this direction, the reflected power corresponds to that part of the incident field that was reflected; the reflection coefficient of the surface determined the latter. This coefficient Γ_{pl} is normalized against the associated penetration loss. The reflected power is given by [26]

$$P_{\text{rp}} = P_{\text{ip}}\Gamma_{\text{pl}}^{2} \tag{5.1}$$

where P_{ip} shows the incident power wave and P_{rp} shows the reflected power wave. Reflections and scattering are key phenomena in THz communication, significantly impacting signal propagation and system performance. Due to the short wavelength of THz waves, reflections occur when the waves encounter surfaces, such as walls, buildings, or objects, causing the signal to bounce in different directions. Depending on the surface material and angle of incidence, reflections can either extend the signal's reach or cause multipath interference, where multiple versions of the signal arrive at the receiver at different times, degrading communication quality.

Scattering, on the other hand, happens when THz waves interact with small particles or rough surfaces, such as dust, aerosols, or foliage. This results in the signal energy being dispersed in multiple directions, causing signal attenuation and reducing the strength of the received signal. Scattering is more pronounced at higher frequencies, like THz, due to the smaller wavelength, making it a significant challenge in outdoor and complex indoor environments.

In THz communication systems, managing reflections and scattering is essential for optimizing performance. Techniques like beamforming, which focuses the signal into narrow beams, and advanced channel modeling, which accounts for these effects, are often employed to mitigate their impact. These factors are especially important in short-range, high-data-rate applications, where line-of-sight communication is typically preferred to minimize reflections and scattering losses [28–32] (Figure 5.1).

5.3.2 Beamforming and Antenna Patterns

Optimal THz communication systems employ beamforming and antenna patterns in an effort to counter high path loss and limited range. Beamforming is therefore the

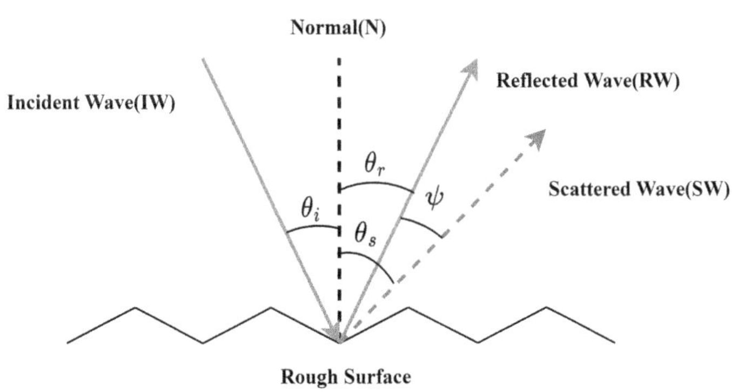

FIGURE 5.1 A diagram illustration of a surface radio wave incident.

transmission or reception of signals using phased arrays of antennas, whereby there is manipulation of both phase and amplitude of individual elements. Consequently, it creates narrow, highly directional beams that can be steered toward the user or device, focusing signal energy to maximize efficiency and minimize interference. Another definition of antenna patterns describes the distribution of the signal in space. In multi-user or multi-stream scenarios, the adaptive antenna pattern can then be specifically designed to serve more than one user at a time by generating different beams to each of them, making spatial multiplexing possible. This contributes to the efficient utilization of the spectrum, especially in high-density environments, because of interference minimization between data streams [16–25].

Together, beamforming and customization of the antenna pattern heighten the performance of THz systems, offering higher data rates, extended coverage, and improved reliability across applications such as small cells, wireless backhaul, and ultra-high-speed wireless networks. These methods are also a key for emerging technologies like 3D holographic video, extended reality, and tactile internet, needing point-to-point communication that relies on precise, high-bandwidth communication.

5.3.2.1 Analog Beamforming Patterns

When using analog beamforming, both receiver and transmitter antenna patterns, which indicate the gain in the various directions surrounding the antenna array, can be used to compute the effective gain of the incoming signal. In order to facilitate the assessment of mmWave systems, several antenna layouts have been presented in the literature. The subsequent subsections provide further details about this.

5.3.2.1.1 Uniform Linear Array (ULA) Model

The most common antenna configuration used in the communication systems, from THz to mmWave technologies, is the Uniform Linear Array (ULA) model. In the context of this model, there are a number of elements aligned in straight line and separated equally by equal distances between two consecutive elements. The average distance between the elements is half the wavelength of signal during transmission and reception. The model of ULA can be used in this manner that it is able to support directional beamforming since signals from each antenna element are combined with specific phase shifts so that a narrow, high-gain beam forms in a desired direction.

The ULA model is better suited for beamforming applications since control over the directions and shapes of a beam is highly precise. This results in increased strength of signals and decreased interference. The model is best for application situations, such as point-to-point communication scenarios, multi-user environments, and high-speed wireless networks. The ULA model is also relatively easy to implement, hence commonly employed both in simulation and real-world deployment for performance evaluation purposes with advanced wireless systems.

An M-array ULA's antenna gain function $F_{\sin}(\varphi_1)$, where d_1 is the antenna element separation and λ_1 is the signal wavelength [33], is

$$F_{\sin}(\varphi_1) = \frac{\sin^2(\pi M \varphi_1)}{M^2 \sin^2(\pi \varphi_1)} \tag{5.2}$$

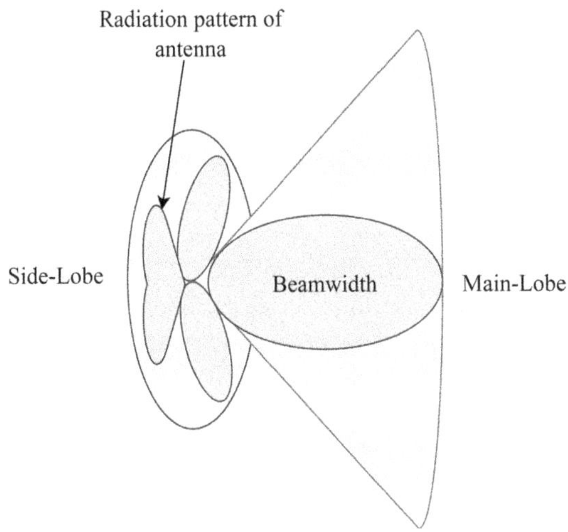

FIGURE 5.2 The sectorized antenna model [14].

where $\varphi_1 = \frac{d_1}{\lambda_1}\cos\delta$ shows the cosine direction that corresponds to the spatial departure angle (δ) of the transmitted signal (Figure 5.2). Hence, the array gain function may be approximated as a squared sinc-function, which is given by

$$F_{\text{sinc}}(\varphi_1) \triangleq \frac{\sin^2(\pi M \varphi_1)}{(\pi M \varphi_1)^2} \tag{5.3}$$

5.3.2.1.2 Sectorized Antenna Model

One of the most common models in wireless communication systems, which covers and manages interference in THz and mmWave networks, is the Sectorized Antenna Model. Here, the coverage area of the antenna can be divided into several sectors, which can be covered using a directional antenna or beam per sector. Each sector focuses on a particular region with a higher gain for efficient transmission of the signal over that area and less interference with other sectors. Sectorized antennas are quite handy in high-capacity, spatial reuse environments, such as cellular networks, where multiple users or devices need to be served simultaneously [34].

By separating various sectors, the system can handle several concurrently functioning data streams, each targeted to a particular sector, thereby enhancing the network overall. The Sectorized Antenna Model is also very effective when users are in fixed or predictable locations.

In such a case, the sectorized directional beams could be optimized for particular areas. This model is applied very often in system design and evaluation where an efficient use of spectrum and reliable performance are needed in dense and high traffic environments.

5.3.2.1.3 Multi-Lobe Antenna Model

The Multi-lobe Antenna Model is designed to generate multiple beams or "lobes" in different directions, allowing for simultaneous communication with multiple users or devices. Each lobe represents a focused beam of signal energy, enabling the antenna to cover several spatial regions at once, enhancing network capacity and efficiency. Unlike single-beam antennas, which concentrate signal power in a single direction, multi-lobe antennas distribute their energy across various lobes, allowing them to serve multiple users located in different directions [35].

This model is particularly useful in scenarios where users or devices are spread out across a wide area, as it allows for efficient spatial multiplexing. By creating distinct lobes, the antenna can communicate with several users at once without significant interference between them. The Multi-lobe Antenna Model is often employed in applications like massive MIMO, mmWave, and THz communication systems, where high data rates and multiple simultaneous connections are required. It is ideal for environments with diverse or unpredictable user locations, as the multiple lobes can dynamically adjust to changes in user positions, ensuring reliable and efficient communication [36].

5.3.2.1.4 Gaussian Antenna Model

Sometimes it is used to represent very directive antennas in high frequency bands like THz and mmWave communication systems. The radiation pattern of an antenna in the case of this model is characterized by a Gaussian function with a main lobe at some angle and, later on, it tails off gradually while deviating further away from the beam axis in terms of strength in signal. This produces a smooth, symmetric, bell-shaped shape that closely models the behavior of real-world antennas with thin, high-gain beams.

One of the attractions of the Gaussian Antenna Model is its simplicity in mathematically capturing the very directional nature of antennas. Its representation provides a good approximation for applications that require precision point-to-point signal transmission, such as those involved in point-to-point links, beamforming and beam steering.

This model features a most significant reduction of side lobes by forcing more energy into the main lobe, ensuring efficient signal transmission in the direction of interest. The model is thus made suitable for performance evaluation in high frequency short range systems such as wireless backhaul and ultra-fast indoor networks, with the fact of capacitive application requiring focused beams and minimizing interference [37,38]. According to this model, the antenna gain is

$$G_{\text{Gau}}(\theta) = \left(g_{\text{mlg}} - g_{\text{slg}}\right)e^{\mu\theta^2} + g_{\text{slg}} \tag{5.4}$$

where g_{mlg} represents the maximum main-lode gain, occurring at $\theta = 0$, g_{slg} shows the side-lobe gain, while the parameter μ regulates the 3 dB beam-width.

5.3.2.1.5 Cosine Antenna Model

The Cosine Antenna Model is a simplified representation of an antenna's radiation pattern, often used in wireless communication system analysis, particularly for omnidirectional or moderately directional antennas. In this model, the antenna's gain is proportional to the cosine of the angle θ from the main axis, with maximum gain at $\theta=0$ (directly in front of the antenna) and gradually decreasing as θ increases.

Mathematically, the gain is typically expressed as

$$G(\theta) = G_{\max} \cos^n(\theta) \tag{5.5}$$

where $G_{\max}$ is the maximum gain in the direction of the main lobe and n is the parameter that controls the directivity of the antenna. A larger n results in a more concentrated beam, while a smaller n yields a wider radiation pattern [34].

This model is useful for approximating antenna behavior in scenarios where high precision is not required, and it provides a simple way to model the attenuation of signal strength as the angle deviates from the main direction. The Cosine Antenna Model is commonly applied in studies of low- to medium-gain antennas used in applications such as base stations, sensor networks, and indoor wireless communications.

5.3.2.2 Multi-User Pattern

Multi-user pattern of antenna dictates the type of directionality for signals transmitted and received. As a general observation, highly directional antennas are usually used in the THz frequency range, which narrows down the energy into very thin beams and reduces interference while maximizing data rates. In the case of multi-element antennas, it is possible to create several beams simultaneously and thus enable spatial multiplexing, where separate data streams are transmitted to different users within the same frequency band without causing interference between them. In multi-stream transmission, techniques of adaptive antennas such as beam forming and beam steering are used in order to change the direction of the beams dynamically according to the position and movement of users or devices. These antenna patterns must be designed so that high-gain directional beams track the user in real time; therefore, signals are made of a good quality and can work on efficient spectrum use. Tailoring antenna patterns for various transmission paths is an important concern for high-density environments, in which a large number of users or devices have to be served simultaneously. Advanced antenna designs, such as phased arrays or metasurface-based antennas, increasingly support the functionality for ensuring robust, high-speed communication in THz systems.

5.3.2.3 THz Beamforming

THz beamforming is a crucial technique in THz communication systems [40], enabling highly focused signal transmission and reception to overcome challenges like high path loss and limited transmission range. Since THz waves have much shorter wavelengths compared to lower frequency bands, they are more susceptible to attenuation and interference, making precise beamforming essential for maintaining strong, reliable links.

Beamforming uses an array of antennas to dynamically control the direction of the THz signal by adjusting the phase and amplitude of the transmitted signals. This allows the creation of narrow, high-gain beams that can be directed toward specific users or devices, concentrating the signal energy where it's needed most. In a multi-user environment, beamforming can support multiple simultaneous connections by steering separate beams to different users, thereby improving overall network efficiency and reducing interference.

THz beamforming is particularly useful in short-range, high-data-rate applications like small cells, wireless backhaul, and indoor wireless networks [42], where the precise direction of the beam can compensate for the limited propagation range. Additionally, it plays a key role in emerging technologies like holographic video conferencing, extended reality, and ultra-HD streaming, which require high-bandwidth and low-latency communication. By leveraging advanced beamforming algorithms and antenna arrays, THz communication systems can achieve improved coverage, capacity, and data rates [39].

5.3.3 Channel Models

The channel models play a very significant role in THz communication; it will provide accurate and reliable predictions and simulations of the behavior of THz waves propagating through various environments. This is because, at THz frequencies, high path losses are combined with molecular absorption and sensitivity to obstacles. For this reason, development becomes very crucial for the proper and exact channel models aimed at the optimization of system performance.

Besides free-space path loss, molecular absorption caused by gases in the atmosphere, reflection, scattering, diffraction, and multipath effects [43] should also be considered in THz channel models. Furthermore, because of the limitation imposed on the maximum transmission distance, such models are used to simulate both LOS and NLOS scenarios where the propagation path of the radio signal may be obstructed or reflected while traveling from a transmitter to a receiver.

The channel models for indoor scenarios should include the effects of the surface materials and objects contributing toward signal reflection and scattering. In outdoor environments, the models need to capture atmospheric conditions also, because the THz wave propagation is highly affected by these. Such channel models are obviously of great importance to design efficient THz communication systems by optimizing the signal processing techniques and the challenges arising from high frequencies in practical applications.

5.3.4 THz Channel

The THz channel refers to the medium through which THz-frequency electromagnetic waves propagate, typically between 0.1 and 10 THz. While the THz band offers vast bandwidth, enabling ultra-high data rates up to the terabit-per-second (Tbps) range, it also presents significant challenges. High path loss, due to free-space propagation, and molecular absorption by atmospheric gases, such as water vapor, greatly limit transmission distance [44]. As a result, THz communication is often confined to short ranges, typically within a few meters. Additionally, THz waves are highly sensitive to reflections, scattering, and diffraction when encountering surfaces or small particles, which can degrade signal quality. Given these characteristics, THz communication typically relies on line-of-sight (LOS) transmission, making it suitable for applications like short-range wireless links, indoor networks, and chip-to-chip communication. To overcome these challenges and maximize performance, techniques such as beamforming and adaptive modulation are often employed in THz systems.

5.4 THz COMMUNICATION SYSTEMS

THz communication systems represent a disruptive approach to the transfer of high-speed data. They will harness the THz frequency band, delivering unprecedented bandwidth and capacity. Indeed, the systems are intended to support data rates in the Tbps range and will be used to enable an extended list of applications including ultra-HD video streaming, real-time holographic conferencing, and the tactile internet. However, THz communication systems face many problems with high path loss, atmospheric absorption, and very limited distance of transmission. To handle all these problems, advanced technologies like beamforming and, in particular, beam tracking and innovative transceiver designs are used. Such systems prove well suited for short-range, high-density applications, like small cells in indoor networks, as well as high-speed backhaul for 5G and beyond. The envisioned THz communication systems are candidates to be realized at both the macroscale, such as in the context of future data centers and vehicular networks, and the micro/nanoscale, such as inter-core communication within chips and nano-networking for medical devices [43].

Smaller Range: THz bands have smaller communication range as compared to mmWave transmission due to high propagation losses and absorption by molecules. For example, in a small cell network, the THz band would cover an extent of about 10 m only. Furthermore, molecular absorption is frequency dependent, resulting in band-splitting and decreased bandwidth availability within the THz frequency range.

THz Transceiver Design: The development of a THz transceiver represents an important step for the realization of more efficient THz communication systems. A transmitter-receiver device operating within the THz frequency domain is required to support extremely high data rates while coping with power efficiency, signal attenuation, and shrinking devices. The major components in the design of a THz transceiver include developing high-frequency oscillators, mixers, amplifiers, and antennas capable of operation in the THz band. Advanced materials, in particular, such as graphene, are pursued to enhance performance. Moreover, the compact and scalable integration of the components is of paramount importance for the exploitation of the technologies in areas such as wireless communication [41], imaging, and sensing at both macro and nanoscales [44].

THz Beam Tracking: THz beam tracking is essentially the necessity for stable and reliable communication links in systems that work at THz frequency bands, as THz signals are by nature very directional. THz waves are much more sensitive to blockages, atmospheric absorption, and alignment deviations at their very high operating frequency. Beam tracking is, therefore, the continuous adjustment of the THz beam trajectory to match the movement of the communicating devices or due to environmental changes. This technology is pretty crucial in applications such as high-speed wireless communication, and proper alignment of the beam will ensure maximum rates for data transfer and reduce signal attenuation.

5.4.1 THz Communications' Possible Uses in 6G

Terahertz communication offers a diverse range of applications at both the macro and micro/nano ranges. This section explores some of these applications.

5.4.1.1 Macroscale THz Communication

THz macroscale communication uses the terahertz frequency range for wireless communication over long distances that are usually hundreds of meters or several kilometers. Terahertz frequency promises ultra-high data rates with very low latency and high capacities, making it apt for applications like 6G networks, wireless backhaul, and high-speed data transfer in data centers. However, there are significant path losses and atmospheric absorption as well as only shallow penetration of obstacles in macroscale THz communication. In this regard, high-gain antennas, advanced beamforming techniques, and intelligent reconfigurable surfaces are being developed to alleviate some of these limitations [45,46]. Indeed, macroscale THz communication is the critically important area of research for future wireless systems. T-WLAN offers the ability to provide seamless connectivity bridging high-speed wired networks such as optical fiber links to personal devices: mobile phones, laptops, and smart TVs. T-WPAN offers ultra-fast communication between close vicinity devices. Kiosk downloading is a very important WPAN application in which big multimedia files like video are downloaded to mobile phones located near the fixed kiosks. Similar to the examples above for mmWave networks, other possible applications of THz communication such as improved vehicular and aerial communications, security and wireless connections in data centers, to name a few, are also quite similar and are not repeated here [47,48].

5.4.1.2 Microscale or Nanoscale THz Communication

Nanomachine communication is another use for the THz spectrum. Such nanomachines are likely to be capable of undertaking relatively primitive operations, such as computation, actuation, data storage, and sensing. Transmission ranges vary between a few micrometers to many meters, depending on the application. The following list includes some significant uses for nanomachine communications [45].

i. **Tracking Health Condition**: The human body has nanosensors and nanomachines that can track glucose levels, cholesterol, ion concentrations, and biomarkers from cancer. All this information can then be sent wirelessly with THz communication through an external device—for example, a mobile phone or smart band—to be further processed in some external medical equipment or by the healthcare provider.

ii. **Nuclear, Biological, and Chemical Defenses**: The molecules of dangerous chemical and biological weapons can be highly effectively detected by nanosensors. Unlike traditional macroscale chemical sensors, nanosensors identify extremely low concentrations of such molecules at the level of a single molecule. Thus, nanodevices that communicate in the THz frequency range can be used in defense for detection purposes to dangerous nuclear, biological, and chemical agents.

iii. **Internet of Bio-Nano-Things (IoBNT) and Internet of Nano-Things (IoNT)**: The IoBNT is defined as the integration of nanomachines with the current networks of communication. The interconnected nanodevices can be utilized in various applications, which may include environmental monitoring, health status, and real-time tracking. Almost any device may be equipped with antennae and nano-transceivers to enable Internet connectivity.

   Although the concept of IoBNT is very much like the concept of IoNT, the use of biological nanomachines in the case of IoBNT is instead of silicon-based nanomachines. The biomedical field can benefit from the diverse uses of these biological nanomachines, which can be created using genetically modified cells or synthetic biological components [49].

iv. **Wireless Communication within Network-on-Chip Systems**: THz waves can be applied for inter-core communication between tightly embedded processing cores on a single chip using planar nano-antennas with sizes in the order of a few micrometers. This method allows for ultra-high-speed inter-core communication that is appropriate when not much space is available. The use of graphene in nano-antennas allows for the design of completely flexible and scalable chip-based wireless networks [50].

5.5 CONCLUSION

A significant success of 5G communications has been the practical realization of using mmWave frequencies in commercial wireless networks, which was considered impossible only a few years ago because of the problematic propagation characteristics of such frequencies. Although such speeds are still being rolled out, concerns are already sprouting regarding whether gigabit-per-second speeds supported by 5G mmWave systems are enough to support new uses including expanded reality and 3D gaming. These applications are anticipated to require low latency, excellent dependability, and data speeds that vary between hundreds of gigabits up to several terabits per second—the main design objectives for next-generation 6G networks. THz systems for communication are seen as the next big thing in wireless communication studies because of their capacity to transmit such high data rates across small distances. This chapter provides a cohesive review of both the mmWave and THz bands, acknowledging their critical roles in 6G and beyond. It focuses on channel models, propagation characteristics, design and implementation issues, and possible applications. It concludes with a brief summary of ongoing standardization efforts in these frequency bands.

REFERENCES

1. Rappaport, T. S., Sun, S., Mayzus, R., Zhao, H., Azar, Y., Wang, K., … & Gutierrez, F. (2013). Millimeter wave mobile communications for 5G cellular: It will work! *IEEE Access*, 1, 335–349.
2. Sarieddeen, H., Saeed, N., Al-Naffouri, T. Y., & Alouini, M. S. (2020). Next generation terahertz communications: A rendezvous of sensing, imaging, and localization. *IEEE Communications Magazine*, 58(5), 69–75.
3. Khan, F., Pi, Z., & Rajagopal, S. (2012, October). Millimeter-wave mobile broadband with large scale spatial processing for 5G mobile communication. In *2012 50th Annual Allerton Conference on Communication, Control, and Computing (Allerton)*, Monticello, IL (pp. 517–523). IEEE.
4. Boccardi, F., Heath, R. W., Lozano, A., Marzetta, T. L., & Popovski, P. (2014). Five disruptive technology directions for 5G. *IEEE Communications Magazine*, 52(2), 74–80.
5. Hemadeh, I. A., Satyanarayana, K., El-Hajjar, M., & Hanzo, L. (2017). Millimeter-wave communications: Physical channel models, design considerations, antenna constructions, and link-budget. *IEEE Communications Surveys & Tutorials*, 20(2), 870–913.
6. Rangan, S., Rappaport, T. S., & Erkip, E. (2014). Millimeter-wave cellular wireless networks: Potentials and challenges. *Proceedings of the IEEE*, 102(3), 366–385.
7. Petrov, V., Komarov, M., Moltchanov, D., Jornet, J. M., & Koucheryavy, Y. (2017). Interference and SINR in millimeter wave and terahertz communication systems with blocking and directional antennas. *IEEE Transactions on Wireless Communications*, 16(3), 1791–1808.
8. Attiah, M. L., Isa, A. A. M., Zakaria, Z., Abdulhameed, M. K., Mohsen, M. K., & Ali, I. (2020). A survey of mmWave user association mechanisms and spectrum sharing approaches: An overview, open issues and challenges, future research trends. *Wireless Networks*, 26, 2487–2514.
9. Marcus, M., & Pattan, B. (2005). Millimeter wave propagation: Spectrum management implications. *IEEE Microwave Magazine*, 6(2), 54–62.
10. Tharek, A. R., & McGeehan, J. P. (1988, June). Propagation and bit error rate measurements within buildings in the millimeter wave band about 60GHz. In *8th European Conference on Electrotechnics, Conference Proceedings on Area Communication*, Stockholm, Sweden (pp. 318–321). IEEE.
11. Allen, G., & Hammoudeh, A. (1990). 60GHz propagation measurements within a building. In *1990 20th European Microwave Conference*, Budapest, Hungary (Vol. 2, pp. 1431–1436). IEEE.
12. Allen, G., & Hammoudeh, A. (1991, April). Frequency diversity propagation measurements for an indoor 60GHz mobile radio link. In *1991 Seventh International Conference on Antennas and Propagation, ICAP 91 (IEE)*, York (pp. 298–301). IET.
13. Davies, R., Bensebti, M., Beach, M. A., & McGeehan, J. P. (1991, May). Wireless propagation measurements in indoor multipath environments at 1.7GHz and 60GHz for small cell systems. In *1991 Proceedings 41st IEEE Vehicular Technology Conference*, St. Louis, MO (pp. 589–593). IEEE.
14. Yang, H., Herben, M. H., & Smulders, P. F. (2005). Impact of antenna pattern and reflective environment on 60GHz indoor radio channel characteristics. *IEEE Antennas and Wireless Propagation Letters*, 4, 300–303.
15. Allen, G., & Hammoudeh, A. (1991, January). Outdoor narrow band characterisation of millimetre wave mobile radio signals. In *IEE Colloquium on Radiocommunications in the Range 30–60GHz*, London (pp. 4–1). IET.
16. Daniele, N., Chagnot, D., & Fort, C. (1994). Outdoor millimetre-wave propagation measurements with line of sight obstructed by natural elements. *Electronics Letters*, 30(18), 533–534.
17. Wang, F., & Sarabandi, K. (2005). An enhanced millimeter-wave foliage propagation model. *IEEE Transactions on Antennas and Propagation*, 53(7), 2138–2145.

18. Fong, B., Fong, A. C. M., Hong, G. Y., & Ryu, H. (2005). Measurement of attenuation and phase on 26-GHz wide-band point-to-multipoint signals under the influence of rain. *IEEE Antennas and Wireless Propagation Letters*, 4, 20–21.
19. Ben-Dor, E., Rappaport, T. S., Qiao, Y., & Lauffenburger, S. J. (2011, December). Millimeter-wave 60GHz outdoor and vehicle AOA propagation measurements using a broadband channel sounder. In *2011 IEEE Global Telecommunications Conference-GLOBECOM 2011*, Houston, TX (pp. 1–6). IEEE.
20. Rappaport, T. S., Gutierrez, F., Ben-Dor, E., Murdock, J. N., Qiao, Y., & Tamir, J. I. (2012). Broadband millimeter-wave propagation measurements and models using adaptive-beam antennas for outdoor urban cellular communications. *IEEE Transactions on Antennas and Propagation*, 61(4), 1850–1859.
21. Hirata, A., Takahashi, H., Takeuchi, J., Kukutsu, N., Kim, D., & Hirokawa, J. (2012, March). 120-GHz-band antenna technologies for over-10-Gbps wireless data transmission. In *2012 6th European Conference on Antennas and Propagation (EUCAP)*, Prague, Czech Republic (pp. 2564–2568). IEEE.
22. Keusgen, W., Weiler, R. J., Peter, M., Wisotzki, M., & Göktepe, B. (2014, September). Propagation measurements and simulations for millimeter-wave mobile access in a busy urban environment. In *2014 39th International Conference on Infrared, Millimeter, and Terahertz Waves (IRMMW-THz)*, Tucson, AZ (pp. 1–3). IEEE.
23. Weiler, R. J., Peter, M., Keusgen, W., & Wisotzki, M. (2014, September). Measuring the busy urban 60GHz outdoor access radio channel. In *2014 IEEE International Conference on Ultra-WideBand (ICUWB)*, Paris, France (pp. 166–170). IEEE.
24. Guan, K., Ai, B., Peng, B., He, D., Li, G., Yang, J., ... & Kürner, T. (2018). Towards realistic high-speed train channels at 5G millimeter-wave band—Part I: Paradigm, significance analysis, and scenario reconstruction. *IEEE Transactions on Vehicular Technology*, 67(10), 9112–9128.
25. Bas, C. U., Wang, R., Sangodoyin, S., Choi, T., Hur, S., Whang, K., ... & Molisch, A. F. (2019). Outdoor to indoor propagation channel measurements at 28GHz. *IEEE Transactions on Wireless Communications*, 18(3), 1477–1489.
26. Ju, S., Shah, S. H. A., Javed, M. A., Li, J., Palteru, G., Robin, J., ... & Rappaport, T. S. (2019, May). Scattering mechanisms and modeling for terahertz wireless communications. In *ICC 2019-2019 IEEE International Conference on Communications (ICC)*, Shanghai, China (pp. 1–7). IEEE.
27. Järveläinen, J., Haneda, K., Kyrö, M., Kolmonen, V. M., Takada, J. I., & Hagiwara, H. (2012, November). 60GHz radio wave propagation prediction in a hospital environment using an accurate room structural model. In *2012 Loughborough Antennas & Propagation Conference (LAPC)*, Shanghai, China (pp. 1–4). IEEE.
28. Degli-Esposti, V., Fuschini, F., Vitucci, E. M., & Falciasecca, G. (2007). Measurement and modelling of scattering from buildings. *IEEE Transactions on Antennas and Propagation*, 55(1), 143–53.
29. Kulkarni, M. N., Visotsky, E., & Andrews, J. G. (2018). Correction factor for analysis of MIMO wireless networks with highly directional beamforming. *IEEE Wireless Communications Letters*, 7(5), 756–759.
30. Kokkoniemi, J., Rintanen, P., Lehtomaki, J., & Juntti, M. (2016, December). Diffraction effects in terahertz band-measurements and analysis. In *2016 IEEE Global Communications Conference (GLOBECOM)*, Washington, DC (pp. 1–6). IEEE.
31. Federici, J., & Moeller, L. (2010). Review of terahertz and subterahertz wireless communications. *Journal of Applied Physics*, 107(11).
32. Bao, L., Zhao, H., Zheng, G., & Ren, X. (2012, October). Scintillation of THz transmission by atmospheric turbulence near the ground. In *2012 IEEE Fifth International Conference on Advanced Computational Intelligence (ICACI)*, Nanjing, China (pp. 932–936). IEEE.

33. Balanis, C. A. (2005). *Antenna Theory: Analysis and Design*. John wiley & Sons.
34. Yu, X., Zhang, J., Haenggi, M., & Letaief, K. B. (2017). Coverage analysis for millimeter wave networks: The impact of directional antenna arrays. *IEEE Journal on Selected Areas in Communications*, 35(7), 1498–1512.
35. Lu, W., & Di Renzo, M. (2015, November). Stochastic geometry modeling of cellular networks: Analysis, simulation and experimental validation. In *Proceedings of the 18th ACM International Conference on Modeling, Analysis and Simulation of Wireless and Mobile Systems*, Cancun, Mexico (pp. 179–188).
36. Di Renzo, M., Lu, W., & Guan, P. (2016). The intensity matching approach: A tractable stochastic geometry approximation to system-level analysis of cellular networks. *IEEE Transactions on Wireless Communications*, 5(9), 5963–5983.
37. Thornburg, A., & Heath, R. W. (2015, October). Ergodic capacity in mmWave ad hoc network with imperfect beam alignment. In *MILCOM 2015-2015 IEEE Military Communications Conference*, Tampa, FL (pp. 1479–1484). IEEE.
38. Deng, N., & Haenggi, M. (2018). A novel approximate antenna pattern for directional antenna arrays. *IEEE Wireless Communications Letters*, 7(5), 832–835.
39. Sun, S., MacCartney, G. R., & Rappaport, T. S. (2017, May). A novel millimeter-wave channel simulator and applications for 5G wireless communications. In *2017 IEEE International Conference on Communications (ICC)*, Paris, France (pp. 1–7). IEEE.
40. Tekbıyık, K., Ekti, A. R., Kurt, G. K., & Görçin, A. (2019). Terahertz band communication systems: Challenges, novelties and standardization efforts. *Physical Communication*, 35, 100700.
41. Han, C., Bicen, A. O., & Akyildiz, I. F. (2014). Multi-ray channel modeling and wideband characterization for wireless communications in the terahertz band. *IEEE Transactions on Wireless Communications*, 14(5), 2402–2412.
42. Priebe, S., Kannicht, M., Jacob, M., & Kürner, T. (2013). Ultra broadband indoor channel measurements and calibrated ray tracing propagation modeling at THz frequencies. *Journal of Communications and Networks*, 5(6), 547–558.
43. Hossain, Z., Mollica, C., & Jornet, J. M. (2017, September). Stochastic multipath channel modeling and power delay profile analysis for terahertz-band communication. In *Proceedings of the 4th ACM International Conference on Nanoscale Computing and Communication*, New York, NY (pp. 1–7).
44. Priebe, S., & Kurner, T. (2013). Stochastic modeling of THz indoor radio channels. *IEEE Transactions on Wireless Communications*, 12(9), 4445–4455.
45. Kim, S., & Zajić, A. (2015, April). Statistical modeling of THz scatter channels. In *2015 9th European Conference on Antennas and Propagation (EuCAP)*, Lisbon, Portugal (pp. 1–5). IEEE.
46. Kim, S., & Zajić, A. (2016). Statistical modeling and simulation of short-range device-to-device communication channels at sub-THz frequencies. *IEEE Transactions on Wireless Communications*, 5(9), 6423–6433.
47. Elayan, H., Shubair, R. M., Jornet, J. M., & Johari, P. (2017). Terahertz channel model and link budget analysis for intrabody nanoscale communication. *IEEE Transactions on Nanobioscience*, 16(6), 491–503.
48. Elayan, H., Stefanini, C., Shubair, R. M., & Jornet, J. M. (2018). End-to-end noise model for intra-body terahertz nanoscale communication. *IEEE Transactions on Nanobioscience*, 17(4), 464–473.
49. Chen, Y., & Han, C. (2018, April). Channel modeling and analysis for wireless networks-on-chip communications in the millimeter wave and terahertz bands. In *IEEE INFOCOM 2018-IEEE Conference on Computer Communications Workshops (INFOCOM WKSHPS)*, Honolulu, HI (pp. 651–656). IEEE.
50. Heath, R. W., Gonzalez-Prelcic, N., Rangan, S., Roh, W., & Sayeed, A. M. (2016). An overview of signal processing techniques for millimeter wave MIMO systems. *IEEE Journal of Selected Topics in Signal Processing*, 10(3), 436–453.

6 Terahertz Communication and Its Application in Microwave

Swati Nema, Shweta B. Thomas, S. Sangeetha, and Deepika Rani Sona

6.1 INTRODUCTION

Terahertz (THz) communication, which operates in the 0.1–10 THz frequency range, is considered one of 6G's core technologies. With its huge frequency band resources, it can meet 6G's spectrum requirements with a very high data transmission rate of Tbps. The short wavelength of infrared light and the large wavelength of millimeter waves, which are equivalent to the THz frequency range, are shown in Figure 6.1. THz communication effectively addresses the capacity constraints and spectrum scarcity of existing wireless systems by combining the benefits of microwave and optical communication. The THz band can offer an improved data transfer rate and more spectrum resources than millimeter-wave communication. Because of their shorter wavelength and narrower antenna aperture, THz waves are more focused and non-interfering than millimeter waves. THz signals are more stable than visible light signals when there are changes in the surrounding light, air turbulence, cloud dust, light intensity, or flicker. However, in spite of quasi-optical transmission properties of THz communication, some microwave communication features are still present [1].

THz waves can penetrate a wide range of infrared opaque materials, such as paper semiconductors ceramics, and plastics, and provide high-quality image resolution equivalent as optics (about 100 μm). They are used to identify organic molecules

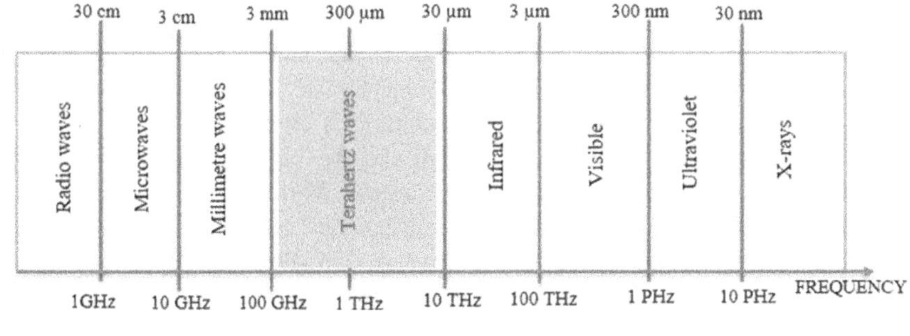

FIGURE 6.1 Position of THz waves in the radio spectrum.

DOI: 10.1201/9781003599111-6

TABLE 6.1
Different Technologies for Wireless Communication [3]

Wireless Communication	Frequency	Range	Power	Network Topology	Noise Source	Security
Millimeter waves (mmW)	30–300 GHz	Short	Medium	Point to multi-point	Thermal	Medium
Terahertz (THz) Waves	100 GHz to 10 THz	Short to medium	Medium	Point to multi-point	Thermal	High
Infrared (IF)	10–430 THz	Short to long	Low	Point to point	Sun light	High
Visible light	430–790 THz	Short	Low	Point to point	Sun light	High
Ultraviolet (UV) waves	790 THz to 30 PHz	Short	Low	Point to multi-point	Sun light	—

spectroscopically and interact with van der Waals forces or molecular hydrogen bonds without emitting any ionizing radiation. THz photons, which have a low energy (1 THz is equivalent to 4 meV), are not hazardous to people, in contrary to high-energy X-rays. Since THz waves include the superconducting and semiconducting materials' fundamental vibrational energy levels and the molecules' rotational and vibrational energy levels, they are extremely useful for the analysis of spectrum and identification of material.

THz band offers extremely high transmission rates and a vast amount of available bandwidth resources. For achieving a Tbps rate, THz communication is regarded as a prominent air interface technique as an alternative. Table 6.1 shows the comparison between various wireless communication technologies. Additionally, it is anticipated to be used in cases like short-distance ultra-high-speed transmission, small-scale communication, holographic communication, and ultra-large-capacity data backhaul. Furthermore, a THz communication application's expansion direction is its exceptionally broad bandwidth, which is used for high-resolution image sensing as well as the high-precision position of a network device or a terminal [2].

6.2 THz SPECTRUM

The THz spectrum has long been considered the radio spectrum's last unexplored region. Particularly in bands above 275 GHz, very few astronomical and scientific services are used in these frequency ranges. Despite having enough spectrums, a tremendous transmission rate, and an anti-interference nature, there are still a lot of real-world technical restrictions. This scenario is changing with the advancement of integrated circuits and components, and the surge of multiple areas that demand extremely high data transmission rates. World Radio Communication Conference 2019 (WRC-19) approved four frequency bands, which are globally harmonized around a bandwidth of 137 GHz (i.e., 275–296, 306–313, 318–333, and 356–450 GHz). The 275–450 GHz frequency range was designated for the deployment of land mobile and fixed service applications based on the study for Agenda Item 1.15 (WRC-19). The addition of the spectrum allocated at the previous WRC has thus resulted in an increase in the number of 230 GHz mobile service spectrums.

It can support current wireless transmission applications with very high data rates, including wireless cellular fronthauling and backhauling, fixed wireless access (FWA), and certain short-range link communications. Their connection speeds may increase from tens of Mbps to several Gbps, and potentially reach hundreds of Gbps or even Tbps, making them comparable to optical fiber networks.

6.3 BUILDING BLOCKS FOR THz COMMUNICATION

The fundamental frequency and cutoff frequency are the primary factors that determine the various process and material properties of THz devices. A large antenna array is typically required to ensure power transmission because the increase in operating frequency and high integration becomes important, and a THz small wavelength is very beneficial for implementing a small size large-scale antenna array. However, the small wavelength creates some challenges [4].

The hardware building blocks include the analog front-ends, the digital back-ends, and the antenna systems.

6.3.1 Front-End (Analog)

The transmitter produces a THz carrier signal, filters out-of-band emissions, amplifies signal power, and modulates the signal with desired information. The receiver identifies the THz signal, removes noise, boosts the signal, and retrieves the sent data. Performance indicators for the analog front-end include power transmission, receiver sensitivity, amplitude and phase noise, modulation bandwidth, and operational frequency bands [5].

6.3.2 Antenna System

At the transmitter front-end, it is responsible for transforming on-chip modulated THz signals into EM waves to propagate into free space. On the receiver side, coupling of free-space radiation onto the on-chip front-ends is done. The antenna system consists of one or more antennas paired with lenses or other reflecting devices. Beam width, directivity gain, and radiation efficiency are the main performance indicators of an antenna system [5].

6.3.3 Back-End (Digital)

It creates control signals for analog front-ends, processes information signals digitally, and works as a link between analog wireless channels and digital computing devices. At the transmitter, it has digital-to-analog converters (DACs), and at the receiver, analog-to-digital converters (ADCs), with sampling frequency and resolution being key performance criteria [5].

6.4 THz TRANSCEIVERS

These waves enable the practical application of technologies based on photonics and electronics to generate them. Presently, there are three primary technological approaches for creating THz systems, the plasmonic, photonic, and electronic approaches, as described in Figure 6.2.

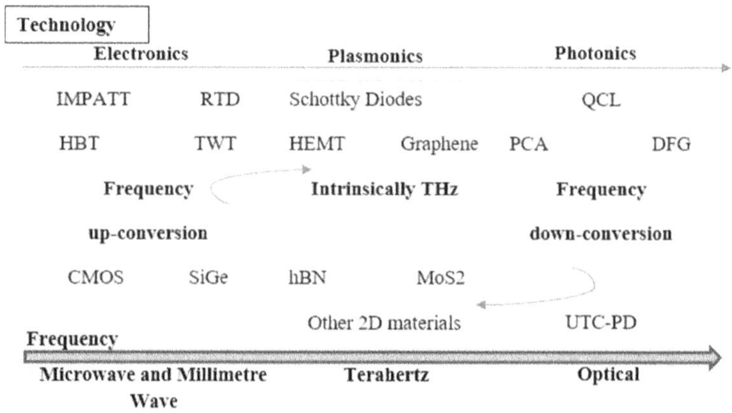

FIGURE 6.2 Technological approaches for creating THz systems [5].

6.4.1 Electronic Approaches

The latest advances in nanofabrication technology have enhanced semiconductor devices for THz waves. These devices include a variety of silicon-based technologies [6], such as Silicon CMOS and Silicon-Germanium (SiGe) BiCMOS, which provide a high-level integration with existing electronic devices [7,8] and electronics made of gallium arsenide and indium phosphate [9], for frequency-multiplying chains high-electron-mobility transistors (HEMT), heterojunction bipolar transistors (HBT), and Schottky diodes to be used [10,11].

IMPACT ionization Avalanche Transit Time (IMPATT) diodes [12,13], Resonant Tunnelling Diodes (RTDs) [14,15], and Traveling Wave Tubes (TWTs) [16,17] are a few cases of electronic systems. The generated signal frequency in each of these situations is a few hundred GHz. RTDs and IMPATT diodes emit power in the sub-mW range, although heavier TWTs emit nearly few Ws. In addition to all of these, the most popular method for producing a THz carrier signal is up-converting a signal of lower frequency (millimeter-wave and microwave bands) via frequency-multiplying chains. Several device technologies can be used to construct frequency multipliers. The transmitting and receiving devices used in the electronics approach are shown in Figure 6.3.

The process of mixing and frequency multiplication is extremely irregular. This presents many difficulties in THz communications, such as tremendous sensitivity for peak-to-average power ratio (PAPR), substantial distortion of the waveform, and exceptional increase in phase noise. Electronic arrays are being used to boost transmission power and enable mobile THz networking.

The authors of [18] introduced a novel architecture of silicon for the generation and shaping of THz wave signals in a single microchip, as well as scalability. This kind of contribution is essential because of the difficulty in THz creation to produce a broad spectrum of wavelengths, especially in a microchip.

6.4.2 Photonic Approaches

Higher data speeds for photonics devices are largely dependent on how the optical fibers are fed into the THz emission circuits. Photonics-based methods offer a unique

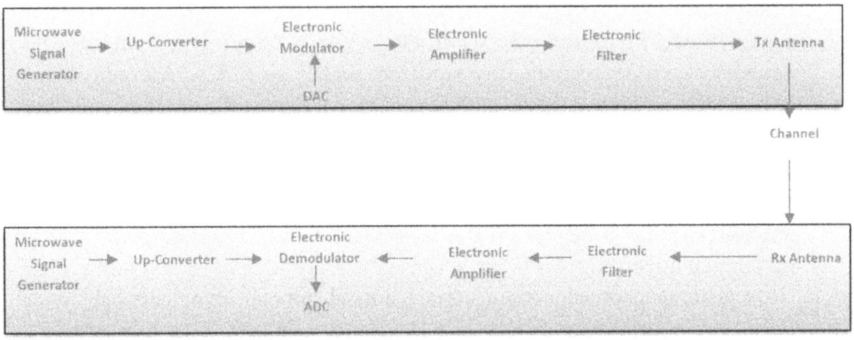

FIGURE 6.3 Electronic approach [5].

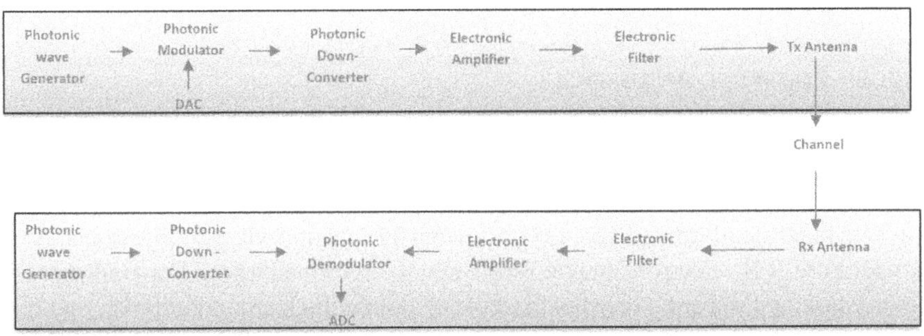

FIGURE 6.4 Photonic approach [5].

opportunity to attract clients or innovators by showcasing technology using readily available lasers, modulators, and photodiodes. Figure 6.4 shows the transmitting and receiving devices used in the photonic approach.

Because of carrier switching properties, for hybrid networks to use a radio/optical interface, these devices also enable the handling of multi-carrier THz channels [19]. The three main methods for photonic THz signal production, modulation, and detection at the moment are as follows:

1. **Quantum Cascade Lasers (QCLs):** The low-energy photons emitted by these semiconductor lasers correspond to THz and far-infrared frequencies [20,21]. QCLs typically operate in the frequency band over 2–3 THz. It is still difficult to determine actual power values when working at room temperature, even though output powers of several hundred milliwatts have been shown at cryogenic temperatures [22,23]. At present, their use in real-world applications is limited by this.
2. **Difference Frequency Generation (DFG):** Two optical carrier signals with wavelengths that are identical to the desired THz waves but varied by an increase can be photo-mixed to create a THz carrier signal [24]. These systems typically comprise a uni-traveling photodiode (UTC), which works as a down-converter for THz frequencies, after two narrow-band lasers [25].

3. **Photoconductive Antennas (PCAs):** Another way to down convert optical signals to the THz band is by using PCAs [26,27]. Typically, THz metallic antenna over a photoconductive substrate makes up a PCA. An optical signal, usually an optical pulse, activates photocarriers at the antenna gap. The carriers are subsequently accelerated through the structure of the antenna under the influence of the DC bias field, which results in the production of THz photons. The main disadvantage of this method is its low conversion efficiency; at hundreds of gigahertz, it receives a low power output, even less than a milliwatt.

   In general, photonic transmitters are used in heterodyne networks with frequency-multiplying chain-based electronic receivers. This is because of the absence of any optical mechanism process that up-converts THz signals, which carry information to the optical spectrum, where they are processed using already available optical technology.

6.4.3 Plasmonic Approaches

Furthermore, most traditional materials that are used for lower frequencies (microwave and millimeter-wave ranges) are not suitable for high frequencies and usually show significant losses at the THz range. Therefore, graphene is used as a substitute for THz band small transceivers [28], allowing THz nanodevices to function at room temperature. THz electromagnetic waves may be reformed using thin graphene layers because of graphene's benefits in terms of atomic thickness, tuneability, and high kinetic inductance [29].

THz signals have been developed by utilizing the characteristics of surface plasmon polariton (SPP) waves [28,30]. It is used to construct ultra-broadband modulators to achieve the 1 Tbps rate of data to anticipate a 6G network. Graphene can be mixed with III–V semiconductors or other 2D materials like molybdenum or hexagonal boron nitride (hBN) to improve some of these characteristics. As a result, graphene-based THz components' performance in producing, modifying, and detecting THz waves has been extremely beneficial [31].

Plasmonic THz devices operate above hundreds of GHz and a few THz with 10% of their exceeding carrier frequency bandwidths. Their small size (from hundreds of nanometers to a few micrometers) allows for low power consumption and integration in nanoscale applications and dense compact arrays. THz waves radiate through a nano-dipole and nano-patch antennas. Transmitting and Receiving devices used in the plasmonic approach are shown in Figure 6.5. Graphene, for its small electric size, is considered to perform excellent for THz radiation as compared to its metal counterpart. This leads to the manufacturing of new devices with unparalleled functionalities, including light concentrators, nano-antennas, superlenses, hyperlenses, and sub-wavelength guiding structures. Other uncommon THz communications devices include liquid crystals and phase-change materials like vanadium dioxide (VO_2) [32,33]. For high-speed modulation, phase change substances may be used as tunable components for both the transmitter side and the receiver end. Additionally, tunable plasmonic THz sources are made possible using superconductors such as $Bi_2Sr_2CaCu_2O_8$ + (BSCCO) [34].

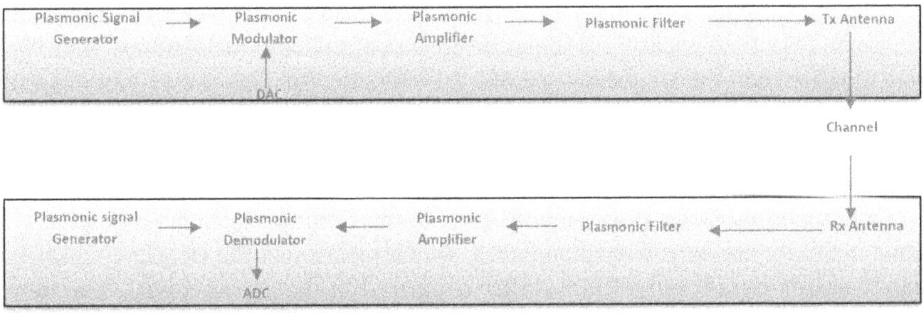

FIGURE 6.5 Plasmonic approach [5].

6.5 ANTENNA SYSTEMS

An intelligent antenna system is needed for transmission, reception, and reflection. Generally, plasmonic antennas are used for plasmonic front-ends, and normal antennas are typically used alongside electronic and photonic front-ends. THz signals' short wavelength enables the production of very small-sized conventional THz antennas. Generally, two primary directional antennae are used:

6.5.1 Fixed Directional Antennas

These include more complex designs like multi-reflector antennas [78], which give gains around 30–35 dBi; horn antennas, which typically have gains ranging from 20 to 25 dBi (or gain in dB for isotropic antenna); and cascade-dish reflector antennas, which have gains of about 40–55 dBi. Leaky wave antennas are another type of non-traditional fixed directional radiating structure. These antennas allow for new spatial and frequency multiplexing techniques because their radiation direction is dependent on the excitation frequency [35,36].

6.5.2 Beamforming Antenna Arrays

These can be made using new nanomaterials or traditional metals, and they can be either fixed or dynamic. For instance, at frequencies lower than 300 GHz, metallic antenna arrays with a combination of electronic front-ends have been demonstrated in [37, 38] having a maximum of 16 adjustable components. A vast graphene-based plasmonic antenna array consisting of 1,024 elements has been proposed combined with plasmonic front-ends [40,41].

THz frequencies permit small antenna sizes and wide antenna array components to be placed over terminal devices. This result, diversity gain, and directivity gain for antenna over Multiple-Input-Multiple-Output (MIMO) techniques are accomplished. To improve communication distance in THz systems, the Ultra-Massive MIMO (1,024 × 1,024) concept was introduced [39].

Dielectric lenses enhance the gain and dynamic abilities of directional THz antennas, providing up to 30 dBi of gain on on-chip quasi-omnidirectional antennas [42] and 40–55 dBi in lens-integrated horn antennas. Metamaterials and metasurfaces are engineered structures with unique EM properties that can be used for programmable lenses at THz frequencies. These lenses consist of metallic meta-atoms, with their size smaller than the frequency-specific wavelength. Antenna arrays and metasurfaces can be used for reflection, enabling reflective intelligent surfaces (RISs), which are crucial for intelligent propagation environments in 6G systems, overcoming LoS propagation requirements [43].

One important factor is the antenna bandwidth. Commercial communication systems typically use narrowband antennas, which means that the frequency response bandwidth is significantly 10% smaller compared to the carrier signal. The transmitted waveforms may get distorted if they surpass the antenna bandwidth because antenna systems function as filters [44]. Other prospects for THz antenna systems include the creation of structures that allow wavefronts other than the conventional Gaussian beams to be constructed. For instance, bending Airy beams [45] or self-healing Bessel beams [46] can expand their possibilities for intelligent THz propagation settings. Furthermore, orbital angular momentum (OAM) can be defined on top of the wavefronts [47,48]. Frequently contrasted with MIMO, OAM does not require complex digital signal processing and generates intrinsically orthogonal parallel spatial channels. By using these OAM modes, THz systems' capacity can be increased without requiring an increase in system bandwidth.

6.6 CHANNEL MEASUREMENT METHODS

It is important to determine the features of the spectrum and design channel models. Numerous measurement techniques have long been used in practice. Even with high-directivity antennas, THz channel measuring systems need to detect wideband properties, enable long-distance sounds because of significant path loss, and take a fair amount of time while scanning the spatial domain [5]. Three types of channel measurement are used:

6.6.1 Vector Network Analyzer Method

To analyze a linear system's frequency response, one frequency at a time, Vector Network Analyzer (VNA) is a common technology to be used, as shown in Figure 6.6. VNAs are frequently used to analyze the wireless channel by connecting antennas at both the transmitting end and receiving side. They are used to determine discrete electrical or RF components. The VNA has several drawbacks, including long measurement periods, low power output, increased noise, and specific range. Low-noise amplifiers and external power can be used to overcome the first two problems, which are low output power and high noise. By using directional broadcasts or optical fiber, the distance can be increased [49].

6.6.2 Sliding Correlation Method

An alternative method of measuring a channel involves sending a signal with autocorrelation expressed as a Dirac delta function, in which case the channel is thought to be time-invariant and the same is described in Figure 6.7. Increased power transmission

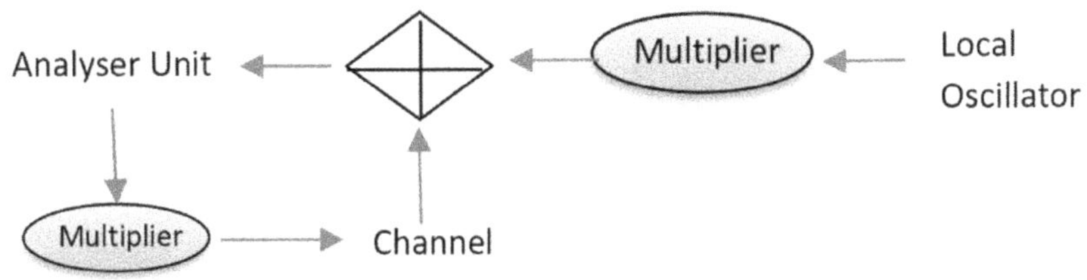

FIGURE 6.6 Vector network analyses [5].

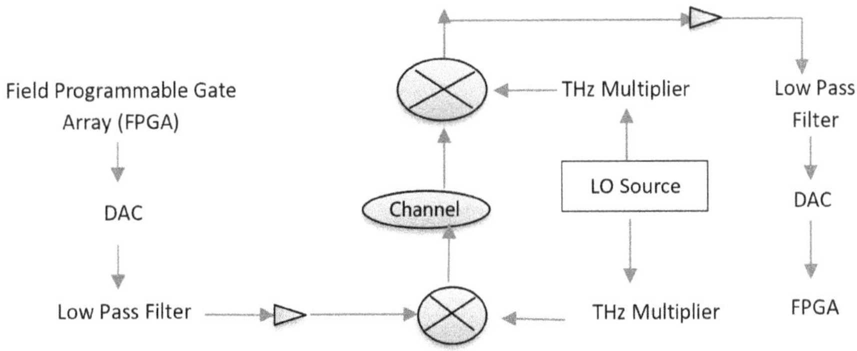

FIGURE 6.7 Sliding correlation method [5].

with a low PAPR can be exploited using this correlation-based method. This approach has the benefit of enabling real-time measurements. Nevertheless, this approach has two major shortcomings. The power spectral density tends to be unevenly distributed, on the one hand. Conversely, ultra-high sample rate DACs and ADCs are required. The sliding correlation method allows for the resolution of the second issue and improves the SNR by increasing the measurement period since various chip speeds can be used for the correlation between the receiver and the emitter [50].

6.6.3 Direct Pulse Method

It is also referred to as THz time-domain spectroscopy (THz-TDS). It is the simplest method that trains extremely narrow pulses in the time domain with a period greater than the maximum excess delay, or the relative delay between the multipath component whose energy drops below a specific threshold from the strongest one and the first-arriving component, as shown in Figure 6.8. By using this measuring technique, it is simple to determine the channel impulse response by looking at the received signals. This can be achieved by considering both the delay between the transmission and sampling times, as well as the amplitude of the channel impulse response. This technique works well for THz waves with ultra-broad bandwidth [51], but it has drawbacks like low power consumption, big equipment size, short range, and a limited range of applications beyond measuring the narrow beams' effects on reflection, scattering, and diffraction.

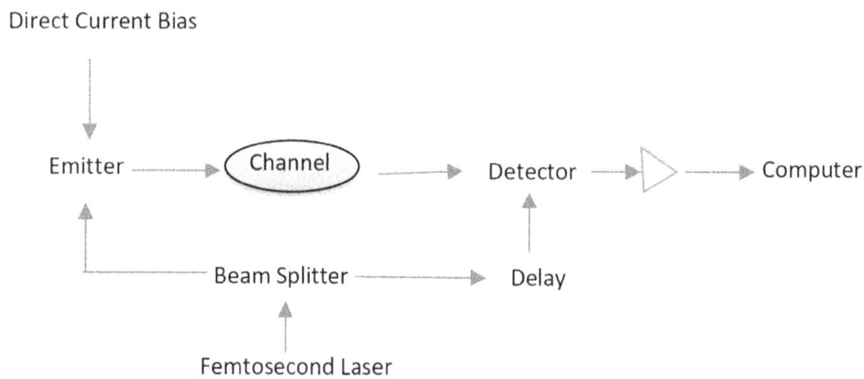

FIGURE 6.8 Direct pulse method [5].

6.7 THz CHANNEL MODELING

In general, many models describe in the literature survey fall into one of three groups:

6.7.1 Deterministic Channel Modeling

The propagation of waves is accurately described by deterministic channel models. In general, numerical findings match measurements, but precise transceiver positions, material properties, and full information of the geometry of the wireless system environment are needed. Two types of channels are commonly utilized in THz communications:

1. **Ray-Tracing (RT):** The Resonance Transmitter (RT) is a trustworthy method to model wireless environments for specific locations, reducing computational complexity for huge systems. It involves a stationary transmitter to generate signals for different objects for different surface situations, analyzing wavefronts as simple particles to explore electromagnetic wave characteristics. The RT uses backtracking methods to create a visibility tree for deducing paths, while another way is to use ray launching, which is a grid of rays to explore the trajectory [52].
2. **Finite-Difference Time-Domain (FDTD):** This channel model is obtained numerically through Maxwell's equations. In particular, the whole area is separated into grids that are independently examined to speed up the computation. After that, both the magnetic and electric fields were examined and updated continuously in spatial and temporal regions. With careful theoretical calculations, the FDTD can accurately model rough surfaces and small, intricate scatterers. Nevertheless, achieving complete geometrical precision comes at a significant computational cost in terms of memory and time characteristics during propagation [53].

6.7.2 Statistical Channel Modeling

This approach is better because it has less computational complexity than deterministic channel modeling. Instead of using narrowband models, which typically assume Rayleigh or Rician fading, customized models are required as THz channel is thought to be relatively limited with manageable multipath elements. Certain SCMs use fitted distributions to describe the temporal and spatial channel properties, including complex amplitudes, directions of departure/arrival (DoD/DoA), and time of arrival (ToA). While certain SCM explains the mathematical properties of channel and antenna impulse response without taking propagation into account [5].

6.7.3 Hybrid Channel Modeling

Deterministic channel modeling uses less time and resources while achieving excellent accuracy. Statistical channel modeling sacrifices accuracy but operates with a little computing load. Therefore, the design of hybrid models looks appealing and trustworthy to resolve hard THz channel models for minimum latency by purposely combining two or more approaches.

Due to water vapor molecules, the THz channel experiences notable absorption from the atmosphere. Along with air attenuation, free-space path loss is regarded as a natural physical phenomenon. The THz band channel is therefore regarded as having good frequency selectivity [54]. As a result, transmission distance is limited by the attenuation, and application determines the proper carrier frequency.

However, even if there are absorption peaks centered at particular frequencies, there are transmission windows that make it possible to establish effective communication at the THz frequency spectrum. Since there are currently no models that simulate outdoor conditions, the research for modeling channels in the THz range is less or limited to indoor settings only. In [55], for THz channels, the first statistical model was presented using frequencies in the range of 275 and 325 GHz. For D2D scatter channels at the sub-THz band, a geometrical statistical model was proposed [55] to overcome these problems. With the help of these models, scattering and reflection patterns in a sub-THz D2D environment are simulated.

For kiosk applications, [56] introduced a stochastic model for the THz region, which ranges between 220 and 340 GHz. To extract the channel properties of three distinct kiosk applications, 3D ray-tracing simulator was carried out for channel measurements. At THz frequencies, near-proximity communication systems may be designed using the proposed channel model. In [57], the THz scenario's statistical channel characterization is presented. Although the research focuses on 240–300 GHz frequency bands, it is the first to depict single-sweep measurement results of THz waves. To develop sustainable transceiver systems, especially those that deal with antenna misalignment issues, measured data allow the acquisition of finer temporal information.

6.8 APPLICATIONS

6.8.1 Automotive

Unmanned autonomous vehicles (UAVs) have become widely available, enabling applications in various fields such as weather control, traffic detection, and emergency search and rescue. To achieve these applications, UAVs require reliable communication links. THz frequency band is a reliable solution for this, offering high-capacity wireless backhaul and can mitigate the fast-paced environment for UAVs. These links are suitable for swapping safety-critical data and serve as a gateway in the sky between UAVs and aircraft to provide internet access during missions [58].

6.8.2 Sensing and Detection

Applications that use mmWave and THz frequencies benefit from sub-millimeter wavelengths and the environment's frequency-dependent features to obtain comprehensive data and insights. These applications can utilize wider channel bandwidths, frequency-selective resonances, and absorption of materials. Antennas with a high gain even in small form factors allow directional sensing, with finer spatial resolution as frequencies increase, enabling sub-millimeter spatial differentiation. Beam scanning allows for the creation of images of physical spaces by monitoring signal signatures at various angles. Real-time implementation of electrical beam steering algorithms and small radio propagation distances make it possible to measure properties in less than seconds. This opens a new way for wireless possibilities, enabling reality sensing and detailed 3-D maps.

The use of spectroscopy in smartphones for 6G and beyond could enable the detection of chemicals, allergens, and defects in food, drink, and air. This technology could lead to new sensing applications such as miniaturized radars, touchless smartphones, explosion detection, gas detection, scanning of body security, detection of air quality, health monitoring, precision time or frequency transfer, and wireless connectivity. Real-time maps could also predict channel characteristics, aid in antenna alignment, and adapt wireless capabilities. This is used in transportation and shopping type of commercial applications [59].

6.8.3 Hybrid Wireless Communication Link

The hybrid radio frequency/free space optical (FSO) system is gaining interest in further wireless communication technology. There are two hybrid methods: parallel and serial. Parallel hybrids use two communication links, one-way or two-way, depending on weather conditions. THz is useful in fog and wind, while FSO uses minimal wind for strong beam-tracking. Serial hybrids use multi-hop links optimized for system performance [58].

6.8.4 Imaging and Scanning

Millimeter-wave (mmWave) and THz radar offer significant advantages over Light Detection and Ranging (LIDAR) in various applications due to their superior resilience to adverse weather conditions such as fog, rain, and clouds, which severely degrade LIDAR performance. Although LIDAR is excellent in high-resolution imaging, its operational capabilities are significantly limited in challenging atmospheric conditions. mmWave and THz radar can effectively support driving, flying, and other critical operations in bad weather, military, and national security scenarios. Furthermore, high-resolution video-based radar systems can complement lower-frequency radar technologies. Dual-frequency radar systems, incorporating both lower and higher frequencies, can enhance performance in extremely challenging conditions such as heavy fog or rain. THz waves possess unique capabilities that extend beyond conventional imaging. They can penetrate certain materials, enabling "through-wall" imaging and the detection of Near-Local Objects (NLOS) that are otherwise obscured. This capability has significant implications for applications such as search and rescue, surveillance, autonomous navigation, and localization. While optical NLOS imaging offers high resolution, it faces significant challenges, including complex hardware requirements, computationally demanding image reconstruction algorithms, and limitations in field of view and signal strength. Lower-frequency NLOS radar systems, while less susceptible to signal attenuation, often require precise knowledge of the surrounding environment and are primarily limited to object detection rather than detailed imaging. THz waves uniquely combine the advantages of both microwave and visible light, offering both small wavelengths for high spatial resolution and wide bandwidths for high data rates. This enables the creation of high-resolution 3D images using relatively compact imaging systems. THz radar systems achieve this by illuminating the scene with THz waves and precisely measuring the time of flight (ToF) of the reflected signals [59].

6.8.5 Data Centers

The escalating demand for cloud services has intensified competition among data centers to enhance performance. However, the inherent limitations of static wired connections, such as finite bandwidth and the inability to dynamically adapt to traffic surges, pose significant challenges. While wireless networking offers greater flexibility and efficiency, its applicability within data centers is hindered by short operational ranges and susceptibility to signal blockage. To overcome these limitations, integrating THz links as a complementary technology within data center infrastructures presents a promising solution. This approach can potentially improve performance, reduce cabling costs, and maintain high throughput levels [58].

6.8.6 Wireless Cognition

Real-time communication links enable machines or devices to perform large computations remotely, enabling complex tasks and fast data rates. This allows drone fleets to operate without a power budget, as fixed base stations support real-time cognition and wireless connection. This allows autonomous vehicles to perform cognitive processing remotely, even without local cognition platforms [59].

6.8.7 Mobile Heterogeneous Networks (HetNets)

THz waves, despite their increased path fading and directional antenna requirements, can be transformed into a more efficient femtocell regime. Femtocells reduce the distance between base stations and users, maintaining a high signal-to-noise ratio (SNR). These base stations may be used for Low-Power Nodes for various applications, including in-home services, automation, and traffic lights. Machine learning (ML) can enhance the THz femtocell experience by modeling problems like large-scale MIMOs, D2D networks, and HetNets. Reinforcement learning is crucial for autonomous spectrum identification and parameter tuning [58].

6.8.8 3D Beamforming for Antenna

3D MIMO technology is used for revolutionizing THz wireless, enhancing communication range, reducing interference, and enabling directional beam formation. This technology can improve signal strength at the end of the receiver by fixing the vertical main lobe precisely, thereby enhancing diversity and spatial separation, potentially boosting signal quality or concurrent user support [58].

6.8.9 Space

The environment attenuation, which is critical to space communications in the THz band, does not affect space applications in the free atmosphere, in contrast to THz communications on the ground. Additionally, compared to laser communication systems, THz wave systems have less attenuation in hostile environments including rain, fog, haze, and battlefields, and compared to traditional millimeter-wave and microwave systems, they can provide significantly larger bandwidth resources. THz systems offer inter-satellite connections between low earth orbit (LEO) and geostationary (GEO) satellites. THz, microwave, millimeter wave, and visible waves are utilized to create links within satellites and the ground for high communication reliability that depends on several variables, including location and weather [60].

6.9 MACHINE LEARNING-BASED TERAHERTZ COMMUNICATION

Models based on mathematical calculations explain wireless signal transmission through the air, which is the foundation of modern wireless communications. To describe the physical systems, these mathematical models are typically imperfect. Furthermore, we generally do not obtain their mathematical equations in many common communication situations, such as indoor THz communications and the dense building blocks of wireless networks. The ML/DL-enabled approach has been shown to address high-dimensional, complicated electromagnetic environments, model-free, and unsolvable non-linear problems. A great deal of research is focused on ML/DL for upcoming 6G THz Communications systems to address complicated issues, such as estimation of the channel, antenna design-beamforming, beam tracking, prediction of blockage, and interference mitigating, due to the inherent difficulties in THz communications, such as high size, transmission vulnerability, and model-free.

- **For Channel Estimation:** Channel estimate is a crucial and difficult challenge in THz communications systems because of the extremely high-frequency operation and small RF chains number. Several ML/DL-empowered methods have been proposed to address this problem. Utilizing the inherent sparsity in the THz communications channel estimation problem, a method may be put forth that makes use of a traditional neural network architecture to both train the sensing matrix and reconstruct the sparse signal. An end-to-end autoencoder is capable of channel estimation in addition to other modules, such as coding, modulation, and demodulation [61].

- **For Intelligent Beamforming and Smart Antenna:** Throughput, spectral efficiency, and energy efficiency may all be substantially boosted by intelligent beamforming and smart antenna steering systems, which can also reduce interference, expand coverage area, and support very challenging applications for mobile devices. The use of ML and deep learning (DL) techniques can lead to more intelligent beamforming systems for 6G communications in the THz band. Depending on the volume of training data and the complexity of the learned models, the differences between digital and analog beamforming within the communication chain will be defined by either differing degrees of cognition or flexibility [61].
- **For Predication of Blockage and Intelligent Beam Tracking:** Strengthening reliable communication lines and overcoming path loss can be implemented by strategies of prediction of blockage and intelligent beam tracking. Beam tracking often starts with tracking the user's movement in highly mobile THz systems, Future mobile THz communication systems will employ beam tracking and blockage prediction techniques based on user trajectory, made possible by ML/DL-empowered approaches for learning user mobility and blockage. To solve the LoS blockage issue, DL can motivate intelligently prediction of blockage and improve Relay Network efficiency through agile scheduling [61].
- **For Interference Mitigation:** When many users access the transmissions in dense THz networks, the uncoordinated directional transmissions will result in significant interference. This interference is typically received randomly during a brief period, and the received strength is also high. Using ML/DL-based techniques, like reinforcement learning, to effectively identify and intelligently cancel out sporadic interference from disorganized directions while requiring significantly less computation time [61].

6.10 CHALLENGES

Significant efforts have been made to address several issues or discover new information about THz communications, encompassing device development, fundamental building blocks, packaging solutions, antenna design, channel modeling, mitigating interference with passive services, and identifying suitable applications. Despite the promise of exceptionally high bandwidth, the THz communications community faces significant hurdles in realizing practical and energy-efficient beamforming transceivers capable of achieving Tbps data rates at these frequencies. Large-scale phased-array antennas' high signal loss necessitates the use of additional amplifiers, which results in significant power dissipation [62].

Channel estimate in THz communication is challenging because of the hybrid beamforming and the enormous number of antennas. In real-world hybrid beamforming systems, direct channel dimension estimation at the digital baseband is not possible due to the quickly changing channel within the coherence period. This limitation hinders the accurate acquisition of channel state information. Beam scanning and search techniques are commonly employed to address this challenge. A significant body of research focuses on characterizing channel propagation in both

Line-of-Sight (LOS) and Non-Line-of-Sight (NLOS), although additional research is necessary to better understand the channels and see how the mobility element influences the behavior of the channels under various circumstances.

Another issue is the creation of effective algorithms and protocols for accurate channel estimation. Positioning is important because the spread angle and delay domain affect performance, and the connection between positioning and communication must also be taken into account. For 3D beamforming schemes, a quick beam scanning technique is employed; however, it is time-consuming and ineffective. Therefore, by forecasting the target's paths, the study can be done to minimize resource usage. The resolution of spatial and enormous variety attained by large-scale MIMO radars requires a great deal of research. Regarding the DL method, there are several issues to be taken into account, such as algorithmic issues, implementation-focused issues, data-related issues, and performance analysis [57].

At THz frequencies, highly directed antennas are needed to reduce path loss. It is still very difficult to design antennas with broad operation, high gain, and small size. Ultra-compact antennas are necessary for portable THz devices, necessitating creative designs such as planar arrays, leaky-wave antennas, or structures based on metamaterial. These technologies are still in their early phases of development. At THz frequencies, conventional antenna materials like metals experience higher ohmic losses. Although they are promising, novel materials like graphene, carbon nanotubes (CNTs), and metamaterials are not yet commercially scalable.

6.11 CONCLUSIONS

A major advancement in wireless communication and sensing systems has been made possible by THz communication and its combination with microwave technology. Ultra-high-speed data transmission, improved spectrum efficiency, and extremely sensitive imaging and sensing applications are all made possible by the THz band. This spectrum increases the potential for next-generation solutions in telecommunications, healthcare, defense, and environmental monitoring when combined with the well-established advantages of microwave technology, such as dependability, penetration capability, and long-range communication.

With the ability to handle large device connectivity, enable data rates in the terabits per second (Tbps) range, and reduce latency to levels close to real-time, THz communication is especially well-suited to transform 6G and beyond. It is perfect for applications such as non-destructive testing, medical diagnostics, and security screening because of its high-frequency functioning, which enables small and accurate imaging systems. When paired with microwave technology's resilience, this hybrid strategy overcomes shortcomings in both areas and provides a balance between performance and adaptability.

There are still many obstacles in the way of this immense potential. Developments in materials research, device downsizing, and economical manufacturing techniques are necessary for the effective creation, modulation, and detection of THz signals. Other challenges include air absorption, propagation losses, and the requirement for sophisticated antennas and beamforming methods. It will take a concentrated effort in research and development, cooperation between government agencies, business, and academia, and large infrastructural investments to overcome these obstacles.

REFERENCES

1. W. Xun, Y. Liu, Y. Ni, H. Zhao, Z. Xu, and X. Ge, A survey on terahertz communication theory assisted by intelligent reflecting surface and device-to-device technologies, *2022 IEEE 22nd International Conference on Communication Technology (ICCT)*, Nanjing, China, pp. 678–682, 2022, doi: 10.1109/ICCT56141.2022.10073033
2. O. Li et al., Integrated sensing and communication in 6G: a prototype of high resolution multichannel THz sensing on portable device, *EURASIP Journal on Wireless Communications and Networking*, vol. 2022, no. 1, p. 106, Oct. 2022, doi: 10.1186/s13638-022-02172-w
3. H. Elayan, O. Amin, R. M. Shubair, and M. S. Alouini, Terahertz communication: The opportunities of wireless technology beyond 5G, *2018 International Conference on Advanced Communication Technologies and Networking (CommNet)*, Marrakech, Morocco, pp. 1–5, 2018, doi: 10.1109/COMMNET.2018.8360286
4. G. Wang et al., Terahertz Sensing and Communication towards Future Intelligence Connected Networks, no. September 2022.
5. I. F. Akyildiz, C. Han, Z. Hu, S. Nie, and J. M. Jornet, Terahertz band communication: An old problem revisited and research directions for the next decade, *IEEE Transactions on Communications*, vol. 70, no. 6, pp. 4250–4285, 2022, doi: 10.1109/TCOMM.2022.3171800
6. M. Fujishima, S. Amakawa, K. Takano, K. Katayama, and T. Yoshida, Terahertz CMOS design for low-power and high-speed wireless communication, *IEICE Transactions on Electronics*, vol. E98.C, no. 12, pp. 1091–1104, 2015, doi: 10.1587/transele.E98.C.1091
7. S. Hara et al., 300-GHz CMOS transceiver for terahertz wireless communication, *2018 Asia-Pacific Microwave Conference (APMC)*, Kyoto, Japan, IEEE, Nov. 2018, pp. 429–431. doi: 10.23919/APMC.2018.8617468
8. K. O. Kenneth et al., Opening terahertz for everyday applications, *IEEE Communications Magazine*, vol. 57, no. 8, pp. 70–76, Aug. 2019, doi: 10.1109/MCOM.2019.1800909
9. T. Nagatsuma, Terahertz technologies: Present and future, *IEICE Electronics Express*, vol. 8, no. 14, pp. 1127–1142, 2011.
10. K. M. K. H. Leong et al., 850 GHz receiver and transmitter front-ends using InP HEMT, *IEEE Transactions on Terahertz Science and Technology*, vol. 7, no. 4, pp. 466–475, Jul. 2017, doi: 10.1109/TTHZ.2017.2710632
11. J. V. Siles, K. B. Cooper, C. Lee, R. H. Lin, G. Chattopadhyay, and I. Mehdi, A new generation of room-temperature frequency-multiplied sources with up to 10× higher output power in the 160-GHz–1.6-THz range, *IEEE Transactions on Terahertz Science and Technology*, vol. 8, no. 6, pp. 596–604, Nov. 2018, doi: 10.1109/TTHZ.2018.2876620
12. A. Biswas et al., 1.0 THz GaN IMPATT source: Effect of parasitic series resistance, *Journal of Infrared, Millimeter, and Terahertz Waves*, vol. 39, no. 10, pp. 954–974, Oct. 2018, doi: 10.1007/s10762-018-0509-z
13. S. Banerjee, THz Solid-state source based on IMPATT devices, in A. Banerjee, B. Chakraborty, H. Inokawa, and J. N. Roy (Eds.) *Terahertz Biomedical and Healthcare Technologies*, Elsevier, 2020, pp. 1–41. doi: 10.1016/B978-0-12-818556-8.00001-X
14. S. Diebold et al., Modeling and simulation of terahertz resonant tunneling diode-based circuits, *IEEE Transactions on Terahertz Science and Technology*, vol. 6, no. 5, pp. 1–8, 2016, doi: 10.1109/TTHZ.2016.2592180
15. K. Kasagi, S. Suzuki, and M. Asada, Large-scale array of resonant-tunneling-diode terahertz oscillators for high output power at 1 THz, *Journal of Applied Physics*, vol. 125, no. 15, Apr. 2019, doi: 10.1063/1.5051007
16. A. Baig et al., Performance of a nano-CNC machined 220-GHz traveling wave tube amplifier, *IEEE Transactions on Electron Devices*, vol. 64, no. 5, pp. 2390–2397, May 2017, doi: 10.1109/TED.2017.2682159

17. P. Hu et al., Demonstration of a Watt-level traveling wave tube amplifier operating above 0.3 THz, *IEEE Electron Device Letters*, vol. 40, no. 6, pp. 973–976, Jun. 2019, doi: 10.1109/LED.2019.2912579
18. X. Wu and K. Sengupta, Dynamic waveform shaping with picosecond time widths, *IEEE Journal of Solid-State Circuits*, vol. 52, no. 2, pp. 389–405, Feb. 2017, doi: 10.1109/JSSC.2016.2616349
19. T. Nagatsuma, Advances in terahertz communications accelerated by photonics technologies, *2019 24th OptoElectronics and Communications Conference (OECC) and 2019 International Conference on Photonics in Switching and Computing (PSC)*, Fukuoka, Japan, IEEE, Jul. 2019, pp. 1–3. doi: 10.23919/PS.2019.8818026
20. B. S. Williams, Terahertz quantum-cascade lasers, *Nature Photonics*, vol. 1, no. 9, pp. 517–525, 2007.
21. X. Wang et al., High-power terahertz quantum cascade lasers with ~0.23 W in continuous wave mode, *AIP Advances*, vol. 6, no. 7, Jul. 2016, doi: 10.1063/1.4959195
22. Q. Lu, F. Wang, D. Wu, S. Slivken, and M. Razeghi, Room temperature terahertz semiconductor frequency comb, *Nature Communications*, vol. 10, no. 1, p. 2403, Jun. 2019, doi: 10.1038/s41467-019-10395-7
23. A. Khalatpour, A. K. Paulsen, C. Deimert, Z. R. Wasilewski, and Q. Hu, High-power portable terahertz laser systems, *Nature Photonics*, vol. 15, no. 1, pp. 16–20, Jan. 2021, doi: 10.1038/s41566-020-00707-5
24. M. Sung et al., Design considerations of photonic THz communications for 6G networks, *IEEE Wireless Communications*, vol. 28, no. 5, pp. 185–191, Oct. 2021, doi: 10.1109/MWC.001.2100002
25. C. C. Renaud, M. Natrella, C. Graham, J. Seddon, F. Van Dijk, and A. J. Seeds, Antenna integrated THz uni-traveling carrier photodiodes, *IEEE Journal of Selected Topics in Quantum Electronics*, vol. 24, no. 2, pp. 1–11, Mar. 2018, doi: 10.1109/JSTQE.2017.2725444
26. N. M. Burford and M. O. El-Shenawee, Review of terahertz photoconductive antenna technology, *Optical Engineering*, vol. 56, no. 1, p. 010901, Jan. 2017, doi: 10.1117/1.OE.56.1.010901
27. N. T. Yardimci and M. Jarrahi, Nanostructure-enhanced photoconductive terahertz emission and detection, *Small*, vol. 14, no. 44, Nov. 2018, doi: 10.1002/smll.201802437
28. J. M. Jornet and I. F. Akyildiz, Graphene-based plasmonic nano-transceiver for terahertz band communication, *The 8th European Conference on Antennas and Propagation (EuCAP 2014)*, The Hague, Netherlands, IEEE, Apr. 2014, pp. 492–496. doi: 10.1109/EuCAP.2014.6901799
29. C. J. Docherty and M. B. Johnston, Terahertz properties of graphene, *Journal of Infrared, Millimeter, and Terahertz Waves*, vol. 33, no. 8, pp. 797–815, Aug. 2012, doi: 10.1007/s10762-012-9913-y
30. J. M. Jornet and I. F. Akyildiz, Graphene-based plasmonic nano-transceiver for terahertz band communication, *The 8th European Conference on Antennas and Propagation (EuCAP 2014)*, The Hague, Netherlands, IEEE, Apr. 2014, pp. 492–496. doi: 10.1109/EuCAP.2014.6901799.
31. M. Hasan, S. Arezoomandan, H. Condori, and B. Sensale-Rodriguez, Graphene terahertz devices for communications applications, *Nano Communication Networks*, vol. 10, pp. 68–78, Dec. 2016, doi: 10.1016/j.nancom.2016.07.011
32. J. Wang et al., Liquid crystal terahertz modulator with plasmon-induced transparency metamaterial, *Optics Express*, vol. 26, no. 5, p. 5769, Mar. 2018, doi: 10.1364/OE.26.005769
33. Y. Jeong, Y. Bahk, and D. Kim, Dynamic terahertz plasmonics enabled by phase-change materials, *Advanced Optical Materials*, vol. 8, no. 3, Feb. 2020, doi: 10.1002/adom.201900548

34. K. Delfanazari, R. A. Klemm, H. J. Joyce, D. A. Ritchie, and K. Kadowaki, Integrated, portable, tunable, and coherent terahertz sources and sensitive detectors based on layered superconductors, *Proceedings of the IEEE*, vol. 108, no. 5, pp. 721–734, May 2020, doi: 10.1109/JPROC.2019.2958810
35. M. Esquius-Morote, J. S. Gomez-Diaz, and J. Perruisseau-Carrier, Sinusoidally modulated graphene leaky-wave antenna for electronic beamscanning at THz, *IEEE Transactions on Terahertz Science and Technology*, vol. 4, no. 1, pp. 116–122, Jan. 2014, doi: 10.1109/TTHZ.2013.2294538
36. N. J. Karl, R. W. McKinney, Y. Monnai, R. Mendis, and D. M. Mittleman, Frequency-division multiplexing in the terahertz range using a leaky-wave antenna, *Nature Photonics*, vol. 9, no. 11, pp. 717–720, Nov. 2015, doi: 10.1038/nphoton.2015.176
37. K. Sengupta and A. Hajimiri, A 0.28 THz power-generation and beam-steering array in CMOS based on distributed active radiators, *IEEE Journal of Solid-State Circuits*, vol. 47, no. 12, pp. 3013–3031, Dec. 2012, doi: 10.1109/JSSC.2012.2217831
38. A. Tang et al., A 65nm CMOS 140 GHz 27.3 dBm EIRP transmit array with membrane antenna for highly scalable multi-chip phase arrays, *2014 IEEE MTT-S International Microwave Symposium (IMS2014)*, Tampa, FL, IEEE, Jun. 2014, pp. 1–3. doi: 10.1109/MWSYM.2014.6848244
39. I. F. Akyildiz and J. M. Jornet, Realizing ultra-massive MIMO (1024 × 1024) communication in the (0.06–10) terahertz band, *Nano Communication Networks*, vol. 8, pp. 46–54, Jun. 2016, doi: 10.1016/j.nancom.2016.02.001
40. A. Singh, M. Andrello, N. Thawdar, and J. M. Jornet, Design and operation of a graphene-based plasmonic nano-antenna array for communication in the terahertz band, *IEEE Journal on Selected Areas*, vol. 38, no. 9, pp. 2104–2117, Sep. 2020, doi: 10.1109/JSAC.2020.3000881
41. F. Shen, J. Qin, and Z. Han, Planar antenna array as a highly sensitive terahertz sensor, *Applied Optics*, vol. 58, no. 3, p. 540, Jan. 2019, doi: 10.1364/AO.58.000540
42. N. Llombart, G. Chattopadhyay, A. Skalare, and I. Mehdi, Novel terahertz antenna based on a silicon lens fed by a leaky wave enhanced waveguide, *IEEE Transactions on Antennas and Propagation*, vol. 59, no. 6, pp. 2160–2168, Jun. 2011, doi: 10.1109/TAP.2011.2143663
43. S. Nie and I. F. Akyildiz, Beamforming in intelligent environments based on ultra-massive MIMO platforms in millimeter wave and terahertz bands, *ICASSP 2020-2020 IEEE International Conference on Acoustics, Speech and Signal Processing (ICASSP)*, Barcelona, Spain, IEEE, May 2020, pp. 8683–8687. doi: 10.1109/ICASSP40776.2020.9053786
44. C. Han, L. Yan, and J. Yuan, Hybrid beamforming for terahertz wireless communications: Challenges, architectures, and open problems, *IEEE Wireless Communications*, vol. 28, no. 4, pp. 198–204, Aug. 2021, doi: 10.1109/MWC.001.2000458
45. C. Liu, L. Niu, K. Wang, and J. Liu, 3D-printed diffractive elements induced accelerating terahertz Airy beam, *Optics Express*, vol. 24, no. 25, p. 29342, Dec. 2016, doi: 10.1364/OE.24.029342
46. S. Liu et al., Anomalous refraction and nondiffractive bessel-beam generation of terahertz waves through transmission-type coding metasurfaces, *ACS Photonics*, vol. 3, no. 10, pp. 1968–1977, Oct. 2016, doi: 10.1021/acsphotonics.6b00515
47. L. Zhu et al., Experimental demonstration of basic functionalities for 0.1-THz orbital angular momentum (OAM) communications, *Proceedings of the IEEE Optical Fiber Communication (OFC) Conference*, San Francisco, CA, pp. 1–3, 2014.
48. X. Su et al., Multipath and receiver aperture effects in a THz wireless communications link using OAM multiplexing, *2020 IEEE Globecom Workshops (GC Wkshps)*, Taipei, Taiwan, IEEE, Dec. 2020, pp. 1–6. doi: 10.1109/GCWkshps50303.2020.9367533

49. A. W. Mbugua, W. Fan, K. Olesen, X. Cai, and G. F. Pedersen, Phase-compensated optical fiber-based ultrawideband channel sounder, *IEEE Transactions on Microwave Theory and Techniques*, vol. 68, no. 2, pp. 636–647, Feb. 2020, doi: 10.1109/TMTT.2019.2948842
50. H. Cox, Spatial correlation in arbitrary noise fields with application to ambient sea noise, *The Journal of the Acoustical Society of America*, vol. 54, no. 5, pp. 1289–1301, Nov. 1973, doi: 10.1121/1.1914426
51. H. Harde, R. A. Cheville, and D. Grischkowsky, Terahertz studies of collision-broadened rotational lines, *The Journal of Physical Chemistry A*, vol. 101, no. 20, pp. 3646–3660, May 1997, doi: 10.1021/jp962974c
52. J. S. Lu et al., A discrete environment-driven GPU-based ray launching algorithm, *IEEE Transactions on Antennas and Propagation*, vol. 67, no. 2, pp. 1180–1192, Feb. 2019, doi: 10.1109/TAP.2018.2880036
53. Y. Zhao, Y. Hao, and C. Parini, FDTD characterization of UWB indoor radio channel including frequency dependent antenna directivities, *IEEE Antennas and Wireless Propagation Letters*, vol. 6, pp. 191–194, 2007, doi: 10.1109/LAWP.2007.891963
54. J. M. Jornet and I. F. Akyildiz, Channel modeling and capacity analysis for electromagnetic wireless nanonetworks in the terahertz band, *IEEE Transactions on Wireless Communications*, vol. 10, no. 10, pp. 3211–3221, Oct. 2011, doi: 10.1109/TWC.2011.081011.100545
55. S. Priebe and T. Kurner, Stochastic modeling of THz indoor radio channels, *IEEE Transactions on Wireless Communications*, vol. 12, no. 9, pp. 4445–4455, Sep. 2013, doi: 10.1109/TWC.2013.072313.121581
56. D. He et al., Stochastic channel modeling for kiosk applications in the terahertz band, *IEEE Transactions on Terahertz Science and Technology*, vol. 7, no. 5, pp. 502–513, Sep. 2017, doi: 10.1109/TTHZ.2017.2720962
57. A. R. Ekti et al., Statistical modeling of propagation channels for terahertz band, *2017 IEEE Conference on Standards for Communications and Networking (CSCN)*, Helsinki, Finland, IEEE, Sep. 2017, pp. 275–280. doi: 10.1109/CSCN.2017.8088634.
58. H. Elayan, O. Amin, R. M. Shubair, and M.-S. Alouini, Terahertz communication: The opportunities of wireless technology beyond 5G, *2018 International Conference on Advanced Communication Technologies and Networking (CommNet)*, Marrakech, Morocco, IEEE, Apr. 2018, pp. 1–5. doi: 10.1109/COMMNET.2018.8360286.
59. T. S. Rappaport et al., Wireless communications and applications above 100 GHz: Opportunities and challenges for 6G and beyond, *IEEE Access*, vol. 7, pp. 78729–78757, 2019, doi: 10.1109/ACCESS.2019.2921522
60. T. Nagatsuma and A. Kasamatsu, Terahertz communications for space applications, *2018 Asia-Pacific Microwave Conference (APMC)*, Kyoto, Japan, 2018, doi: 0.23919/APMC.2018.8617598
61. Z. Chen et al., Terahertz wireless communications for 2030 and beyond: A cutting-edge frontier, *IEEE Communications Magazine*, vol. 59, no. 11, pp. 66–72, Nov. 2021, doi: 10.1109/MCOM.011.2100195
62. H.-J. Song and N. Lee, Terahertz communications: Challenges in the next decade, *IEEE Transactions on Terahertz Science and Technology*, vol. 12, no. 2, pp. 105–117, Mar. 2022, doi: 10.1109/TTHZ.2021.3128677

7 A Review on State-of-the-Art Implementations in Optically Transparent Antennas

Natarajan Rajesh, Madurakavi Karthikeyan, Ramprabhu Sivasamy, and Kanniyappan Vinayagam

7.1 INTRODUCTION

Recent developments in wireless communication demand different antenna implementations to cater the need. The antennas are fabricated using metallic layers as radiator and ground plane along with the substrate [1,2]. Also, advancements in metasurface antennas are a promising technology for future communication systems and other applications requiring high efficiency and performance [3–5]. The performances of these antennas are good in terms of radiation efficiency since it used a continuous layer as a conductor. Such implementations will obstruct the human vision, and it cannot be used for applications like vehicular implementations, solar panel integration, etc. To alleviate this problem, metallic layers are printed on vehicular glasses, glass windows and solar panel as substrate materials, which obtained 50% of transparency. Further to improve the transparency, metal mesh films are used instead of metallic layers as radiators. The efficiency of these antennas is less compared to conventional antennas since the impedance matching is difficult. Off late, metal films are replaced by transparent conductive layers, which are made up of materials like ITO, doped fluorine tin oxide (FTO), AgHT, AgGL, etc. These are some of the materials for improving transparency. These conducting oxides exhibit high optical transparency by the same time less radiation efficiency. Radiation efficiency is governed by the material conductivity, surface resistance and substrate permittivity. In a transparent antenna, the conductivity mainly depends on electron mobility, permittivity and metal oxide characterization process. Inherently, it is a material property and should be done properly to have a control over the antenna performance. This review is focused upon the way of obtaining the optical transparency, improving radiation efficiency and characterization of such metal oxides.

DOI: 10.1201/9781003599111-7

Antennas made up of ITO as the conducting layer with its controlling parameters on radiation efficiency is given in reference [6]. Recently, the glass antennas are reported from initial papers [7–10] when efforts were made to improve the transparency without any compromise in performance. The optical transparency of an ITO film depends on thickness (t) and skin depth of the film (δ), which is given by equation (7.1). Surface resistance of the film is given in equation (7.2) and electrical conductivity is computed from skin depth.

$$T(t) = e^{-2t/\delta} \tag{7.1}$$

$$R_s 1/(N_e q \mu_e t) \tag{7.2}$$

It is obvious from equation (7.2) that more electron mobility (μ_e) and high electron density (N_e) will reduce the surface resistance of the metal film. In material characterization, electron mobility (μ_e) is the only parameter to control the optical and electrical characteristics of the transparent layer. It is proved that higher electron mobility and lower metal thickness result in high efficiency due to reduced surface resistance. Multi-layer film has been made of ITO/Cu/ITO by *r. f* sputtering method and the formula attributed to radiation resistance were computed in [11]. Resistivity of the deposited material (ρ) may be computed from sheet resistance (R_s) of the sample and thickness of the film (d) using the relationship (7.3). Multi-layer film has been formed with two ITO layers and Cu as an intermediate layer without compromise in transparency. The equivalent surface resistance can be calculated similar to parallel resistors and is given in equation (7.4).

$$R_s = \frac{\rho}{d} \tag{7.3}$$

$$\frac{1}{R_s} = \frac{1}{R_{s\,\mathrm{ITO}}} + \frac{1}{R_{s\,\mathrm{CU}}} + \frac{1}{R_{s\,\mathrm{ITO}}} \tag{7.4}$$

The conductive ITO films are produced with the aid of sputtering process and its performance is measured in the UV-VIS-IR spectral region 350–2,500 nm [12]. These ITO films are used as art protection coatings used in the visible spectrum. ITO film deposition has been prepared on MRC 8620 deposition plant with the following conditions. The chamber is supplied with the base pressure of 10^{-4} to 10^{-5} Pa, RF power of 200 W and a substrate temperature ranging from 240°C to –290°C. The films are annealed for 2 hours in air at different temperatures of 100°C, 200°C and 300°C. In reference [13], conductive FTO (fluorine doped tin oxide) coated with ITO (indium tin oxide) is proposed. The double layer deposition with FTO is so chosen that its heat resistance is more compared with single layer ITO films. Optical transmittance of the different films weas compared and given in Figure 7.1. These layers are deposited on glass substrate using spray pyrolysis deposition (SPD) method.

The base material is put forth under atomization and compressed air at 0.06 MPa and deposited on a 7059 glass corning substrate. The substrate temperature is

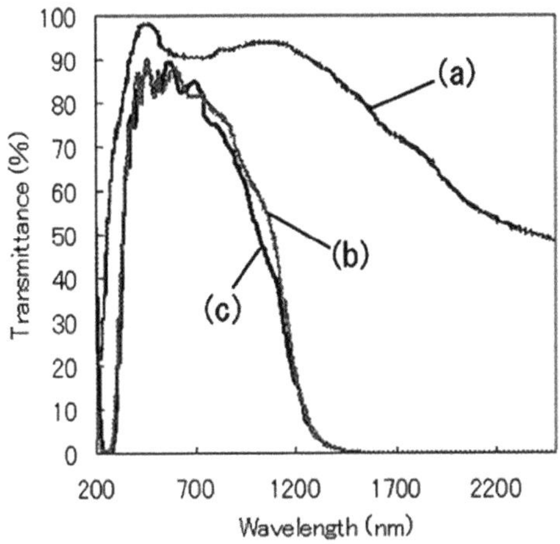

FIGURE 7.1 Transmittance of conductive coatings: (a) FTO, (b) ITO and (c) FTO/ITO [13].

maintained at 350°C and 400°C for ITO films and FTO films, respectively. This deposition leads to a new film FTO/ITO with a lesser resistivity of 1.4×10^{-4} Ω Cm and more than 80% transmittance. The lowest resistivity is possible due to post-annealing of the films for 1 hour at 600°C in air.

The chapter is divided into six different sections, and details are mentioned below. Section 7.2 outlines different methods to improve the radiation efficiency of transparent antennas. Section 7.3 explains ultra-wideband implementations and Section 7.4 discussed on the solar panel integration of transparent antennas. High frequency implementations are explicated in Section 7.5. The challenges in designing optical transparent antennas are discussed and concluded finally in Section 7.6.

7.2 RADIATION EFFICIENCY IMPROVEMENT IN TRANSPARENT ANTENNAS

7.2.1 Water Patch Antenna

In reference [14], optically transparent monopole antenna is designed with sea water as a conduction medium. Sea water is filled with a clear acrylic tube with transparency of 90% and fed with Sub Miniature—a connector at bottom of the tube—which is clear from Figure 7.2. To improve the radiation efficiency, short metal tube with metal disc is used in the bottom of the acrylic tube. The placement of the metal increased the excitation and exhibits the efficiency of 71% in the entire WLAN bandwidth.

Similar kind of water monopole antenna is proposed for maritime wireless communication (VHF band) in reference [15]. A clear acrylic tube is loaded with sea water, and the feed probe is loaded with aluminium disc to excite the TM mode radiation. The resonant frequency of the antenna is tuned by adjusting the level of vertical acrylic tube. The bandwidth is proportional to the diameter of the acrylic vertical cylinder. The antenna efficiency was varied between 50.2% and 72.3% in

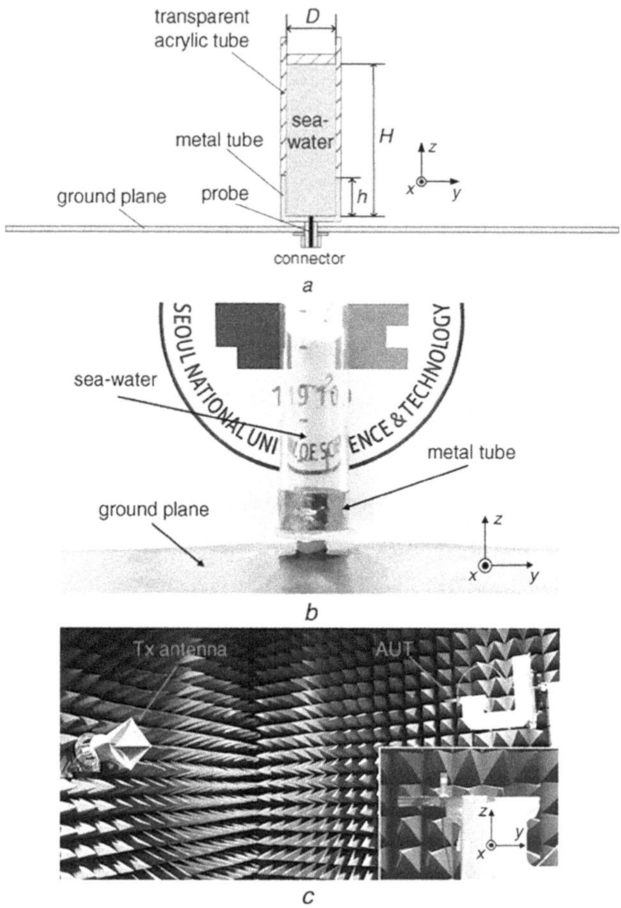

FIGURE 7.2 Cross-section view of the monopole antenna made up of sea water [14].

the band of 40–200 MHz. Surface resistance of the reported antenna improved the radiation efficiency.

In reference [16], water-based monopole antenna is designed and fed with shunt element that exhibits improved radiation efficiency. Conducting tube added to the acrylic cylinder setup is fed with shunt feed assembly in which Gamma shaped element is connected at the conductor top, as displayed in Figure 7.3. Transmission line analogy is utilized on the feeding structure to calculate the input impedance of the antenna. From (7.5), the impedance mainly depends on geometrical dimensions of a_1 and a_3 dimensions.

$$Z_c = \frac{1}{\Pi}\sqrt{\frac{\mu_0}{\varepsilon_0}}\ln\frac{d+\sqrt{d^2-4a_1a_3}}{2\sqrt{a_1a_3}} \tag{7.5}$$

Here, the feeding height (h_1) and the distance (d) directly affect/increased the antenna efficiency. This calculation is more appropriate for the static case alone.

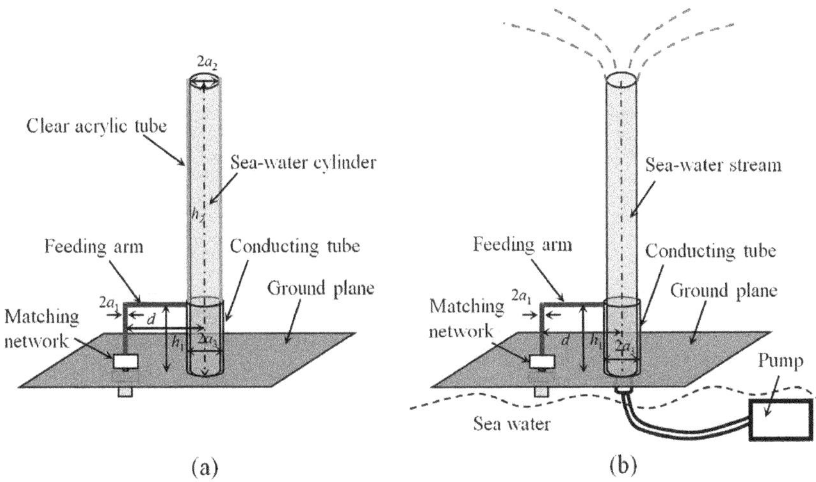

FIGURE 7.3 Water monopole antenna with shunt feed: (a) static arrangement and (b) dynamic arrangement [16].

7.2.2 Radiation Efficiency in Other Antennas

In reference [17], a reflect array with improved radiation efficiency is proposed. Most of the transparent antenna designs utilized the ITO material whose conductor loss is the major challenge for optical transparency. Power loss and the energy of the antenna can be correlated using the following equation:

$$P_c \approx \frac{w \times W_t}{h\sqrt{\Pi f \mu \sigma}} \tag{7.6}$$

The conductor loss can be effectively reduced by reducing the total energy. Thick substrate with low dielectric constant will exhibit low conductor loss at the cost of phase variation. This is mitigated by the novel sub-wavelength element. The antenna exhibits more than 73% radiation efficiency with 25.8 dB measured gain.

Conductive transparent films are introduced on the high current density areas which improved the radiation efficiency significantly without disturbing the optical transparency in [18]. A CPW patch antenna was made with copper having surface resistance of 0.02 Ω/S, which served as a reference antenna. The same antenna is made with Silver Coated Polyester Film whose surface resistance is 4.5 Ω/S. The radiation efficiency of the AgHT antenna is less due to Ohmic loss attributed to the surface resistance. To improve the efficiency, a 1 mm wide conductive strip is applied on the edges of the AgHT surface and it improved the efficiency considerably.

In reference [19], transparent silver/titanium bilayer is printed on a glass substrate, and optical transparency of 81.3% with radiation efficiency of 85% is presented. A conductive Ag mesh has been formed, and the periodic array was implemented on photo etching. The pitch of the aperture is 300 μm and the strip width is close to 20 μm, as shown in Figure 7.4. This reduced the surface resistance of the antenna and attributed to high radiation efficiency.

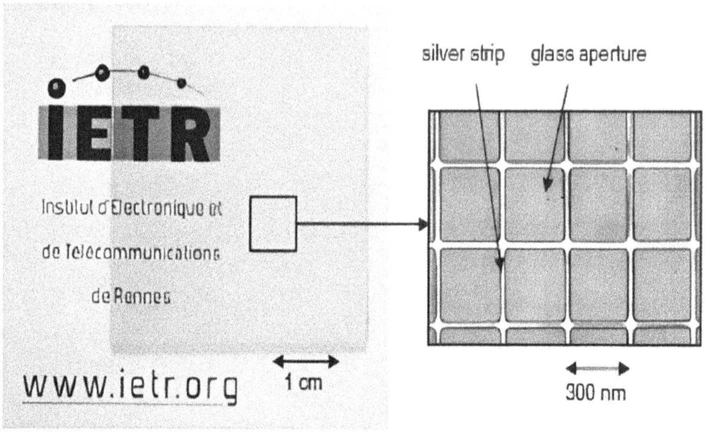

FIGURE 7.4 Geometry of meshed silver/titanium bilayer grid [19].

7.3 ULTRA-WIDEBAND TRANSPARENT ANTENNAS

In [20], a novel optically transparent ultra-wideband water patch antenna is designed with Omni directional radiation characteristics. A linearly polarized antenna reported with water as the conducting medium, which is enclosed by a transparent plexi glass, as shown in Figure 7.5. This is cost-effective implementation and ensures optical transparency. A disc probe feed is used on air substrate, which exhibits wide impedance bandwidth of 35% from 2 to 2.85 GHz with a maximum gain of 4 dBi.

The dimensions of disc loaded with probe, patch thickness and the ground plane influencing the performance of the antenna. To inject water smoothly into the disc, two valves must be realized on the container. Antenna performance was measured with salt water and compared against distilled water. It is proved that the performance is far better for distilled water.

In reference [21], circular monopole antenna is made up of highly conductive transparent AgHT-4 as conductor. The antenna exhibits a circular radiation pattern in the H–plane and a monopole radiation pattern in the E–plane in the lower part of UWB spectrum in the band (1–8.5 GHz). Measurements have been made with coaxial cable and optical fibre for both the reflection antenna and the conductor antenna made of aluminium. The gain of the transparent antenna is 5 dB less than the aluminium antenna.

A lightweight, compact CPW fed transparent antenna for extended UWB application is proposed in [22]. Rectangular radiator is modified to staircase patched to accommodate multiple resonant frequencies. A quarter circle slot with two rectangular stubs is introduced onto the partial ground plane. The schematic is given in Figure 7.6

Silver coated with polyester film (AgHT-8) is utilized as a conductive element and polyethylene terephthalate (PET) is used as a substrate. The reported antenna covered the super wideband between 3.15 and 32 GHz.

In reference [23], transparent antenna prepared on water gives broadside radiation pattern. Similar to reference [24], plexi glass is used to fill the distilled water and

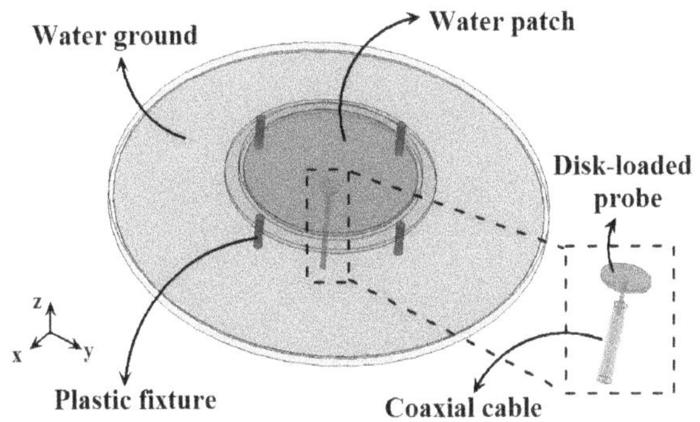

FIGURE 7.5 Geometry of the 3D water-filled patch element [20].

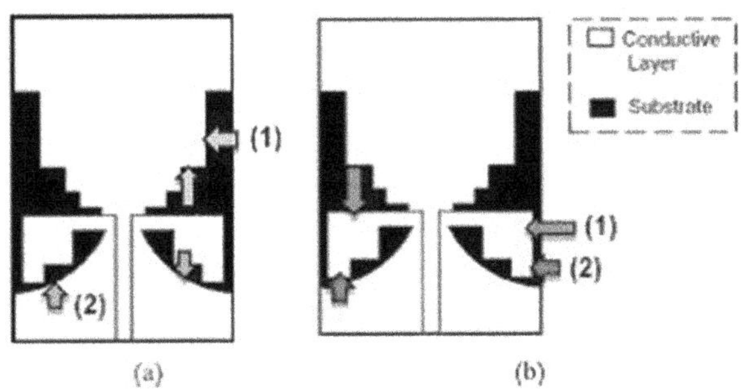

FIGURE 7.6 (a) Staircase enabled patch with semi-circle slot (major slot) and (b) minor slot enabled configuration [22].

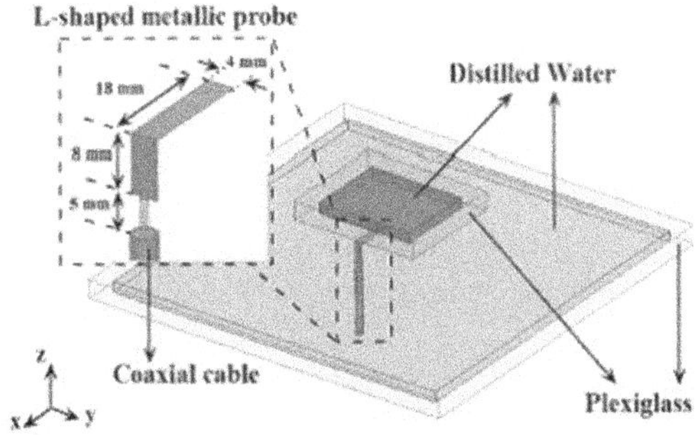

FIGURE 7.7 3D view of water patch antenna with broadside radiation pattern.

served as a radiator and ground plane. An air substrate with L-shaped probe is used to give the input to the antenna, as shown in Figure 7.7. Plexi glass and air substrate offer maximum optical transparency to the antenna with more than 75% radiation efficiency.

In [24], a compact monopole water patch antenna is designed to operate in wide impedance bandwidth. The water patch with disc loaded probe is used to design the

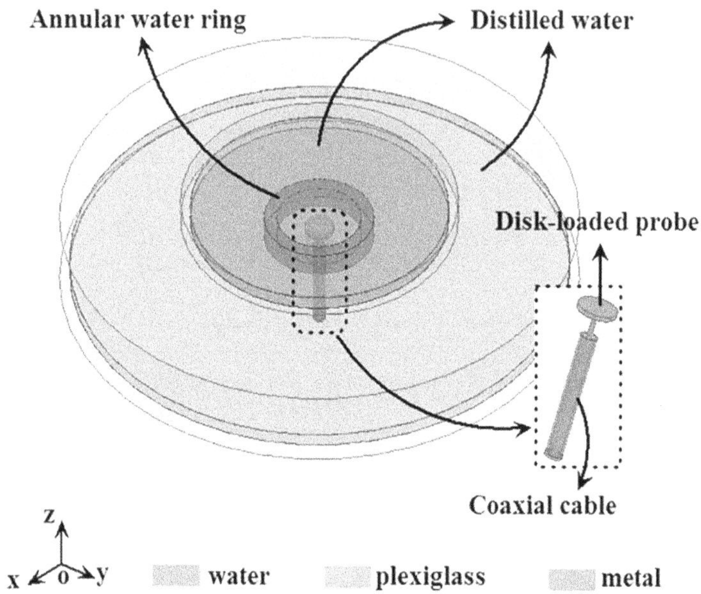

FIGURE 7.8 Perspective view of compact water patch showing annular ring beneath the patch [20].

antenna. Implementing an annular ring underneath the water patch reduced the lower operating frequency 1.95 GHz. Since there is no current existing within the water medium, the E field vectors are accumulated beneath the conductor and above the ground (Figure 7.8).

The distilled water acts as a conducting medium, termed as PEC, for the electromagnetic waves to confine within the boundary. The antenna performance is explained with the help of transmission line theory with its equivalent circuit, as shown in Figure 7.9.

The edges of the antenna act as the open-ended transmission line and modelled as parallel G and C. The water patch modelled as a transmission line with characteristic impedance of Z_{water}. When the annular ring is inserted beneath the water patch, the impedance is changed toZ_{ring}. This antenna exhibits 100% optical transmittance and 41.8% impedance bandwidth with VSWR less than 2.

Bandwidth enhancement of meshed patch antenna is obtained through proximity coupling in reference [25]. The solar panel of any satellite consists of solar cell array with a specific gap (G). Microstrip lines are placed on top of the glass panel confined within a gap (G), which is used to feed the meshed resonant structures.

Each structure is placed vertical to the stripline such that the voltage maxima coincide with the centre of the patch. The bandwidth enhancement was done by adding multiple mesh structure with slightly increasing resonant structures. The proposed method induced a bandwidth more than 2.5 times the bandwidth of a single mesh patch (Figure 7.10).

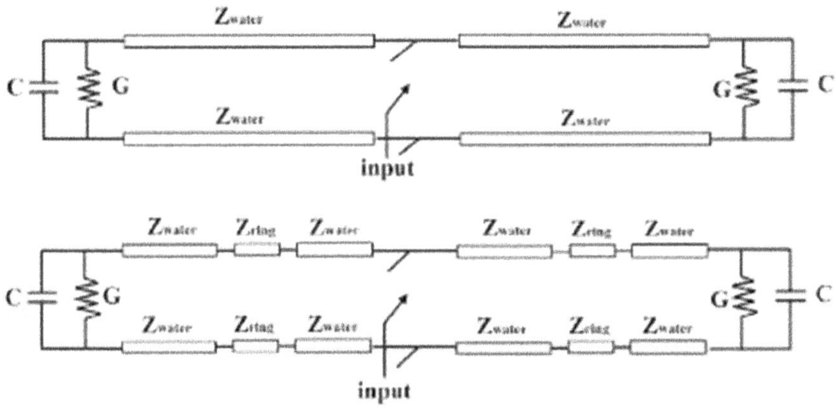

FIGURE 7.9 Transmission line equivalent circuit of a water patch with (a) and without (b) an annular ring [24].

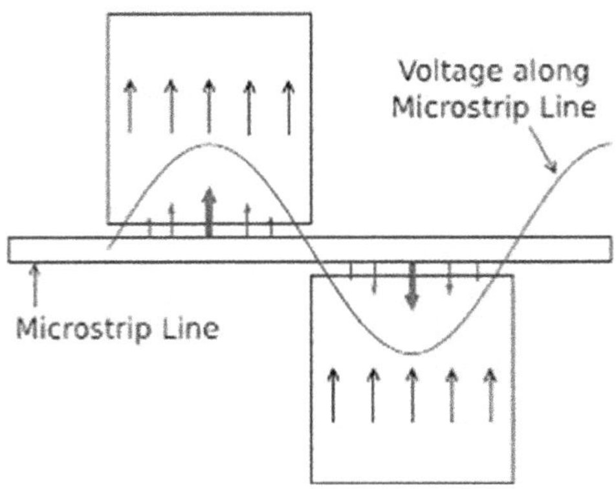

FIGURE 7.10 In-phase alignment of patches in proximity feed [25].

Feasibility study of transparent dipole and patch antenna for ultra-high definition (UHD) TV application is conducted in [26]. Metal mesh films were formed with very thin copper having thickness of 5 μm deposited on a PET substrate. The optical transparency of the metal films mainly depends on mesh width (*w*) and mesh strip (*s*). Experiments verified that mesh dimensions directly affect the optical transparency. Properties of the films are summarized in Table 7.1 [26].

$$\mathrm{OT} = \left(\frac{s}{w+s}\right)^2 \tag{7.7}$$

The surface resistance can be calculated using the below formula:

$$R = \frac{M}{M+1} r \tag{7.8}$$

TABLE 7.1
Properties of Conducting Films Like ITO, MMF and CS

Electrode	Thickness	OT_{AVG} (%)	R_s (Ω/Sm)	FoM (10^{-3} Ω^{-1})
ITO	0.2	72	15	2.5
MMF	5	70.2	0.04	686.3
CS	34	—	0.0005	—

where each resistor r is expressed as

$$r = \rho \frac{s + w}{w \times t} \tag{7.9}$$

MMF surface resistance is theoretically computed by the abovementioned formula since the mesh films are not made from a copper sheet. The efficiency of transparent patch antenna is measured to be 83.8% and the reported maximum gain is 6.2 dBi. Optically transparent mesh film based dipole has 72.1% efficiency and 2.4 dBi gain in the working band 470–771 MHz. The next section will discuss about the antennas integrated on solar panel and glass substrates.

7.4 SOLAR PANEL INTEGRATION OF TRANSPARENT ANTENNAS

A multi-layer electrode-based flexible, transparent antenna is proposed for wearable glasses in reference [27]. The conducting electrode is made of an Indium–Zinc-Tin oxide combination and is used as a conducting material. Flexible polyimide substrate with 100 μm is used to place the antenna on glasses. Since these are wearables, the specific absorption rate (SAR) has been calculated for the proposed IAI antenna when it is placed with different locations. The IAI antenna is placed on (1) the top of the lens, (2) the side of the lens and (3) the side of the frame so as not to interrupt the vision of the human. Measured SARs of the antennas are 0.417, 0.246 and 0.839, respectively, for the three antennas.

Another implementation is reported in [28]. A Square conducting patch is made of AgHT-4 film whose surface resistance is 4.5 Ω^{-2} with 75% transparency. The patch is placed on the clear Perspex substrate and is placed on a glass panel of the solar cell. Microstrip line is used to feed the antenna, and the feed line is sandwiched between Perspex and glass panel. The antenna exhibits a peak gain of 3.96 dBi on operating band between 3.4 and 3.8 GHz.

In recent times, there has been a significant interest in integrating TCO material with solar panel [29]. Such materials could introduce conductor loss to the reflect array implementations, which is not discussed in any of the reported papers. The reflect array is deployed at 26 GHz, and the unit cell is designed with 1 μm ITO deposition on a 1 mm thick quartz substrate. It is observed that the periodicities of the elements are directly proportional to the phase range, which is 272° for the antenna.

In [30], circularly polarized subarray solar panel antenna is designed for CubeSat applications. Conventionally, the Fabri-Perot resonator with a rotated feeding element network is prominent in generating circular polarization. The multi-layer antennas had several components acting as multifunction one. First, the antenna array is designed as 2 * 2 elements with ITO coating as the conductor fed with a parallel sequential feed network.

To improve the directivity of the antenna, another layer of partial reflecting surface is designed with circular FSS unit cells. A solar glass panel is inserted between substrate layers TMM 10i and PET-G. The conductive coating of 5 nm CU and 200 nm ITO is chosen to provide better transparency. This structure has low Radar Cross section (RCS) suitable for satellite implementations.

Meshed patch antenna integrated on solar cells provides an optical transparency of 93%, and antenna performance is discussed in reference [31]. The solar panel is used as a substrate for the mesh antenna. Substrate thickness is proportional to antenna gain, so the thickness of the substrate and the line width are chosen to provide maximum peak gain. Beneath the solar panel, there is a photovoltaic cell that is made up of gold served as a ground for the mesh antenna. Here, the photocells conductivity is different upon suppliers. To improve the conductivity of the layer, solar cell is moulded with silicon, and conductivity varies between 0 and 100 S/m.

An ultra-wideband Vivaldi antenna is designed on the amorphous silicon operating in 0.95–2.45 GHz in [32]. The dimension of each panel is 71 * 18 mm^2, and it is separated by 2 mm with other cells. Each cell contains seven electrodes with each dimension to be 0.5 * 16.5 mm^2 separated by 9.5 mm. The energy scavenging ability of the antenna is tested by eight–stage cascaded circuit using Villard multiplier. The performance of the solar panel Vivaldi antenna is degraded by only about 10% less compared with the copper-made antenna.

A Ka band reflect array is integrated on solar cells for satellite applications in [33]. The electromagnetic properties of multi-layer structure have been studied, and an X-shaped sub-wavelength array was augmented to improve its performance. The reported array gain is 27.3 dBi with an aperture efficiency of 40%. Table 7.2 summarized the antenna schematic and its performance comparison.

7.5 HIGH FREQUENCY IMPLEMENTATIONS

A patch antenna array is designed with meshed AuGL material and implemented for 60 GHz applications in [34]. Two antennas are fabricated, one with reflective Au/Ti served as a reference and another one with transparent Au/Ti bilayer. The antennas are fabricated on silica substrate, and the conducting layers are made up of meshed Au/Ti bilayers, which ensured optical transparency, as shown in Figure 7.11.

$$T(\%) = \frac{(p-s)^2}{p^2} \tag{7.10}$$

Transparency (T) of meshed cells is controlled by the pitch (p) and width (w) of the printed layer. A large cell size will attribute high optical transparency at the cost of high Ohmic loss. To alleviate the Ohmic loss, a highly transparent large mesh pattern is used as a ground, and a small cell size is used for the feeding network. The

TABLE 7.2
Comparison between the Solar Panel Integrated Antennas

Ref. No.	Antenna	Conductive Coating	Substrate	Frequency/ Bandwidth
[23]	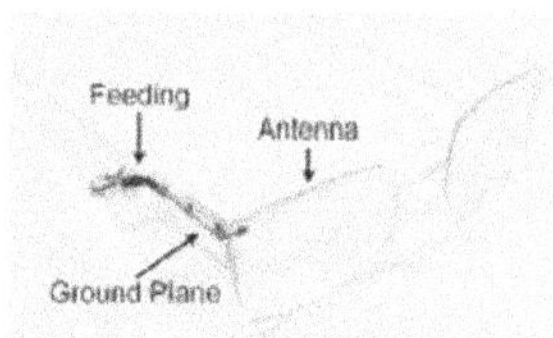	IZTO/Ag/IZTO with 45 nm/10 nm/45 nm thickness	Polyethylene terephthalate 100 μm thick with 4.99 Ω^{-2}	2.4–2.5 GHz
[24]	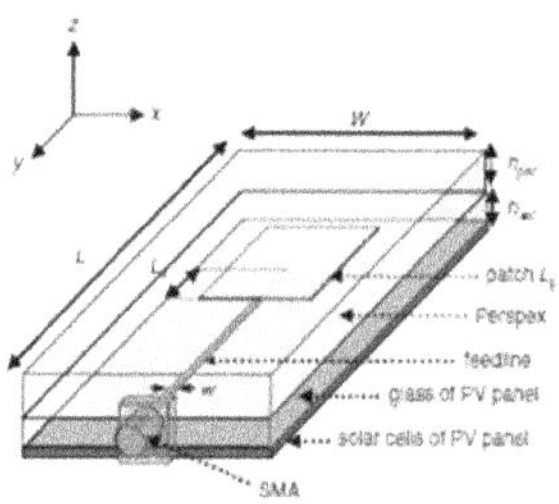	AgHT-4 170 μm thick polyester film coated with tin and aluminium 4.5 Ω^{-2}	Glass ($\varepsilon_r = 6.4$) and Perspex ($\varepsilon_r = 2.6$)	3.4–3.8 GHz
[25]	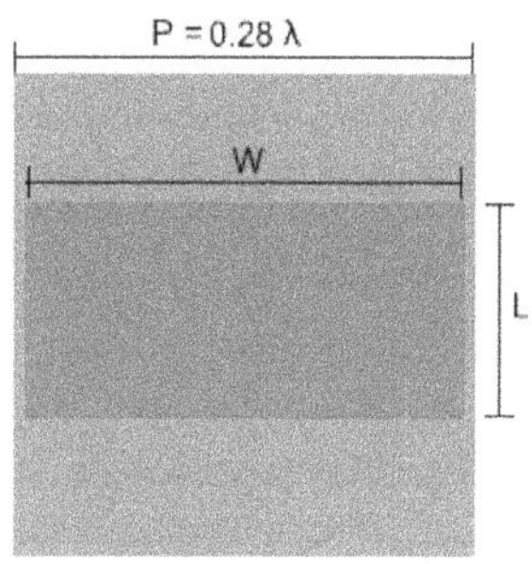	Indium tin oxide (ITO) $\sigma = 2.4 \times 10^5$ S/m $R_s = 4.2\,\Omega^{-2}$	Quartz 1 mm thick, ($\varepsilon_r = 3.8$)	26 GHz
[26]	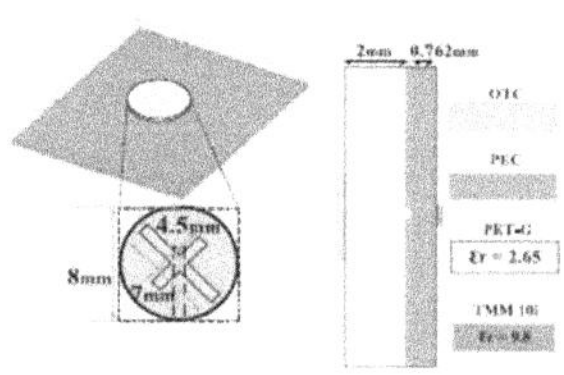	Indium Tin Oxide (ITO) and Cu $\sigma = 10^6$	TMM 10i ($\varepsilon_{r1} = 9.8$) PET-G ($\varepsilon_{r2} = 2.65$)	10 GHz
[27]	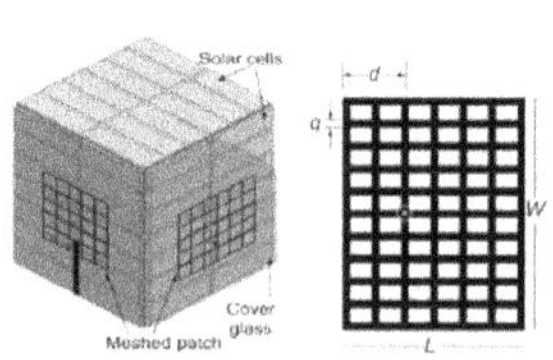	Conductive ink Electroformed meshed conductor up to 100 S/m	PET-G Plastic substrate	2.6 GHz
[28]	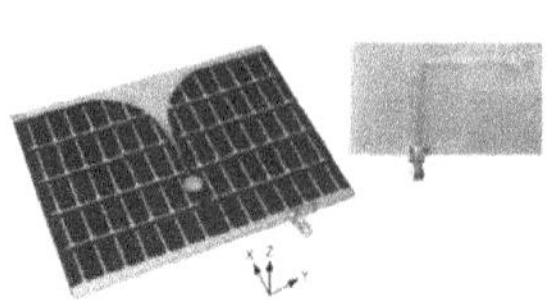	Tin coated Cu	Amorphous silicon 400 nm thick	0.95–2.45 GHz

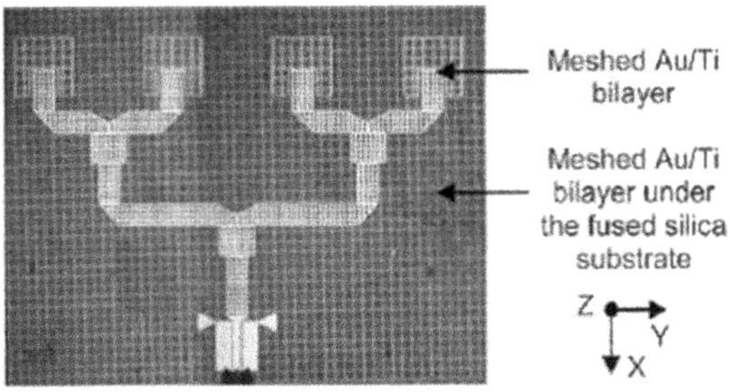

FIGURE 7.11 Meshed transparent patch antenna array printed on corning 7980 silica substrate [34].

proposed antenna exhibits optical transparency up to 74.6%. The overall transparency of the microstrip antenna can be calculated from the following equation:

$$T_{\text{ant}} = T_{\text{sub}} - T_{\text{sub}} \sum_{1}^{n} \frac{A_n}{A_{\text{sub}}} + \sum_{1}^{n} \frac{A_n T_n}{A_{\text{sub}}} \tag{7.11}$$

An array implementation by using transparent elements is implemented for 60 GHz applications [35]. Two antenna arrays are developed, one with reflective double-sided silver coatings, which served as a reference. The proposed antenna developed with a meshed radiating plane and a meshed ground plane, which exhibits a maximum gain of 13.6 dBi. The surface resistance of the antenna is less than 0.5 Ω with the transparency of 80% in the entire visible spectrum. Microwave performance of the antenna is validated in terms of radiation efficiency, optical transparency and other antenna parameters.

Yet another implementation of CPW fed lozenge antenna is reported for high frequency implementation in [1]. The antenna is fabricated on a fused silica 7980 corning substrate, with an Au/Ti meshed bilayer as a conduction patch, as shown in Figure 7.12. Pitch of the grid is 200 μm, with an Au width of 10 μm. It possesses a surface resistance of 0.44 Ω^{-2} and an optical transmittance of 83%. Corner fed square patch antennas are convenient for the array implementation, which exhibits low cross polarization (Table 7.3).

A millimeter-wave transmit array is proposed with beam steering applications in Ka band [2]. Meshed double ring structure elements are used as a unit cell for the transmit array, and it is fed with aperture coupled microstrip patch antennas as the primary radiator. A transparent PMMA substrate is used as a base for transmit array and as a mount. The measured scanning range is reported to be 30°, with a gain more than 19.5 dBi for the bandwidth ranging from 27.5 to 29.5 GHz.

In reference [3], meshed microstrip antenna is proposed to operate at the licensed and also unlicensed spectrum spanning from 5 to 6 GHz. The challenge here is to provide impedance match between the meshed patch and the strip line. The relationship between number of mesh lines and impedance is shown in Figure 7.13. It shows that the feed line impedance is increased for reduced number of mesh lines. Therefore, lowest impedance with reduced mesh cells is the solution to better impedance

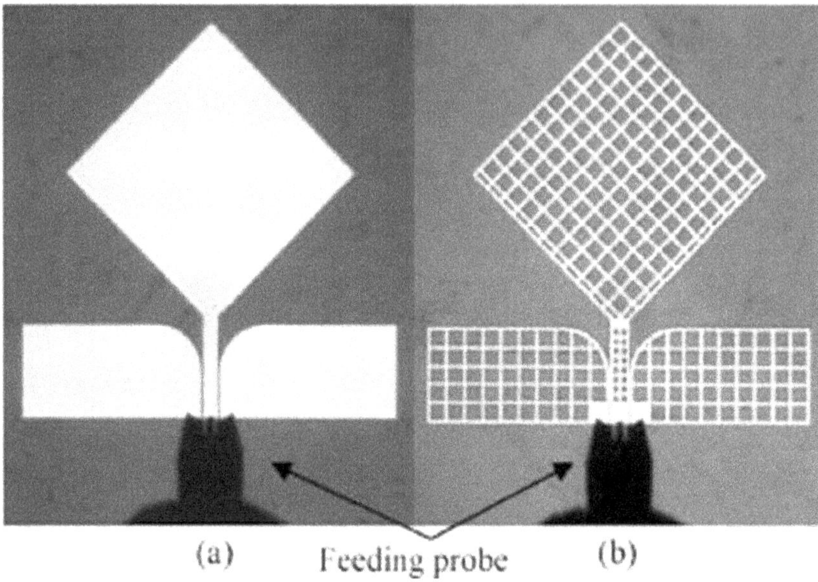

FIGURE 7.12 Coplanar fed monopole lozenge antenna printed on 7980 corning silica substrate [1].

TABLE 7.3
Surface Resistance and Transparencies of Different Conducting Materials

Theoretical Values	AuGL	AgGL	ITO	AgHt
Sheet resistance (Ω/S)	0.44	0.054	8.6	4.5
Max transparency (%)	83	81.3	86	75

matching. In the case of inset feed, the length of the patch is adjusted for the match due to the lower impedance.

In reference [4], challenges in transparent antenna design are explicitly discussed in view of (1) transparent materials property, (2) different losses accumulated due to the film depositions and (3) the impact of surface resistance on antenna efficiency. Antennas are developed for Wi-Max and Wi-Fi applications, and metamaterial-based antennas developed for RCS are developed and listed [36–39].

The following important observations are derived from the experiment. To maximize the optical transmittance, the plasma frequency should be less than the lowest visible wavelength. The frequency is much less than the plasma frequency, and the transmission coefficient tends to be zero. Alternatively, it increases above the plasma frequency, which depends on the permittivity of the conductivity oxide (Figure 7.14).

In view of understanding the correlation between electron mobility, electron density and transparency, an antenna is developed using ITO. From observations, the optical transparency of the patch was not very affected by electron density,

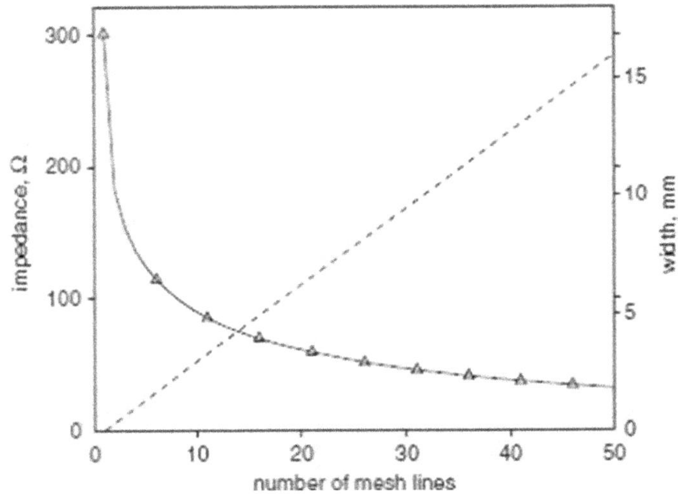

FIGURE 7.13 Impedance (Δ) and line width (- -) of the feed line against the number of mesh lines [3].

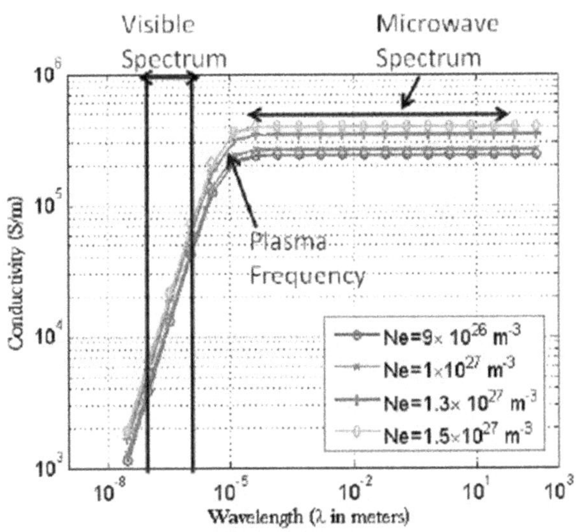

FIGURE 7.14 Conductivity vs. wavelength of standard conducting oxides [4].

but it slightly depends on the electron mobility. Increasing electron mobility resulted in high conductivity and transparency.

7.6 CONCLUSION

This paper presented antenna implementations in terms of radiation efficiency, ultra-wideband antennas and solar panel implementations. To ensure full optical transparency, transparent conducting oxides are used for the antenna design.

The performance of the antennas mainly depends on the material property that is decided by manufacturing itself. Those characterizations are discussed briefly, and the discussions are extended on high frequency implementations and the design challenges.

REFERENCES

1. Hautcoeur, J., Talbi, L. and Hettak, K., 2012. Feasibility study of optically transparent CPW-fed monopole antenna at 60-GHz ISM bands, *IEEE Transactions on Antennas and Propagation*, 61(4), 1651–1657.
2. Liu, G., Kodnoeih, M.R.D., Pham, K.T., Cruz, E.M., Gonzalez-Ovejero, D. and Sauleau, R., 2019. A millimeter-wave multibeam transparent transmitarray antenna at Ka-band, *IEEE Antennas and Wireless Propagation Letters*, 18(4), 631–635.
3. Escobar, E.R., Kirsch, N.J., Kontopidis, G. and Turner, B., 2015. 5.5 GHz optically transparent mesh wire microstrip patch antenna, *Electronics Letters*, 51(16), 1220–1222.
4. Saberin, J.R. and Furse, C., 2012. Challenges with optically transparent patch antennas, *IEEE Antennas and Propagation Magazine*, 54(3), 10–16.
5. Karthikeyan, M., Sitharthan, R., Ali, T., Pathan, S., Anguera, J. and Shanmuga Sundar, D., 2022. Stacked T-shaped strips compact antenna for WLAN and WiMAX applications, *Wireless Personal Communications*, 123, 1523–1536 (2022).
6. Yasin, T., Baktur, R. and Furse, C., 2011. A study on the efficiency of transparent patch antennas designed from conductive oxide films. *IEEE International Symposium on Antennas and Propagation (APSURSI)*, Spokane, WA, pp. 3085–3087.
7. Morimoto, Y., Shiu, S., Huang, I.W., Fest, E., Ye, G. and Zhu, J., 2023. Optically transparent antenna for smart glasses, *IEEE Open Journal of Antennas and Propagation*, 4, 159–167.
8. Liu, Y. and Luk, K.M., 2024. An optically transparent magneto-electric dipole antenna design by integrating flexible transparent metal mesh film on glass, IEEE Transactions on Antennas and Propagation, 72(7), 5569–5577.
9. Chen, B., Wu, B., Zu, H.R., Xie, H.Y., Liu, Y. and He, D.P., 2024. Wideband optically transparent transmitarrays with fine metal line structure, *IEEE Antennas and Wireless Propagation Letters*, 23(6), 1–5.
10. Sayem, A.S.M., Lalbakhsh, A., Esselle, K.P., Buckley, J.L., O'Flynn, B. and Simorangkir, R.B., 2022. Flexible transparent antennas: Advancements, challenges, and prospects, *IEEE Open Journal of Antennas and Propagation*, 3, 1109–1133.
11. Colombel, F., Castel, X., Himdi, M., Legeay, G., Vigneron, S. and Cruz, E.M., 2009. Ultrathin metal layer, ITO film and ITO/Cu/ITO multilayer towards transparent antenna, *IET Science, Measurement & Technology*, 3(3), 229–234.
12. Boycheva, S., Sytchkova, A.K. and Piegari, A., 2007. Optical and electrical characterization of rf sputtered ITO films developed as art protection coatings, *Thin solid films*, 515(24), 8474–8478, 2007
13. Kawashima, T., Matsui, H. and Tanabe, N., 2003. New transparent conductive films: FTO coated ITO, *Thin Solid Films*, 445(2), 241–244.
14. Phan, D.T. and Jung, C.W., 2019. Optically transparent sea-water monopole antenna with high radiation efficiency for WLAN applications, *Electronics Letters*, 55(24), 1269–1271.
15. Hua, C., Shen, Z. and Lu, J., High-efficiency sea-water monopole antenna for maritime wireless communications, *IEEE Transactions on Antennas and Propagation*, 62(12), 5968–5973.
16. Hua, C. and Shen, Z., 2015. Shunt-excited sea-water monopole antenna of high efficiency, *IEEE Transactions on Antennas and Propagation*, 63(11), pp.5185–5190, 2015

17. Peng, J.J., Qu, S.W. and Xia, M., 2020. Optically transparent reflectarray based on indium tin oxide with improved efficiency, *IEEE Transactions on Antennas and Propagation*, 68(4), 3289–3294.
18. Song, H.J., Hsu, T.Y., Sievenpiper, D.F., Hsu, H.P., Schaffner, J. and Yasan, E., 2008. A method for improving the efficiency of transparent film antennas. *IEEE Antennas and Wireless Propagation Letters*, 7, 753–756.
19. Hautcoeur, J., Colombel, F., Castel, X., Himdi, M. and Cruz, E.M., 2009. Optically transparent monopole antenna with high radiation efficiency manufactured with silver grid layer (AgGL), *Electronics Letters*, 45(20), 1014–1016.
20. Sun, J. and Luk, K.M., 2017. A wideband low cost and optically transparent water patch antenna with omnidirectional conical beam radiation patterns, *IEEE Transactions on Antennas and Propagation*, 65(9), 4478–4485.
21. Katsounaros, A., Hao, Y., Collings, N. and Crossland, W.A., 2009. Optically transparent ultra-wideband antenna, *Electronics Letters*, 45(14), 722–723.
22. Hakimi, S., Rahim, S.K.A., Abedian, M., Noghabaei, S.M. and Khalily, M., 2014. CPW-fed transparent antenna for extended ultrawideband applications, *IEEE Antennas and Wireless Propagation Letters*, 13, 1251–1254.
23. Sun, J. and Luk, K.M., 2009. A wideband and optically transparent water patch antenna with broadside radiation pattern, *IEEE Antennas and Wireless Propagation Letters*, 19(2), 341–345.
24. Sun, J. and Luk, K.M., 2019. A compact-size wideband optically-transparent water patch antenna incorporating an annular water ring, *IEEE Access*, 7, 122964–122971.
25. Yasin, T. and Baktur, R., 2017. Bandwidth enhancement of meshed patch antennas through proximity coupling, *IEEE Antennas and Wireless Propagation Letters*, 16, 2501–2504.
26. Tung, P.D. and Jung, C.W., 2019. Optically transparent wideband dipole and patch external antennas using metal mesh for UHD TV applications, *IEEE Transactions on Antennas and Propagation*, 68(3), 1907–1917.
27. Hong, S., Kang, S.H., Kim, Y. and Jung, C.W., 2016. Transparent and flexible antenna for wearable glasses applications, *IEEE Transactions on Antennas and Propagation*, 64(7), 2797–2804.
28. Roo-Ons, M.J., Shynu, S.V., Ammann, M.J., McCormack, S.J. and Norton, B., 2011. Transparent patch antenna on a-Si thin-film glass solar module, *Electronics Letters*, 47(2), 85–86.
29. Kocia, C. and Hum, S.V., 2016. Design of an optically transparent reflectarray for solar applications using indium tin oxide, *IEEE Transactions on Antennas and Propagation*, 64(7), 2884–2893.
30. Zarbakhsh, S., Akbari, M., Farahani, M., Ghayekhloo, A., Denidni, T.A. and Sebak, A.R., 2019. Optically transparent subarray antenna based on solar panel for CubeSat application, *IEEE Transactions on Antennas and Propagation*, 68(1), 319–328.
31. Turpin, T.W. and Baktur, R., 2009. Meshed patch antennas integrated on solar cells, *IEEE Antennas and Wireless Propagation Letters*, 8, 693–696.
32. O'Conchubhair, O., Yang, K., McEvoy, P. and Ammann, M.J., 2015. Amorphous silicon solar Vivaldi antenna, *IEEE Antennas and Wireless Propagation Letters*, 15, 893–896.
33. An, W., Xiong, L., Xu, S., Yang, F., Fu, H.P. and Ma, J.G., 2018. A ka-band high-efficiency transparent reflectarray antenna integrated with solar cells, *IEEE Access*, 6, 60843–60851.
34. Hautcoeur, J., Talbi, L., Hettak, K. and Nedil, M., 2014. 60 GHz optically transparent microstrip antenna made of meshed AuGL material, *IET Microwaves, Antennas & Propagation*, 8(13), 1091–1096.
35. Martin, A., Lafond, O., Himdi, M. and Castel, X., 2018. Improvement of 60 GHz transparent patch antenna array performance through specific double-sided micrometric mesh metal technology, *IEEE Access*, 7, 2256–2262.

36. Priyadharshini, A.S., Arvind, C. and Karthikeyan, M., 2023. Novel ENG metamaterial for gain enhancement of an off-set fed CPW concentric circle shaped patch antenna, *Wireless Personal Communications*, 130, 2515–2530.
37. Naveen, M., Chandra, A., Kumar, R., Karthikeyan, M., Pareek, P. and Singh, L., 2025. A review on distinct microwave metamaterial structures and their uses, *Materials Science and Engineering:* B, 313, 117917.
38. Sukumar, V. and Karthikeyan, M., 2024. Ultra-wide band and high efficient polarization converting metasurface employing bell-shaped unit cell for RCS reduction, *Results in Engineering*, 24, 103342.
39. Karthikeyan, M., Pradeep, J., Harikrishnan, M., Sitharthan, R., Rajesh, M., Chandra, A. and Kumar, R., 2024. Absorbers as biosensors: Leveraging absorption phenomena for enhanced biosensing. *In Advances in All-optical Communication*, pp. 11–1. Bristol, UK: IOP Publishing, 2024.

8 Graphene-Based Antennas and Absorbers for Terahertz Applications

Anoushka Kumari, Saurabh Kumar, and Amit Bage

8.1 INTRODUCTION

"Terahertz" is an unexplored region having a frequency span of 0.1–10 THz, which falls in between the regions of microwave and infrared and has recently gained significant interest from researchers [1]. This region exhibits several unique characteristics, such as smaller single-photon energy, which make it a less harmful alternative to X-rays. Additionally, it possesses high spectral resolution and exceptional penetration capabilities. The terahertz frequency band offers immense potential due to its abundant bandwidth and ability to support power-efficient systems [2–4]. However, its implementation is hindered by challenges such as significant propagation and scattering losses, along with the absence of well-developed devices for terahertz applications [5,6]. Despite these hurdles, terahertz technology remains a promising area for innovative scientific and engineering solutions.

The choice of materials is a critical parameter for achieving highly efficient and high-performing terahertz devices. Traditional materials, including metals like copper, have been widely used but exhibit limitations due to energy losses and restricted tunability at terahertz frequencies [7,8]. In response to these challenges, graphene has emerged as a revolutionary material. Graphene is a hexagonal single atomic layer, that has unique optical, mechanical, and electrical properties. By altering the graphene's chemical potential accordingly, its conductivity can be changed. Additionally, in terahertz regime the Surface Plasmon Polariton (SPP) is supported by the graphene material which lead to high confinement [9–11]. These features have positioned graphene as an emerging material for developing fast, compact, and reconfigurable terahertz devices.

In a system, when discussing devices, antennas and absorbers are two crucial components that underpin numerous applications. Antennas, which transmit and receive electromagnetic waves, and absorbers, which provide precise control over these waves, facilitate many real-time applications such as imaging, biomedical diagnostics, security, and wireless communication [12,13]. Therefore, the development of these two devices is vital for the overall advancement of terahertz systems. The incorporation of graphene into these devices opens new avenues for innovation by

DOI: 10.1201/9781003599111-8

offering enhanced performance, dynamic reconfigurability, and miniaturization—capabilities that were challenging to achieve with conventional materials in the terahertz regime.

This chapter explores graphene's role in advancing terahertz technology, particularly in the context of the two major devices: antennas and absorbers. Firstly, the fundamental properties of graphene and its conductivity model are examined. Antennas are then discussed, focusing on Photonic Band Gap (PBG) and Surface Integrated Waveguide (SIW) techniques integrated with terahertz antennas, highlighting their advantages and relevance. Subsequently, graphene-based terahertz absorbers are explored, with detailed discussions on narrowband and broadband configurations. This chapter investigates the impact of graphene integration into antennas and absorbers, which will ultimately unlock new possibilities in the terahertz domain.

8.2 GRAPHENE

Carbon, one of nature's abundant elements, is the building block of life and a multipurpose resource. Because of its tetravalent nature—that is, its capacity to create four bonds—it can react with a wide variety of elements, leading to a broad range of practical uses. One of the naturally occurring carbon's allotropes is graphite. From this three-dimensional layered structure, a layer of atoms of carbon which are configured in a hexagonal shape, i.e., graphene, is extracted.

8.2.1 Graphene's Fundamental Properties

Since the first successful isolation from graphite in 2004, graphene material has gained special consideration both academically and industrially [14]. The atoms of carbon in graphene material are sp_2 bonded into a honeycomb-shaped lattice, as demonstrated in Figure 8.1. The hexagonal-shaped lattice consists of two interleaving triangular lattices, which provide stability to the material [15,16]. This zero-gap

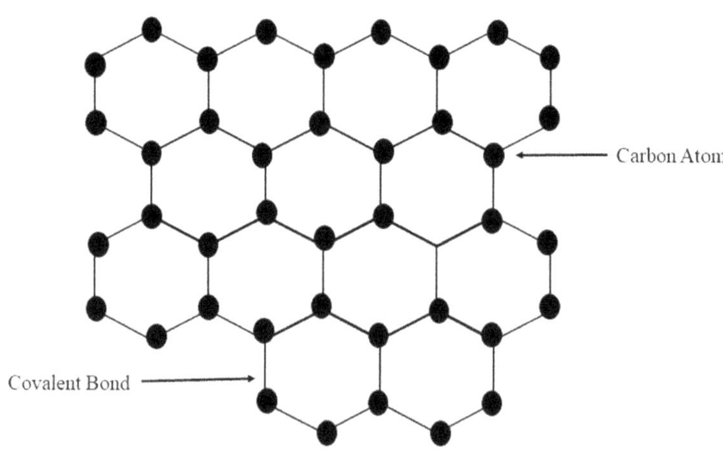

FIGURE 8.1 Structure of carbon allotrope—graphene.

TABLE 8.1
Some Properties of Graphene

Properties	Values
Conductivity (S/m)	10^8
Melting point (K)	3,800
Thermal conductivity (W/m/K)	$3\times10^3 - 5\times10^3$
Mean free path at room temperature (nm)	1×10^3
Current density (A/cm^2)	$>10^8$

semiconductor contains the highest percentage of edge atoms of any allotrope, with a chemical reactivation 100 times greater than that of thicker sheets [17,18]. Comparing it to conventional semiconductors, due to the presence of a Dirac point—where its valence band and conduction bands meet—graphene has several intriguing characteristics, including variable carrier densities and predictable high nonlinearities. In addition, optical excitation can be used to invert conical electronic band, which would result in a THz gain [19–21]. Because of the acoustic photon scattering, it exhibits electron mobilities up to 15,000 cm^2/V/s at room temperature. Also, fermi energy and the charge carrier density can be adjusted by doping and electrical gating. In contrast to the SPPs in metals like gold and copper, graphene plasmons have a strong light–matter interaction, resulting in significantly lower optical losses [22,23]. Therefore, graphene being the first thin material (atomically) paved the road for two-dimensional materials because of its unique physical, mechanical, electrical, and chemical properties. Some of the properties of the material are listed in Table 8.1 [24,25]. Because of the characteristics discussed, graphene turns out to be a potential material for novel applications in optoelectronics and photonics. For instance, a graphene-based flexible terahertz detector is reported in [26], graphene-based metamaterials as a broadband polarization rotator [27], and terahertz modulation is another area of consideration with respect to the applications for graphene [28]. Although in this chapter, we mainly discuss the graphene-based absorbers and antennas.

8.2.2 Graphene Conductivity: Kubo's Model

Graphene's surface conductivity (σ_g) can be determined by a well-established theoretical model known as Kubo's model [29,30]. σ_g is a function of the following parameters: ω and μ_c, i.e., radian frequency and chemical potential, respectively. It also depends on T, i.e., temperature, and Γ is the rate of scattering that is inversely related to the electron relaxation time, i.e., $\tau = 1/\Gamma$. The Kubo's model is defined as two components summation: intra-band (σ_{intra}) and inter-band (σ_{inter}) conductivities [31].

$$\sigma_g(\omega,\mu_c,T,\Gamma) = \sigma_{\text{intra}} + \sigma_{\text{inter}} \tag{8.1}$$

Inter-band transitions are characterized by the generation and recombination of electron–hole pairs that dominate at higher frequencies, like the visible spectrum, particularly above the Pauli blocking frequency. However, free-carrier conductivity, which dominates in the terahertz and infrared regions, is related to the intra-band transitions. Inter-band transitions are suppressed by Pauli blocking when the fermi level is more than half the photon energy [32].

Therefore, conductivity due to intra-band in equation (8.1) can be calculated as

$$\sigma_{\text{intra}} = \frac{ie^2 Tk_B}{\pi\hbar^2(\omega + i2)}\left[\left(\frac{\mu_c}{Tk_B}\right) + 2\ln\left(e^{-\mu_c/Tk_B} + 1\right)\right] \tag{8.2}$$

where k_B and $\hbar$ are the Boltzmann's constant and reduced Planck's constant, respectively.

If $k_B T \ll |\mu_c|$, $\hbar$, then the conductivity due to inter-band can be approximated as

$$\sigma_{\text{inter}} \simeq \frac{ie^2}{4\pi\hbar}\ln\left[\frac{2|\mu_c| - (\omega + i2\Gamma)\hbar}{2|\mu_c| - (\omega + i2\Gamma)\hbar}\right] \tag{8.3}$$

The product of electric field's tangential component and surface conductivity is the surface current density on graphene sheet, which is stated in the following equation:

$$\hat{n} \times \left(H|_{x=0^+} - H|_{x=0^-}\right) = J_{\text{surf}} = \sigma(\omega)E \tag{8.4}$$

The edge effects of the graphene conductivity can be ignored if the lateral size is larger than 100 nm. The conductivity due to K graphene layers is calculated as the summation of individual graphene layers conductivity, as shown below [33–35]:

$$\sigma_\omega = \sum_{k=1}^{K} \sigma_\omega^{(k)} \quad (k = 1, 2, 3, \ldots, K) \tag{8.5}$$

Therefore, because of the graphene surface conductivity, it possesses more confinement, tunable conductivity, and low losses, which make it distinct from other noble metals in terahertz and mid-infrared frequencies.

8.3 GRAPHENE-BASED TERAHERTZ ANTENNAS

Antenna is one of the crucial components in a system, serving as both a transmitter and a receiver of electromagnetic waves. Significant progress has been achieved in the development of terahertz antennas in recent years. For an antenna to be efficient, it must have a high gain, be compact, and be cost-effective. Besides the perspective of electromagnetism, it is important to understand the perspective of physical and material science to improve the efficiency of an antenna. Several studies have employed metals like gold and platinum as conducting materials in terahertz antennas because of their high value of conductivity and resistance to oxidation in air [36]. Whereas, at terahertz frequencies, metal exhibits significant losses that lead to efficiency degradation [37,38]. Additionally, metallic thin films are prone to cracks [39]. To overcome these challenges, carbon-based materials like graphene have emerged as a better alternative, offering better efficiency in terahertz antennas.

In 2012, graphene material was used as an antenna radiator for the first time in the terahertz band [40]. Graphene THz antennas perform better than traditional metallic antennas in the context of radiation efficiency particularly. Further to increasing the radiation efficiency, the conductivity of graphene can be altered by applying the electric bias. Additionally, graphene THz antennas have features that offer high directivity, stable impedance, and frequency reconfiguration, which make them highly versatile and suitable for a broad range of applications. The plasma frequency of graphene terahertz antenna, with thickness Δ, can be defined by the following relationship [41,42]:

$$\omega_P = \sqrt{\frac{2e^2 k_B T}{\Delta \pi \varepsilon_0 h^2} \ln\left(2\cosh\left(\frac{\mu_c}{2k_B}\right)\right)} \tag{8.6}$$

In the terahertz band, both TE (Transverse Electric) and TM (Transverse Magnetic) modes of SPPs arise because of the imaginary part of conductivity of graphene. For the TM mode, the SPP dispersion relationship in terms of refractive indexes is expressed as follows:

$$\sqrt{n^2 - n_{\text{eff}}^2} + n^2\sqrt{n^2 - n_{\text{eff}}^2} + \frac{4\pi}{c}\sigma_\omega\sqrt{1-n_{\text{eff}}^2}\sqrt{n^2 - n_{\text{eff}}^2} = 0 \tag{8.7}$$

where c represents the speed of light, σ_ω denotes the graphene's frequency-dependent conductivity, and n_{eff} represents the complex refractive index. The resonant properties of graphene are determined by SPP wave propagation, which is induced due to coupling of the incident electromagnetic radiation with SPP modes. The resonant condition depends on L and m, i.e., length and number of resonant modes in the antenna, respectively. The relationship can be expressed as:

$$L \approx m\frac{\lambda_{\text{spp}}}{2} \approx m\frac{\pi}{\text{Re}\{k_{\text{spp}}\}} \tag{8.8}$$

where k_{spp} is the mode wavenumber. The mode wavenumber is calculated using k_0 and the free-space wavenumber as follows:

$$k_{\text{spp}} = n_{\text{eff}} k_0 \tag{8.9}$$

Much literature on graphene terahertz antenna designs has been done, showcasing the versatility and potential of graphene in this domain. In [43], a microstrip patch terahertz antenna utilizing a substrate of polyimide was introduced, having 0.75 THz as resonating frequency, 5.09 dB gain, and a radiation efficiency of 86.58%. Another design operating at 2.60 THz employed a triangular shape on a quartz substrate. When biased with 0.5 eV of chemical potential, this antenna achieved radiation efficiency of 82.7% and a gain of 5.97 dB [44]. Graphene also enables functionalities such as phase control, wide bandwidth, and reconfigurability in reflectarray antennas. For instance, a graphene-based terahertz reflectarray antenna with square patches was proposed for 1.3 THz [45,46]. Furthermore, graphene can be integrated into Multi-Input Multi-Output (MIMO) techniques, enhancing spectral efficiency while addressing challenges like antenna size and separation [47]. These advantages, including high efficiency, reconfigurability, and miniaturization, make graphene-based antennas a leading solution for future technologies with wide-ranging real-time applications. The integration of graphene-based antennas with techniques such as PBG and Substrate Integrated Waveguide (SIW) is discussed in detail in this section.

8.3.1 Photonic Band Gap Based Graphene THz Antenna

A class of artificial periodic dielectric structures called photonic crystals, or PBG materials, are designed to prohibit electromagnetic wave propagation within a range of frequency [48]. It was first introduced in 1888 by Ray and his group through their study of electromagnetic wave propagation in a one-dimensional (1D) periodic medium. Further in 1987, the concept was expanded and introduced the two-dimensional structure and three-dimensional structures [49,50]. These materials consist of periodic implants within a surrounding medium, with configurations such as hollow air cylinders of different shapes, including circular, triangular, and elliptical, embedded in the substrate [51]. The diffraction and scattering characteristics of periodic elements have an impact on the propagation of electromagnetic waves. Frequency Band Gap is one of the unique attributes possessed by the PBG substrate, which can effectively restrict the electromagnetic wave propagation within the specific range of frequency by periodically altering the dielectric constant. This feature has potential applications in reducing return losses, enhancing the gain of the antenna, and suppressing surface wave losses. A photonic crystal substrate whose forbidden gap includes the antenna excitation frequency is used in place of the substrate in order to reduce surface modes and reflect power that was previously emitted into the substrate toward the air side. The graphene THz antenna performance in terms of radiation is expected to be much improved by the use of photonic crystals. TM and TE are the two modes of EM wave propagation inside the PBG substrate. A mathematical representation of TM and TE modes is given below [52]:
For the Transverse Magnetic mode:

$$\frac{\partial^2 E_z}{\partial x^2} + \frac{\partial^2 E_z}{\partial y^2} + k^2 n^2 (x, y) E_z = 0 \tag{8.10}$$

For the Transverse Electric mode:

$$\frac{\partial}{\partial x}\left(n^{-2}(x, y)\frac{\partial H_z}{\partial x}\right) + \frac{\partial}{\partial y}\left(n^{-2}(x, y)\frac{\partial H_z}{\partial y}\right) + k^2 H_z = 0 \tag{8.11}$$

The electric field or magnetic field can be expressed as follows:

$$E_z = e^{i\vec{\beta}\vec{r}} \sum_{m=-\infty}^{+\infty} \sum_{n=-\infty}^{+\infty} c_{m,n} e^{i\left(m\vec{b_1} + n\vec{b_2}\right)\cdot\vec{r}} \tag{8.12}$$

where the vector quantity in the x–y plane is represented by $\vec{\beta}$ and $\vec{r}$, and $\overrightarrow{b_1}$ and $\overrightarrow{b_2}$ are reciprocal lattice vectors. Many researches have been done in the field of

PBG-based terahertz graphene antenna. In [53], a graphene antenna array having dual band, utilizing PBG dielectric grating substrate with parasitic coupled feed, was introduced, achieving 16.4 dBi peak gain and −52.58 dB minimum return loss. In 2020, a 2 × 1 complementary split-ring graphene resonator patch array antenna was developed for frequencies between 0.84 and 0.94 THz. This design incorporated PBG substrates with varying hollow cylinder radii, reaching a peak gain of 14.69 dB and a minimum return loss of −57.052 dB when paired with a dielectric grating [54]. Similarly, [51] presented a broadband PBG-based antenna optimized by utilizing genetic algorithms in order to identify most effective PBG configuration. The study explored variations in PBG shapes and dimensions, highlighting their influence on performance. This optimization led to a 3.74% improvement in gain and a significant 208.94% enhancement in return loss. Another study [55] proposed graphene PBG antenna, as shown in Figure 8.2, that offers ultra-wide bandwidth and high gain, with analysis showing that periodicity and dimensions played a crucial role in performance, while shape of photonic crystal had a negligible impact. The gain and radiation efficiency for different PBG periodicities are depicted in Figure 8.3, where increasing the periodic distance from $S = 10$ to −20 μm slightly improved radiation efficiency, though the gain remained relatively constant at approximately 21.14 and 21.22 dB, respectively. Additionally, as depicted in Figure 8.4, increasing air gap diameter enhanced antenna directivity, resulting in a minor gain improvement. For $a = 4$ μm, the directivity was 21.53 dBi, rising to 21.69 dBi for $a = 8$ μm, with a corresponding gain increase from 21.22 to 21.28 dB. These advancements demonstrate potential of PBG structures in significantly enhancing the efficiency and bandwidth of graphene-based terahertz antennas.

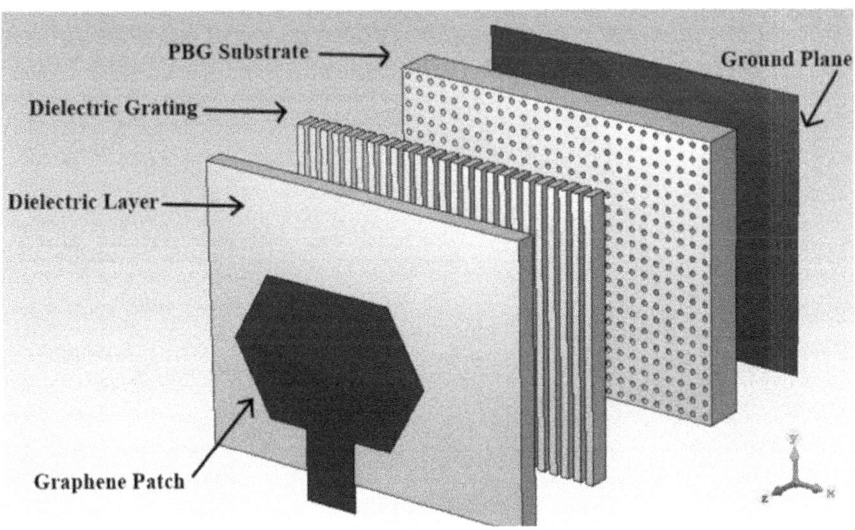

FIGURE 8.2 Proposed antenna's geometry comprised of different layers [55]. (Copyright Permission Taken.)

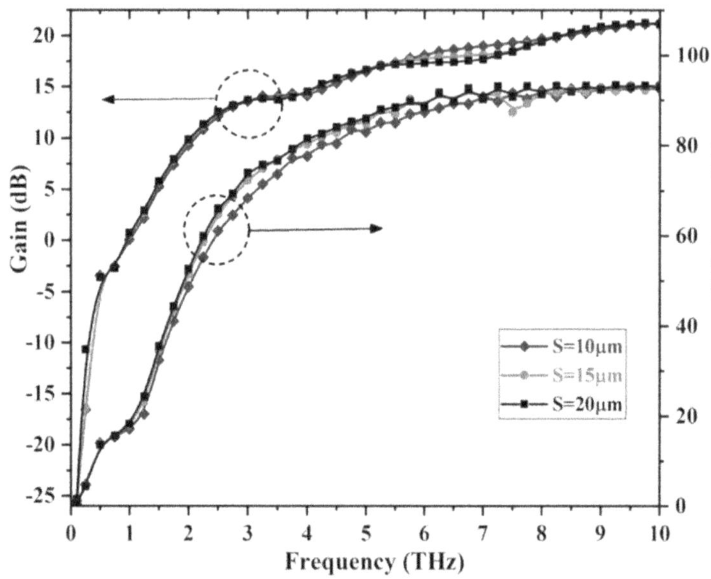

FIGURE 8.3 Radiation efficiency and gain of antenna for different periodicity of PBG [55]. (Copyright Permission Taken.)

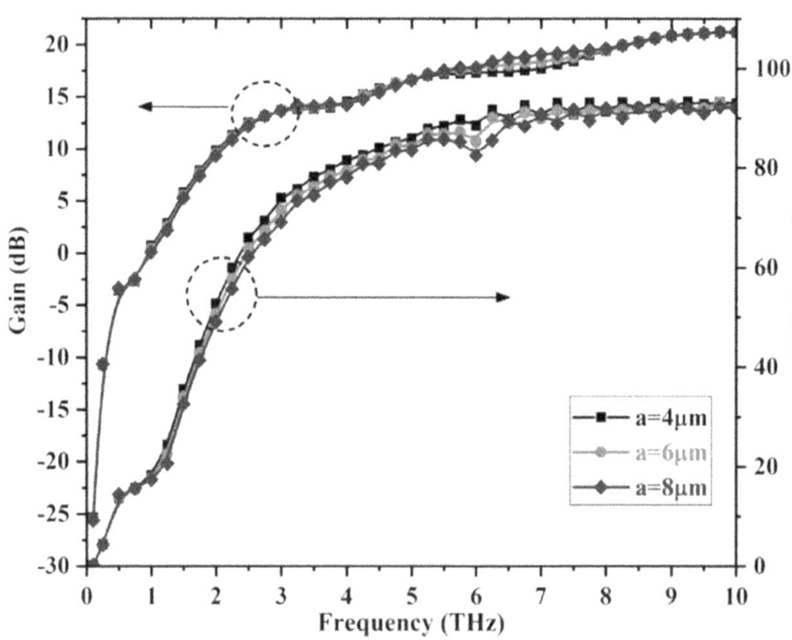

FIGURE 8.4 Radiation efficiency and gain of antenna for different air gap diameters [55]. (Copyright Permission Taken.)

8.3.2 Surface Integrated Waveguide Based Graphene THz Antenna

In the family of substrate integrated circuits (SICs), the SIW, i.e., substrate integrated waveguide, stands out as most popular and mature structure. SIW is an integrated waveguide-like structure synthesized in planar form by embedding two rows of metallic vias or slot trenches into a dielectric substrate [56]. The possibility for full-scale integration of SIW technology is one of its biggest benefits; it allows for the fabrication of active and passive circuits, as well as antennas, on the same substrate using a single processing method. The integration is highly beneficial for the overall development of compact and efficient systems. In addition to its integration capabilities, SIW offers other advantages, including having capacity to handle high power and self-consistent electrical and mechanical shielding, which are crucial for developing beam-forming networks [57,58]. A study conducted in 2021 by Mohd et al. demonstrated the SIW effectiveness as it surpasses surface wave propagation and mitigates mutual coupling [59]. The research proposed a 2×3 antenna array having 0.19–0.20 THz as the range of frequency of operation. The structure is comprised of substrate made up of polyimide, as shown in Figure 8.5. The antenna utilized a 500 nm layer of graphene for metallization and featured square patches surrounded by SIW isolators.

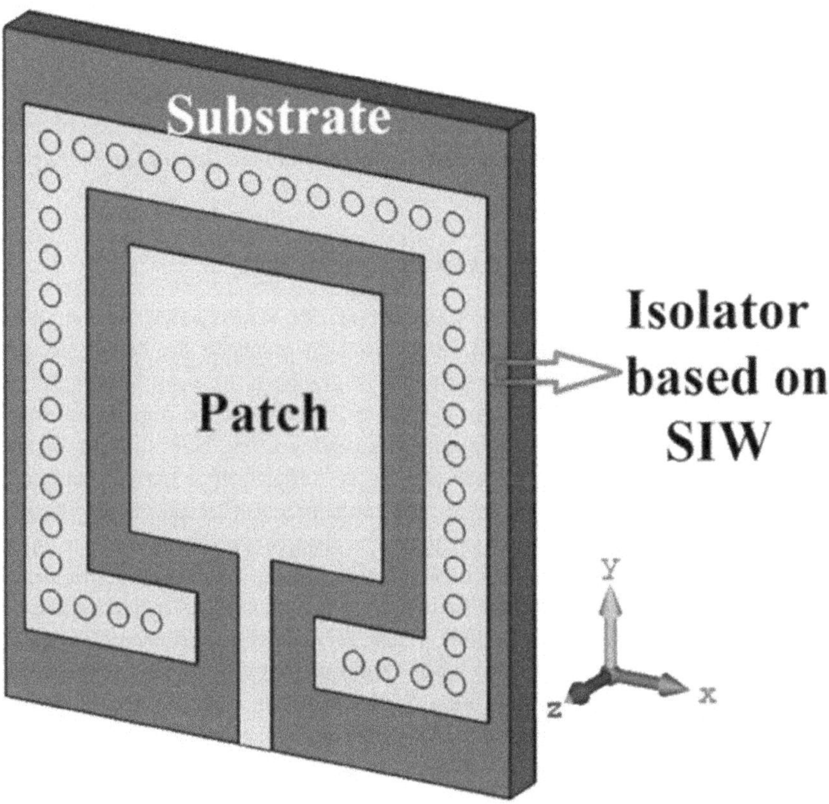

FIGURE 8.5 Isometric view of the proposed antenna [59]. (Copyright Permission Taken.)

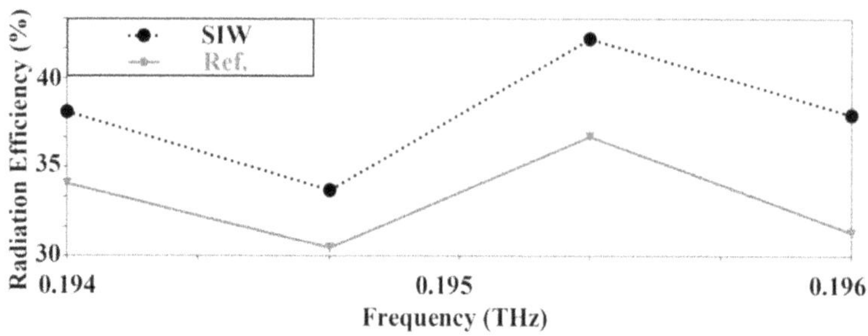

FIGURE 8.6 Radiation efficiency plots of the reference and SIW-loaded antennas [59]. (Copyright Permission Taken.)

The integration of SIW technology suppressed surface waves across the structure, which is a highly desirable feature, particularly in antenna arrays as it helps in reducing the mutual coupling. As a result, isolation between antennas was improved by an average of 22.5 dB compared to arrays without SIW integration. Additionally, the SIW isolators significantly enhanced radiation gain and efficiency due to reduced substrate losses. The radiation efficiency graphs of both antennas are depicted in Figure 8.6. Similarly, [60] proposed graphene SIW antenna having central frequency as 1 THz. Another graphene-based SIW antenna realized on substrate made up of silicon dioxide having 2.3 THz as central frequency is presented in [61]. The structure incorporates interdigital capacitor slots. This reconfigurable antenna achieved a radiation efficiency of 78.7%. These advancements highlight the significant potential of integrating SIW technology with graphene-based antennas for the development of efficient terahertz systems.

8.4 GRAPHENE-BASED TERAHERTZ ABSORBER

Absorbers are responsible for absorbing electromagnetic waves, which is accomplished by reflecting them and stopping their transmission. A graphene absorber is generally composed of three layers. The layer at top is of graphene arranged in periodic patterns. The geometric patterns become crucial as it has impact on impedance matching. A continuous metallic layer is placed at the bottom that blocks all the incident electromagnetic waves. Between these two layers, a dielectric is placed that is used to dissipate the waves. One of the many theoretical models that have been created to comprehend the fundamental physics of graphene absorbers is the interference theory. As stated above, in the three layers of the absorber, the top layer made of graphene is acting as a partial reflector, and the bottom layer is a continuous metallic plate that is considered as a perfect reflector [62]. When THz wave is incident on a graphene absorber, the dielectric spacer reflects waves multiple times, as shown in Figure 8.7.

Therefore, a high level of absorption results from the superposition of multiple reflections canceling out the direct reflection of the air from graphene [63,64]. Since the ground's reflection coefficient (r_{23}) is assumed to be −1, it can be expressed as follows:

$$r = r_{12} - \frac{t_{12}t_{21}}{r_{21} + \exp(-i2\beta)} \tag{8.13}$$

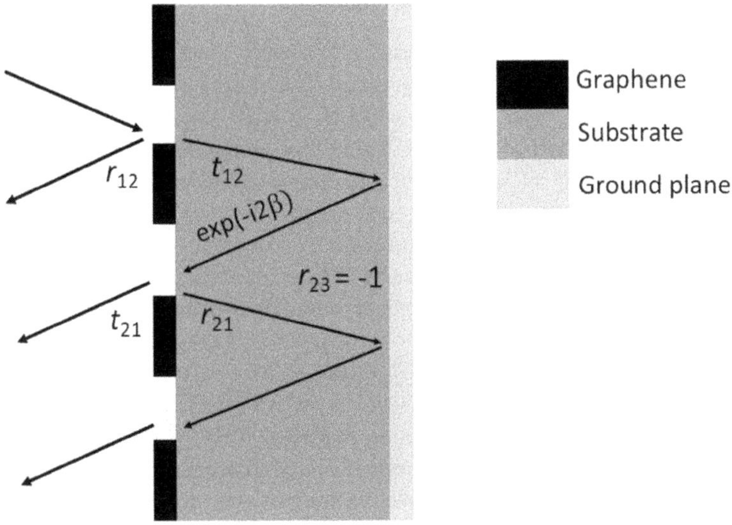

FIGURE 8.7 Schematic of the multiple reflections in the absorber.

In the formula, β is the one-way phase delay in the spacer, which can be defined as $\beta = nkd$, r is the overall reflection, n denotes the refractive index of the dielectric, k represents the wavenumber, and d denotes the thickness of the dielectric. Additionally, the air-side reflection and transmission coefficients are denoted by r_{12} and t_{12}, whereas the substrate side reflection and transmission coefficients are denoted by r_{21} and t_{21}. Reducing the reflection coefficient r_{12} is necessary to achieve complete absorption. The absorber's absorption can be calculated using the formula below [65,66]:

$$A(\omega) = 1 - R(\omega) - T(\omega) = 1 - |S_{12}(\omega)|^2 - |S_{21}(\omega)|^2 \tag{8.14}$$

Here, $S_{21}(\omega)$ is the reflected amplitude, $S_{11}(\omega)$ is the transmitted amplitude, while $T(\omega)$ and $R(\omega)$ correspond to transmission and reflection efficiencies, respectively. The absorption efficiency is at its highest when $T(\omega)$ and $R(\omega)$ are sufficiently minimal. Since the metal layer's thickness is far more than the absorber's skin depth, transmission is prevented. Therefore, $A(\omega)$ of absorber can be defined from the following equation [67,68]:

$$A(\omega) = 1 - R(\omega) \tag{8.15}$$

When the value of $R(\omega)$ is close to zero, the absorber can achieve perfect absorption. The perfect absorption can also be theoretically described using the effective medium approximation and equivalent circuit theory [69,70].

Because of the improvements in graphene fabrication technique brought about by scientific and technological advancements, graphene absorbers have emerged as a research hotspot. Based on the electromagnetic spectral response, these graphene-based absorbers can be majorly classified into two categories: Narrowband and Broadband absorbers that will be briefed in the following sections.

8.4.1 Narrowband Graphene THz Absorber

The purpose of narrowband absorbers is to absorb electromagnetic waves that fall into a certain, limited frequency range. Depending on the need of application, narrowband absorbers can be designed to have single band or multi band. These narrowband absorbers are specifically used in applications that require precise frequency filtering such as sensor and communication devices [71,72]. Many designs have been reported, in context to graphene-based narrowband absorbers. For instance, a terahertz absorber featuring two concentric square rings achieved dual-band absorption at 0.74 THz and 1.71 THz, with absorption rates of 99.8% and 99.9%, respectively [73]. Another graphene-based dual-band absorber with a dumb-bell-shaped structure demonstrated more than 99.6% absorption at 0.2242 THz and 0.5302 THz [74]. Additionally, in [75], ultra-sensitive refractive index-based graphene sensor was proposed using a flower-shaped resonator, as shown in Figure 8.8. This design achieved 99.96% absorption at 10.465 THz, with a quality factor of 15.88. The sensor was polarization-insensitive and tolerant to incident angles of up to 75 degrees. When the testing sample, or analyte, was positioned on the absorber's top, as depicted in Figure 8.9, the sensor achieved 3.923 THz/RIU of sensitivity, a quality factor of 14.92, and a FoM value of 6.111. Furthermore, a tunable graphene-based metamaterial perfect absorber was designed with a non-rotationally symmetric graphene pattern [76]. This design exhibited 99.99% and 99.79% absorption at 4.268 THz and 6.032 THz, respectively, for transverse electric polarized electromagnetic waves. For transverse magnetic polarized waves, the absorber achieved an absorption peak at 4.148 THz. By adjusting the graphene's chemical potential, design demonstrates sensitivity of 1.72 THz/RIU and figure of merit of 86.7. Therefore, narrowband absorbers are a flexible and effective solution for a range of technological developments due to their possibilities in application such as optical switching and sensing.

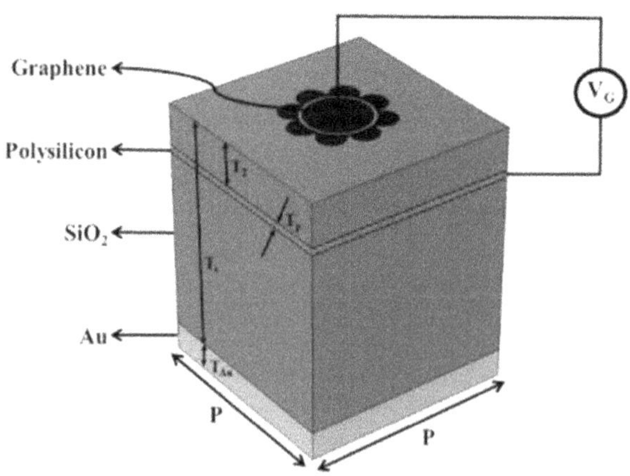

FIGURE 8.8 Proposed refractive index-based flower-shaped absorber for sensing application [75]. (Copyright Permission Taken.)

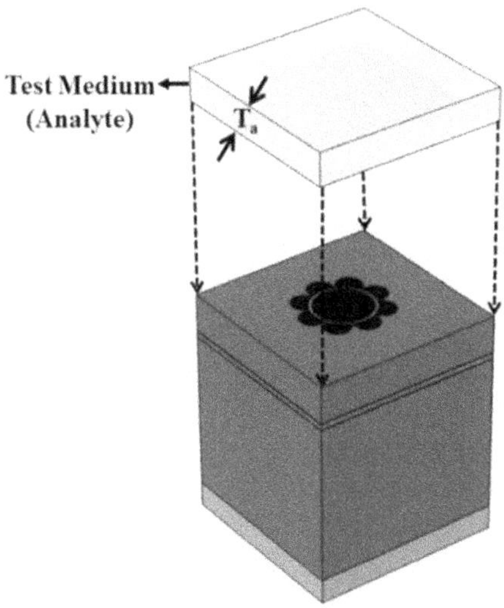

FIGURE 8.9 Absorber structure used for sensing purposes [75]. (Copyright Permission Taken.)

8.4.2 Broadband Graphene THz Absorber

Nowadays, research on absorbers, particularly in context to graphene, has steadily become the focus of study. Although they work well in the situations in which absorbers need to operate within a limited wavelength spectrum, where precise absorption at one or a few wavelengths is wanted. However, these types of absorbers limit in many practical applications like photodetector, THz imaging, solar energy harvesting, etc. [77]. Therefore, research on broadband absorbers with tunable absorptivity behavior is crucial for the development of multifunctional terahertz devices. A broadband response means that the device is able to absorb all the electromagnetic frequencies that are falling on the surface of graphene-based absorber. These broadband absorbers can be duly divided into following classes: single-layer graphene absorber and multi-layer graphene absorbers.

8.4.2.1 Single-Layer Graphene-Based Absorbers

One way to have a broadband response using graphene is based on using a single-layer graphene microstructure. Several researches have been done in context to single-layer graphene absorber. The structure and design method of single-layer graphene absorber are often found to be simple in nature relatively. In 2017, a net-shaped sinusoidal pattern of graphene-based absorber was proposed by Ye et al. that achieved nearly 100% broadband absorption [78]. Additionally, by manipulating the electrostatic doping graphene sheet, the peak absorption may be adjusted between 14% and 100%. The polarization-independent properties in the broadband absorber were achieved by the symmetrical nature of the unit cell and gradient diameter modulation of graphene sheet [79]. Yang et al. proposed a single-layer graphene THz absorber, as shown in Figure 8.10. The absorber structure has a periodical array

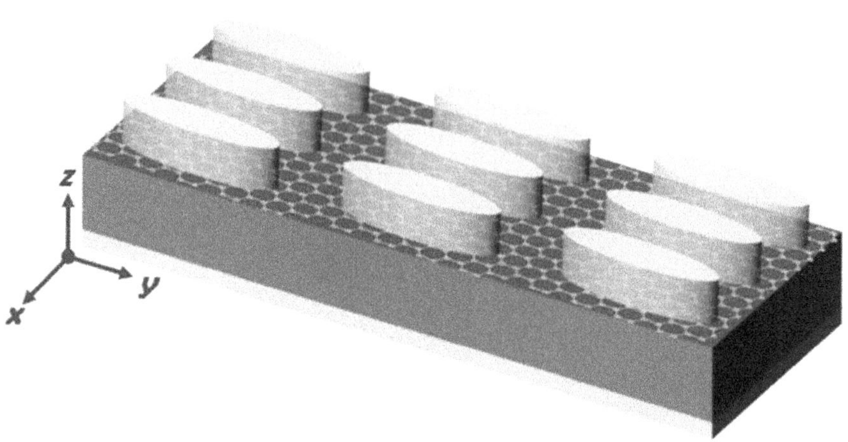

FIGURE 8.10 Schematic representation of the proposed broadband THz absorber [80]. (Copyright Permission Taken.)

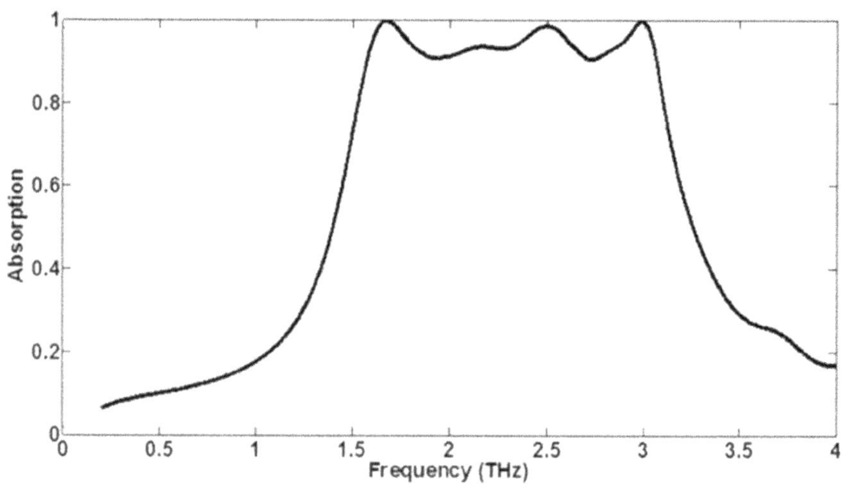

FIGURE 8.11 Absorption spectrum of the proposed broadband THz absorber [80]. (Copyright Permission Taken.)

of dielectric elliptical cylinders loaded with single-layer graphene, which are placed above the substrate that is supported by a metallic sheet [80]. In Figure 8.11, we can see that the design's relative bandwidth for absorption was higher than 90%. There are two main contributing factors for the broadband absorption. The first is due to the monolayer nonstructured graphene that is loaded with elliptical gradient dielectric cylinders, which provides graphene plasmon resonance with large relative frequency intervals. Second, the continuous and gradual variation in the width of the dielectric structures gives rise to a series of continuous plasmon resonances corresponding to each discrete state. Hence, this type of design, which utilizes the combination of dielectric and nonstructured graphene, can facilitate application of broadband terahertz absorbers. Further, fractals—a self-similar repeating pattern—can also be used for having a highly efficient absorber. Nourbaksh et al. designed a Snowflake Koch Fractal single-layer graphene absorber that can have 161% fractional bandwidth over 0.9 absorbances [81]. Additionally, a silicon semi-ellipsoidal absorber with a 95.72% absorption rate from 2 to 10 THz frequency range was proposed. It is separated from the gold bottom by a polydimethylsiloxane layer [82]. Another absorber consists of crosses and concentric circles made up of single-layer graphene, which can achieve absorptivity above 90% and 99% from 1.10 to 1.86 THz and 1.23 to 1.68 THz, respectively [83]. In 2024, Wen.li et al. introduced a bandwidth-tunable single-layer graphene absorber. The design is comprised of a gold sheet used as the bottom, a silicon dioxide layer used as a spacer, and a graphene frame and circle at the center, which is placed at the top of the silicon substrate, as shown in Figure 8.12. Due to the symmetrical nature of the absorber, the absorption spectra do not change with the change in the polarization angle, leading to a polarization-independent absorber. The simulations are done in CST Microwave Studio and are depicted in Figure 8.13. The mechanism of absorption is based on the local surface plasmon theory. Here, the background has been altered by the authors to improve comprehension

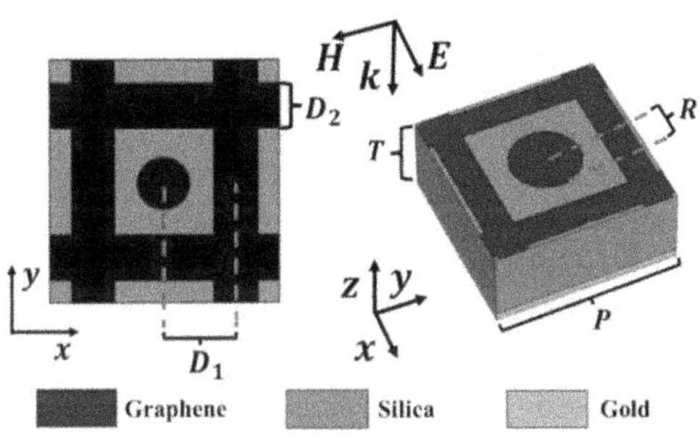

FIGURE 8.12 The unit cell structure of absorber [84]. (Copyright Permission Taken.)

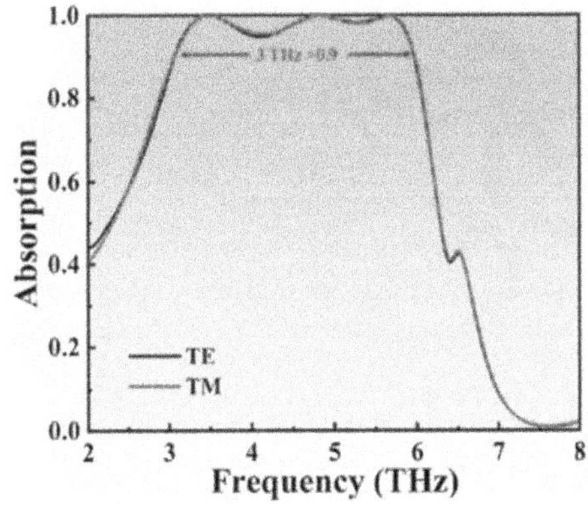

FIGURE 8.13 Absorption spectrum of proposed absorber [84]. (Copyright Permission Taken.)

of the absorption rate's magnitude. The blue and cyan colors signify the high and low absorption, respectively. The absorption rate of greater than 90% is there for the band of 3–6 THz [84]. With their exceptional physical and chemical tunability and optimal absorption performance, the absorbers have great potential for a wide range of terahertz applications.

8.4.2.2 Multi-layer Graphene-Based Absorbers

Another method to design a graphene-based broadband terahertz absorber is to utilize more than one layer of graphene. By stacking multiple layers of graphene, researchers can create a composite material that exhibits improved absorption characteristics, particularly in the terahertz region. These structures take advantage of unique characteristics of graphene material, i.e., its high electronic mobility and the ability to tune its conductivity via the Fermi level. A comprehensive study has been done on broadband metamaterial absorber utilizing multi-layer graphene in [85]. The use of multi-layer approach enhances absorption efficiency while enabling dynamic control over absorption characteristics through Fermi level tuning. With a remarkable absorption efficiency of over 90% for a wide range of frequencies, i.e., from 5.50 to 9.10 THz, the proposed absorber surpasses earlier designs. The absorber performs effectively with the variation in the angle of incidence of waves, i.e., up to 60° for both the TE mode and the TM mode. The structural configuration consists of a periodic graphene pattern layered graphene sheets, dielectric layer made up of silicon dioxide, and a gold bottom layer, as depicted in Figure 8.14. The top layer features a graphene ring with specific inner and outer radii, along with two "T"-shaped graphene bands, arranged periodically in two dimensions. The gold ground plane assures effective wave absorption, while the periodicity and design of the graphene patterns allow for strong interactions between resonators, broadening the frequency response. As shown in Figure 8.15, when $E_f = 0\,\text{eV}$, the reflection is more than 81% and the absorber is considered to be in the off state. However, when $E_f = 0.6\,\text{eV}$, more than 90% absorption is observed at the same frequency band and it is considered in

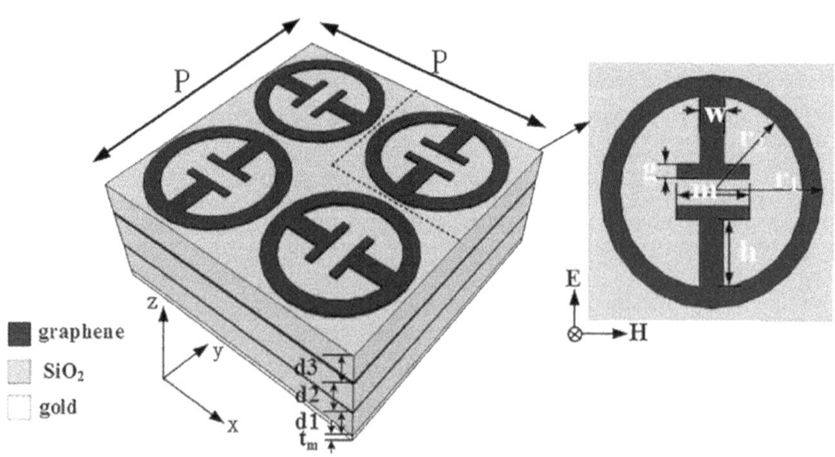

FIGURE 8.14 Schematic of the proposed absorber [85]. (Copyright Permission Taken.)

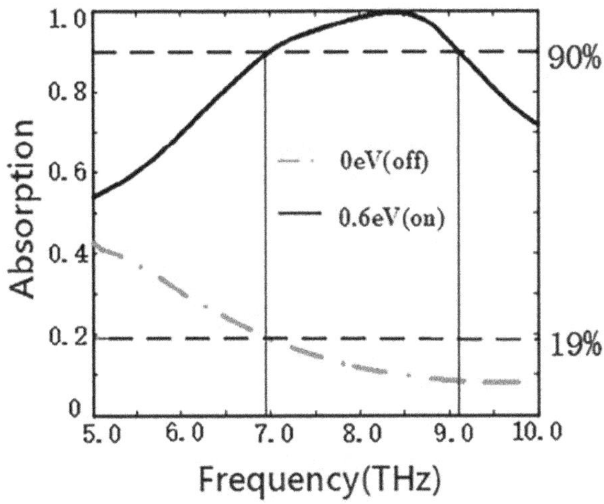

FIGURE 8.15 The proposed absorber is in on state when E_f=0.6eV and in off state when E_f=0eV [85]. (Copyright Permission Taken.)

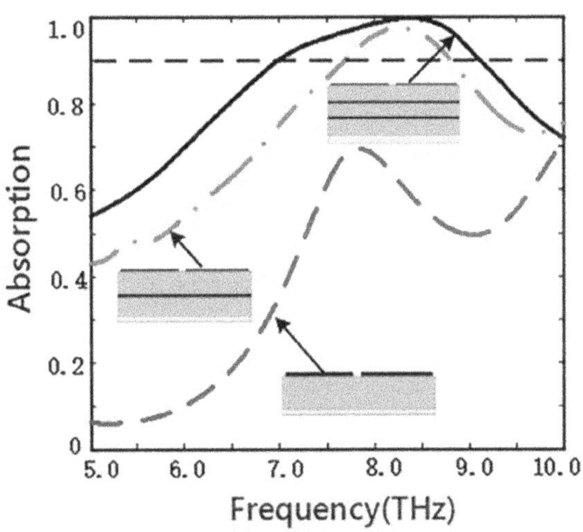

FIGURE 8.16 The absorption spectra of the suggested absorber with different layers graphene film [85]. (Copyright Permission Taken.)

the ON state. In conclusion, when the graphene's potential is varied, the structure can switch between the reflection (OFF state) and absorption (ON state) within the spectrum, which can turn out to be beneficial for the type of applications that require adaptable performance. The absorption spectra of the suggested absorber with different layers of graphene film are shown in Figure 8.16. The absorber is in the ON state when the fermi level is 0.6eV. Therefore, we can infer that high absorption cannot be achieved by using a single layer of graphene. Whenever a graphene's layer is being added, the absorption increases. Further when another layer is added, the band of absorption becomes wider and there is an increase in absorptivity. This shows that the graphene sheets interact strongly, with the neighboring resonators overlapping and spreading the resonance across a broad range of frequency in the terahertz region. Further, the analysis is done on the basis of different top layers. The absorption due to the graphene pattern was more than the graphene sheet, which means that the combination of graphene pattern can result in the increase in absorption. Using a three-periodical array of graphene rings, Iman and Sadegh (2018) proposed a work with a 90% absorption bandwidth that can be increased to 100% of the central frequency of 2.54 THz [86]. Similarly, using three layers of patterned graphene, i.e., combination of graphene disks and ribbons, a bandwidth of 90% absorption was reached, which can be extended up to 131% of the central frequency [87]. Therefore, the multi-layer graphene structures have versatile applications like active camouflage and dynamic light modulators in advanced terahertz technologies.

8.5 SUMMARY

Terahertz technology has emerged as a pivotal area in advanced engineering applications because of its high frequency, wide bandwidth, and superior spatial resolution. However, the realization of terahertz systems remains challenging due to the significant losses exhibited by traditional materials at these frequencies. This chapter investigates how graphene, with its exceptional properties, offers a viable solution to these challenges. Graphene's tunable surface conductivity, high carrier mobility, and ability to support SPP modes make it an ideal material for terahertz systems. The chapter highlights advancements in graphene-based antennas, emphasizing their ability to achieve reconfigurability and polarization control through variations in graphene's chemical potential. The integration of PBG structures and SIWs has further improved the antenna performance, particularly by enhancing gain and reducing mutual coupling in array configurations. Additionally, the chapter explores graphene-based absorbers, detailing their high electromagnetic confinement due to SPP excitation. These absorbers demonstrate remarkable efficiency, with single- and multi-layer designs achieving broader spectral coverage and higher absorption rates. Graphene-based technologies give terahertz systems a revolutionary approach by resolving important issues, opening the door for next-generation imaging, sensing, and telecommunications applications. The insights presented in this chapter underscore the potential of graphene to revolutionize terahertz engineering, enabling innovative, high-performance devices that align with the evolving demands of advanced technologies.

REFERENCES

1. X. Yang, X. Zhao, K. Yang, Y. Liu, Y. Liu, W. Fu, Y. Luo, Biomedical applications of terahertz spectroscopy and imaging, *Trends Biotechnology*, vol. 34, no. 10, pp. 810–824, 2016. https://doi.org/10.1016/j.tibtech.2016.04.008
2. C. M. Armstrong, The truth about terahertz, *IEEE Spectrum*, vol. 49, no. 9, pp. 36–41, 2012.
3. F. Fukunaga, M. Picollo, Terahertz spectroscopy applied to the analysis of artists' materials, *Applied Physics A*, vol. 100, pp. 591–597, 2010.

4. D. M. Mittleman, M. Gupta, R. Neelamani, R. Baraniuk, J. V. Rudd, M. Koch, Recent advances in terahertz imaging, *Applied Physics B*, vol. 68, no. 6, pp. 1085–1094, 1999.
5. J. M. Jornet, I. F. Akyildiz, Channel modeling and capacity analysis for EM wireless nanonetworks in the terahertz band, *IEEE Transactions on Wireless Communications*, vol. 10, no. 10, pp. 3211–3221, Oct. 2011.
6. A. Tamminen, S. Makela, J. Ala-Laurinaho, J. Hakli, P. Koivisto, P. Rantakari, J. Saily, A. Luukanen, A. V. Raisanen, Reflectarray design for 120-GHz radar application: Measurement results, *IEEE Transactions on Antennas and Propagation*, vol. 61, no. 10, pp. 5036–5047, 2013.
7. P. J. Burke, An RF circuit model for carbon nanotubes, *IEEE Transactions on Nanotechnology*, vol. 2, pp. 55–58, 2003.
8. S. Choi, K. Sarabandi, Design of efficient terahertz antennas: Carbon nanotube versus gold, *2010 IEEE Antennas and Propagation Society International Symposium. CNC-USNC/URSI Radio Sci. Meet. - Leading the Wave*, Toronto, ON, AP-S/URSI 2010, 2010.
9. K. S. Novoselov, A. K. Geim, S. V. Morozov, D. Jiang, Y. Zhang, S. V. Dubonos, I. V. Grigorieva, A. A. Firsov, Electric field effect in atomically thin carbon films, *Science*, vol. 306, pp. 666–669, 2004.
10. C. Docherty, M. Johnston, Terahertz properties of graphene, *Journal of Infrared, Millimeter, and Terahertz Waves*, vol. 33, no. 8, pp. 797–815, 2012. https://doi.org/10.1007/s10762-012-9913-y
11. J. S. Gomez-Diaz, J. Perruisseau-Carrier, Microwave to THz properties of graphene and potential antenna applications, *2012 International Symposium on Antennas and Propagation (ISAP)*, Nagoya, Japan, pp. 239–242, 2012.
12. Z. Alipour, S. I. Mirzaei, M. Fardmanesh, Design, fabrication and characterization of a wideband metamaterial absorber for THz imaging, *Engineering Proceedings*, vol. 58, no. 1, p. 92, 2023. https://doi.org/10.3390/ecsa-10-16210
13. T. Hossain, M. Parvez, A. Z. Md. Imran, M. J. Uddin, A. Gafur, S. Z. Rashid, TeraHertz antenna for biomedical application, *2022 International Conference on Innovations in Science, Engineering and Technology (ICISET)*, Chittagong, Bangladesh, 2022, pp. 42–45. https://doi.org/10.1109/ICISET54810.2022.9775884
14. J. W. Jiang, J. S. Wang, B. W. Li, Young's modulus of graphene: A molecular dynamics study, *Physical Review B*, vol. 80, p. 113405, 2009.
15. M. Mirnezhad, M. Modarresi, R. Ansari, M. R. Roknabadi, Effect of temperature on Young's modulus of graphene, *Journal of Thermal Stresses*, vol. 35, pp. 913–920, 2012.
16. L. Zhang, L. J. Long, W. Y. Zhang, D. Du, Y. H. Lin, Study of inhibition, reactivation and aging processes of pesticides using graphene nanosheets/gold nanoparticles-based acetylcholinesterase biosensor, *Electroanalysis*, vol. 24, pp. 1745–1750, 2012.
17. O. Akhavan, E. Ghaderi, A. Esfandiar, Wrapping bacteria by graphene nanosheets for isolation from environment, reactivation by sonication, and inactivation by near-infrared irradiation, *The Journal of Physical Chemistry B*, vol. 115, pp. 6279–6288, 2011.
18. Z. P. Sun, A. Martinez, F. Wang, Optical modulators with 2D layered materials, *Nature Photonics*, vol. 10, pp. 227–238, 2016.
19. F. Bonaccorso, Z. Sun, T. Hasan, A. C. Ferrari, Graphene photonics and optoelectronics, *Nature Photonics*, vol. 4, pp. 611–622, 2010.
20. L. Wang, Z. Sofer, P. Simek, I. Tomandl, M. Pumera, Boron-doped graphene: Scalable and tunable p-type carrier concentration doping, *The Journal of Physical Chemistry C*, vol. 117, pp. 23251–23257, 2013.
21. K. J. A. Ooi, L. K. Ang, D. T. H. Tan, Waveguide engineering of graphene's nonlinearity, *Applied Physics Letters*, vol. 105, p. 111110, 2014.
22. S. E. Hosseininejad, N. Komjani, M. T. Noghani, A comparison of graphene and noble metals as conductors for plasmonic one-dimensional waveguides, *IEEE Transactions on Nanotechnology*, vol. 14, no. 5, pp. 829–836, 2015.

23. J. M. Jornet, I. F. Akyildiz, Graphene-based plasmonic nano-antenna for terahertz band communication in nanonetworks, *IEEE Journal on Selected Areas in Communications*, vol. 31, no. 12, pp. 685–694, 2013.
24. S. Chaudhary, P. K. Luthra, A. Kumar, Use of graphene as a patch material in comparison to the copper and other carbon nanomaterials, *International Journal of Computer Applications in Technology*, vol. 3, no. 3, pp. 272–279, 2013.
25. A. Kausar, Advances in polymer/graphene nanocomposite for biosensor application, *NanoWorld Journal*, vol. 4, 2018. https://doi.org/10.17756/nwj.2018-056
26. X. X. Yang, A. Vorobiev, A. Generalov, M. A. Andersson, J. Stake, A flexible graphene terahertz detector, *Applied Physics Letters*, vol. 111, p. 611, 2017.
27. X. Wen, J. Zheng, Broadband THz reflective polarization rotator by multiple plasmon resonances, *Opt. Express*, vol. 22, pp. 28292–28300, 2014.
28. M. Liu, X. B. Yin, E. Ulin-Avila, B. S. Geng, T. Zentgraf, L. Ju, F. Wang, X. Zhang, A graphene-based broadband optical modulator, *Nature*, vol. 474, pp. 64–67, 2011.
29. B. Wang, X. Zhang, F. García-Vidal, X. Yuan, J. Teng, Strong coupling of surface plasmon polaritons in monolayer graphene sheet arrays, *Physical Review Letters*, vol. 109, p. 073901, 2012.
30. B. Zhao, J. M. Zhao, Z. M. Zhang, Enhancement of near-infrared absorption in graphene with metal gratings, *Applied Physics Letters*, vol. 105, p. 031905, 2014.
31. G. W. Hanson, Dyadic Green's functions and guided surface waves for a surface conductivity model of graphene, *Journal of Applied Physics*, vol. 103, p. 064302, 2008.
32. Y. Wang, H. Mi, Q. Zheng, Z. Ma, S. Gong, Graphene/phase change material nanocomposites: Light-driven, reversible electrical resistivity regulation via form stable phase transitions, *ACS Applied Materials & Interfaces*, vol. 7, pp. 2641–2647, 2015.
33. Y. S. Cao, L. J. Jiang, A. E. Ruehli, An equivalent circuit model for graphene-based terahertz antenna using the PEEC method, *IEEE Transactions on Antennas and Propagation*, vol. 64, no. 4, pp. 1385–1393, 2016.
34. M. Y. Han, B. Özyilmaz, Y. Zhang, P. Kim, Energy band-gap engineering of graphene nanoribbons, *Physicsl Review Letters*, vol. 98, p. 206805, 2007.
35. A. A. Dubinov, V. Ya Aleshkin, V. Mitin, T. Otsuji, V. Ryzhii, Terahertz surface plasmons in optically pumped graphene structures, *Journal of Physics: Condensed Matter*, vol. 23, p. 145302, 2011.
36. T. Niu, W. Withayachumnankul, B. S. Y. Ung, H. Menekse, M. Bhaskaran, S. Sriram, C. Fumeaux, Reflectarray antennas for terahertz communications, *arXiv preprint arXiv:1210.0653*, 2012.
37. M. Walther, D. Cooke, C. Sherstan, M. Hajar, M. Freeman, F. Hegmann, Terahertz conductivity of thin gold films at the metal-insulator percolation transition, *Physical Review B*, vol. 76, p. 125408, 2007.
38. G. W. Hanson, Radiation efficiency of nano-radius dipole antennas in the microwave and far-infrared regimes, *IEEE Antennas and Propagation Magazine*, vol. 50, pp. 66–77, 2008.
39. S. P. Lacour, D. Chan, S. Wagner, T. Li, Z. Suo, Mechanisms of reversible stretchability of thin metal films on elastomeric substrates, *Applied Physics Letters*, vol. 88, p. 204103, 2006.
40. M. Tamagnone, J. S. Gómez-Díaz, J. R. Mosig, J. Perruisseau-Carrier, Analysis and design of terahertz antennas based on plasmonic resonant graphene sheets, *Journal of Applied Physics*, vol. 112, p. 114915, 2012.
41. K. Anusha, D. Mohana Geetha, A. Amsaveni, Overview of THz antenna design methodologies, in *Advances in Terahertz Technology and Its Applications*, S. Das, N. Anveshkumar, J. Dutta, A. Biswas, Eds. Singapore: Springer, 2021. https://doi.org/10.1007/978-981-16-5731-3_19

42. T. A. Elwi, H. M. Al-Rizzo, D. G. Rucker, E. Dervishi, Z. Li, A. S. Biris, Multi-walled carbon nanotube-based RF antennas, *Nanotechnology*, vol. 21, p. 045301, 2009. https://doi.org/10.1088/0957-4484/21/4/045301
43. S. Anand, D. S. Kumar, R. J. Wu, M. Chavali, Graphene nanoribbon based terahertz antenna on polyimide substrate, *Optik International Journal for Light and Electron Optics*, vol. 125, pp. 5546–5549, 2014.
44. R. Bala, A. Marwaha, Development of computational model for tunable characteristics of graphene based triangular patch antenna in THz regime, *Journal of Computational Electronics*, vol. 15, pp. 1–6, 2015.
45. E. Carrasco, J. Perruisseau-Carrier, Reflectarray antenna at terahertz using graphene, *IEEE Antennas and Wireless Propagation Letters*, vol. 12, pp. 253–256, 2013.
46. E. Carrasco, J. Perruisseau-Carrier, Tunable graphene reflective cells for THz reflectarray and generalized law of reflection, *Applied Physics Letters*, vol. 102, pp. 104–103, 2013.
47. Z. Xu, X. Dong, J. Bornemann, Design of a reconfigurable MIMO system for THz communications based on graphene antennas, *IEEE Transactions on Terahertz Science and Technology*, vol. 4, pp. 609–617, 2014.
48. G. Guida, A. Priou, An introduction to photonic band gap (PBG) materials, *Progress in Electromagnetics Research, PIER*, vol. 41, pp. 1–20, 2003. https://doi.org/10.2528/PIER02010801
49. H.-Y. D. Yang, N. G. Alexopoulos, E. Yablonovitch, Photonic band-gap materials for high-gain printed circuit antennas, *IEEE Transactions on Antennas and Propagation*, vol. 45, no. 1, pp. 185–187, Jan. 1997. https://doi.org/10.1109/8.554261
50. B. M. Rai, Modeling transmission of photonic crystal waveguide modes enhanced by Kerr nonlinearity, *Proceedings of the 12th International Conference on Nonlinear Optics Photonics*, 2012.
51. D. Khezzar, D. Khedrouche, T. A. Denidni, New design of a broadband PBG-based antenna for THz band applications, *Photonics and Nanostructures: Fundamentals and Applications*, vol. 46, p. 100947, 2021. https://doi.org/10.1016/j.photonics.2021.100947
52. P.-C. Hoang, *Applications of Photonic Crystals in Communications Engineering and Optical Imaging*, Ph.D. dissertation, Technische Universität Kaiserslautern, 2009.
53. R. K. Kushwaha, P. Karuppanan, Parasitic-coupled high-gain graphene antenna employed on PBG dielectric grating substrate for THz applications, *Microwave and Optical Technology Letters*, vol. 62, pp. 439–447, 2020. https://doi.org/10.1002/mop.32033
54. R. K. Kushwaha, P. Karuppanan, Enhanced radiation characteristics of graphene-based patch antenna array employing photonic crystals and dielectric grating for THz applications, *Optik*, vol. 200, p. 163422, 2020. https://doi.org/10.1016/j.ijleo.2019.163422
55. M. A. K. Khan, M. I. Ullah, M. A. Alim, High-gain and ultrawide-band graphene patch antenna with photonic crystal covering 96.48% of the terahertz band, *Optik*, vol. 227, p. 166056, 2021. https://doi.org/10.1016/j.ijleo.2020.166056
56. K. Wu, D. Deslandes, Y. Cassivi, The substrate integrated circuits: A new concept for high-frequency electronics and optoelectronics, *6th International Conference on Telecommunications in Modern Satellite, Cable and Broadcasting Service, 2003. TELSIKS 2003*, Nis, Yugoslavia, vol. 1, pp. 3–5, Oct. 2003.
57. Y. J. Cheng, K. Wu, W. Hong, Power handling capability of substrate integrated waveguide interconnects and related transmission line systems, *IEEE Transactions on Advanced Packaging*, vol. 31, no. 4, pp. 900–909, Nov. 2008.
58. K. Wu, Y. J. Cheng, T. Djerafi, W. Hong, Substrate-integrated millimeter-wave and terahertz antenna technology, *Proceedings of the IEEE*, vol. 100, no. 7, pp. 2219–2232, Jul. 2012. https://doi.org/10.1109/JPROC.2012.2190252

59. M. Alibakhshikenari, B. S. Virdee, S. Salekzamankhani, S. Aïssa, C. H. See, N. Soin, S. J. Fishlock, A. A. Althuwayb, R. Abd-Alhameed, I. Huynen, J. A. McLaughlin, F. Falcone, E. Limiti, High-isolation antenna array using SIW and realized with a graphene layer for sub-terahertz wireless applications, *Scientific Reports*, vol. 11, 10218, 2021. https://doi.org/10.1038/s41598-021-87712-y
60. N. Kiani, F. Tavakol Hamedani, P. Rezaei, Implementation of a reconfigurable miniaturized graphene-based SIW antenna for THz applications, *Micro and Nanostructure*, vol. 169, p. 207365, 2022. https://doi.org/10.1016/j.micrna.2022.207365
61. N. Kiani, F. Tavakkol Hamedani, P. Rezaei, Design of an ultra-compact reconfigurable IDC graphene-based SIW antenna in THz band with RHCP, *Telecommunication Systems*, vol. 85, pp. 401–413, 2024. https://doi.org/10.1007/s11235-023-01088-0
62. W. Li, Y. Yi, H. Yang, S. Cheng, W. Yang, H. Zhang, Z. Yi, Y. Yi, H. Li, Active tunable terahertz bandwidth absorber based on single layer graphene, *Communications in Theoretical Physics*, vol. 75, p. 045503, 2023.
63. Z. Jia, L. Huang, J. Su, B. Tang, Tunable electromagnetically induced transparency-like in graphene metasurfaces and its application as a refractive index sensor, *Journal of Lightwave Technology*, vol. 39, pp. 1544–1549, 2021.
64. Q. Shangguan, Z. Chen, H. Yang, S. Cheng, W. Yang, Z. Yi, X. Wu, S. Wang, Y. Yi, P. Wu, Design of ultra-narrow band graphene refractive index sensor, *Sensors*, vol. 22, p. 6483, 2022.
65. X. Wu, Y. Li, C. Li, Z. He, Y. Xiang, L. Xiong, D. Chen, Y. Yu, K. Sun, Z. He, P. Chen, The electrochemical performance improvement of $LiMn_2O_4$/Zn based on zinc foil as the current collector and thiourea as an electrolyte additive, *Journal of Power Sources*, vol. 300, pp. 453–459, 2015.
66. Q. Shangguan, Y. Zhao, Z. Song, J. Wang, H. Yang, J. Chen, C. Liu, S. Cheng, W. Yang, Z. Yi, High sensitivity active adjustable graphene absorber for refractive index sensing applications, *Diamond and Related Materials*, vol. 128, p. 109273, 2022.
67. Y. Zhang, Y. Yi, W. Li, S. Liang, J. Ma, S. Cheng, W. Yang, Y. Yi, High absorptivity and ultra-wideband solar absorber based on Ti-Al_2O_3 cross elliptical disk arrays, *Coatings*, vol. 13, p. 531, 2023.
68. B. Tang, Z. Li, E. Palacios, Z. Liu, S. Butun, K. Aydin, Chiral-selective plasmonic metasurface absorbers operating at visible frequencies, *IEEE Photonics Technology Letters*, vol. 29, pp. 295–298, 2017.
69. B. Tang, N. Yang, L. Huang, J. Su, C. Jiang, Tunable anisotropic perfect enhancement absorption in black phosphorus-based metasurfaces, *IEEE Photonics Journal*, vol. 12, p. 4500209, 2020.
70. F. Zhao, J. Lin, Z. Lei, Z. Yi, F. Qin, J. Zhang, L. Liu, X. Wu, W. Yang, P. Wu, Realization of 18.97% theoretical efficiency of 0.9 μm thick c-Si/ZnO heterojunction ultrathin-film solar cells via surface plasmon resonance enhancement, *Physical Chemistry Chemical Physics*, vol. 24, pp. 4871–4880, 2022.
71. A. Rastgordani, Z. Ghattan Kashani, High-sensitive refractive index sensors based on graphene ring metasurface, *Optics Communications*, vol. 474, p. 126164, 2020.
72. Z. Vafapour, W. Troy, A. Rashidi, Colon cancer detection by designing and analytical evaluation of a water-based THz metamaterial perfect absorber, *IEEE Sensors Journal*, vol. 21, no. 17, pp. 19307–19313, 2021.
73. D. X. Yan, J. S. Li, Tuning control of dual-band terahertz perfect absorber based on graphene single layer, *Laser Physics*, vol. 29, p. 046203, 2019.
74. D. X. Yan, M. Meng, J. S. Li, X. Li, Graphene-assisted narrow bandwidth dual-band tunable terahertz metamaterial absorber, *Frontiers in Physics*, vol. 8, p. 306, 2020.

75. A. H. Norouzi Razani, P. Rezaei, P. Zamzam, S. A. Khatami, O. M. Daraei, Absorption-based ultra-sensitive RI sensor based on the flower-shaped graphene resonator for early detection of cancer, *Optics Communications*, vol. 524, p. 128775, 2022. https://doi.org/10.1016/j.optcom.2022.128775
76. Z. W. Wu, J. P. Tian, R. C. Yang, A graphene-based dual-band metamaterial absorber for TE polarized THz wave, *Micro and Nanostructures*, vol. 168, p. 207331, 2022. https://doi.org/10.1016/j.micrna.2022.207331
77. S. Han, J. H. Shin, P.-H. Jung, H. Lee, B. J. Lee, Broadband solar thermal absorber based on optical metamaterials for high-temperature applications, *Advanced Optical Materials*, vol. 4, no. 8, pp. 1265–1273, 2016.
78. L. Ye, Y. Chen, G. Cai, N. Liu, J. Zhu, Z. Song, Q. H. Liu, Broadband absorber with periodically sinusoidally-patterned graphene layer in terahertz range, *Optics Express*, vol. 25, pp. 11223–11232, 2017.
79. X. Huang, W. He, F. Yang, J. Ran, B. Gao, W.-L. Zhang, Polarization-independent and angle-insensitive broadband absorber with a target-patterned graphene layer in the terahertz regime, *Optics Express*, vol. 26, pp. 25558–25566, 2018.
80. J. Yang, Z. Zhu, J. Zhang, J. Zhang, C. Guo, W. Xu, K. Liu, X. Yuan, S. Qin, Broadband terahertz absorber based on multi-band continuous plasmon resonances in geometrically gradient dielectric-loaded graphene plasmon structure, *Scientific Reports*, vol. 8, p. 3239, 2018. https://doi.org/10.1038/s41598-018-21705-2
81. M. Nourbakhsh, E. Zareian-Jahromi, R. Basiri, Ultra-wideband terahertz metamaterial absorber based on snowflake Koch fractal dielectric-loaded graphene, *Optics Express*, vol. 27, pp. 32958–32969, 2019.
82. R. Zhong, L. Yang, Z. Liang, Z. Wu, Y. Wang, A. Ma, Z. Fang, S. Liu, Ultrawideband terahertz absorber with a graphene-loaded dielectric hemi-ellipsoid, *Optics Express*, vol. 28, pp. 28773–28781, 2020.
83. F. He, Z. Xu, K. Li, M. Wang, W. Xie, Q. Luo, B. Chen, W. Kong, M. Yun, Tunable polarization-independent and angle-insensitive broadband terahertz absorber with graphene metamaterials, *Optics Express*, vol. 29, pp. 7158–7167, 2021.
84. W. Li, M. Liu, S. Cheng, H. Zhang, W. Yang, Z. Yi, Q. Zeng, B. Tang, S. Ahmad, T. Sun, Polarization independent tunable bandwidth absorber based on single-layer graphene, *Diamond and Related Materials*, vol. 142, p. 110793, 2024. https://doi.org/10.1016/j.diamond.2024.110793
85. P. Fu, F. Liu, G. J. Ren, F. Su, D. Li, J. Q. Yao, A broadband metamaterial absorber based on multi-layer graphene in the terahertz region, *Optics Communications*, vol. 417, pp. 62–66, 2018. https://doi.org/10.1016/j.optcom.2018.02.034
86. I. Chaharmahali, S. Biabanifard, Ultra-broadband terahertz absorber based on graphene ribbons, *Optik*, vol. 172, pp. 1026–1033, 2018. https://doi.org/10.1016/j.ijleo.2018.07.137
87. S. Biabanifard, M. Biabanifard, S. Asgari, S. Asadi, M. C. E. Yagoub, Tunable ultra-wideband terahertz absorber based on graphene disks and ribbons, *Optics Communications*, vol. 427, pp. 418–425, 2018. https://doi.org/10.1016/j.optcom.2018.07.00800

9 Reconfigurable THz MIMO Antenna Based on Graphene
An Overview

Durai Pandurangan and Naveen Mishra

9.1 INTRODUCTION

The utilization of terahertz (THz) band, which ranges from 0.1 to 10 THz in frequency, is highly desirable for wireless communications because of its vast bandwidth and rapid data transfer speeds [1]. Compared to mmWave and sub-6 GHz frequencies, THz wireless communications provide the advantages of greater data rates, greater directionality, and larger bandwidth [2]. The 6G wireless communication is expected to be operational by 2030 [3]. The THz spectrum will be a critical component in medical cancer treatment, security screening, biomedical imaging, material identification, and radio astronomy. Large bandwidth and extremely high frequency make this spectrum suitable for different applications [4,5]. However, terahertz bands suffer due to high propagation loss and atmospheric absorption loss [6]. Therefore, designing antennas for the terahertz spectrum is highly challenging.

Conventional conductors like copper and aluminum are no longer suitable for the THz spectrum due to skin effect and decreased conductivity. In these materials, the electric charges flow on the surface due to decreased skin depth. Graphene is emerging as a promising nanomaterial for the THz applications due to high conductivity, conductivity tunability, and ability to support surface plasmon polaritons (SPP) waves [7]. Especially, graphene has the advantage of frequency tuning by varying its chemical potential, which further can be controlled by bias voltage.

The addition of multiple-input multiple-output (MIMO) antennas along with THz spectrum will further improve data rate, communication reliability, and increased channel capacity in a signal fading environment compared to single antenna system [8–10]. In MIMO, multiple antenna elements are used for transmitting and receiving the signal. One of the major MIMO antenna design constraints is compactness. The radiating components of a small MIMO antenna should be positioned in close proximity to one another. This results in mutual coupling augmentation, which lowers the performance of MIMO antennas. Otherwise, a minimum gap of $\lambda/2$ or greater should be maintained between the ports to ensure better isolation and reduced coupling.

Mutual coupling reduction techniques are used to reduce the electromagnetic interaction between the antenna elements of the MIMO systems. Defective ground

DOI: 10.1201/9781003599111-9

structure (DGS), Neutralization method (NL), Electromagnetic band gap structure (EBG), Frequency selective surface (FSS), Metamaterials, Parasitic element (PE) method, Artificial magnetic conductor (AMC), and self-decoupling via different antenna orientations are the common coupling reduction techniques widely used in MIMO antennas [11–17]. Each method comes with its own pros and cons, and therefore, appropriate method should be selected [18].

Numerous graphene-based THz antenna works are published in the literature. A graphene-based dual-polarized MIMO antenna has been reported in [19], which offers inter-element isolation of −45 dB due to the presence of two orthogonal modes in the patch, i.e., TM_{10} and TM_{01}. Graphene-based ultra-wideband Yagi-Uda shaped antenna is presented in [20], which achieves orthogonal beam direction with a peak gain of 14dBi and an impedance bandwidth of 120%. In article [21], Metamaterial-inspired THz MIMO antenna for 6G application is explained. The antenna measures $130\times85\,\mu m^2$ in size and has a maximum isolation of −55 dB between radiating elements. In manuscript [22], a two-port MIMO antenna operating at 6 THz is reported. To improve isolation between the ports, graphene-based band stop filter is introduced in the ground plane. This antenna offers −47.56 isolation, 3.99 dBi gain, and 90.17% radiation efficiency. A tunable graphene two-port MIMO antenna with simultaneous transmit and receive mode operation has been presented in [23]. This antenna can be operated in either self-duplexing mode or MIMO mode and provides isolation of −40 dB. MIMO antennas working at THz without tenability features are also reported in [24,25].

In this chapter, graphene-based dual band operating at 1.3 and 2.5 THz, reconfigurable MIMO antenna is designed and presented for THz application. The MIMO antenna consists of quad-ports with self-decoupling in nature, offering isolation of −15 dB over the entire band width. Re-configurability is achieved by changing graphene's chemical potential. The rest of the chapter is organized as follows: Section 9.2 describes the theoretical background of the graphene material. Section 9.3 describes the design of patch antenna to be used as MIMO element. Section 9.4 discusses the design of a dual-port MIMO and quad-port MIMO antenna. Section 9.5 explains the simulated results and discussions.

9.2 GRAPHENE MATERIAL

Graphene is zero-band gap, two-dimensional carbon crystal of one atom thick material with high electrical conductivity. Graphene has unique material properties as compared to other conventional materials such as copper and aluminum. Graphene surface conductivity is tunable, as defined by Kubo's formula [26–28]. The intra-band and the inter-band surface conductivity are mathematically expressed by the following equations:

$$\sigma_{\text{intra}} = -j\frac{e^2 k_B T}{\pi h^2(\omega - j\Gamma)}\left[\frac{\mu_c}{k_B T} + 2\ln\left(e^{\frac{-\mu_c}{k_B T}} + 1\right)\right] \tag{9.1}$$

$$\sigma_{\text{inter}} = -\frac{je^2}{4\pi h} \ln\left(\frac{2|\mu_c| - (\omega - j\Gamma)h}{2|\mu_c| + (\omega - j\Gamma)h}\right) \tag{9.2}$$

$$\sigma_{\text{total}} = \sigma_{\text{intra}} + \sigma_{\text{inter}} \tag{9.3}$$

The sum of the intra and inter surface conductivity is the graphene's total conductivity. Here, 'e' is the charge of the electron (1.602×10^{-19} Coulombs), 'ω' is the angular frequency of operation, 'h' is the Planck's constant ($6.62607015 \times 10^{-34}$ joule-hertz^{-1}), 'T' is the temperature in Kelvin, 'k_B' is the Boltzmann's constant (1.380649×10^{-23} J/K), 'μ_c' is the chemical potential, and Γ is the scattering rate. Figure 9.1 shows the conductivity of a graphene sheet with $\mu_c = 1$ eV, $\tau = 1$ ps (relaxation time), $T = 300$ K, and thickness $= 0.001$ μm.

From Figure 9.1a and b, it is clear that the conductivity of σ_{intra} is higher than σ_{inter} at lower THz frequency. Hence, the effect of inter conductivity at lower THz frequency is neglected. Therefore, the total conductivity of the graphene is approximately equal to intra surface conductivity, as shown in Figure 9.1c. As seen from equation (9.1), intra conductivity is a function of chemical potential, and the chemical

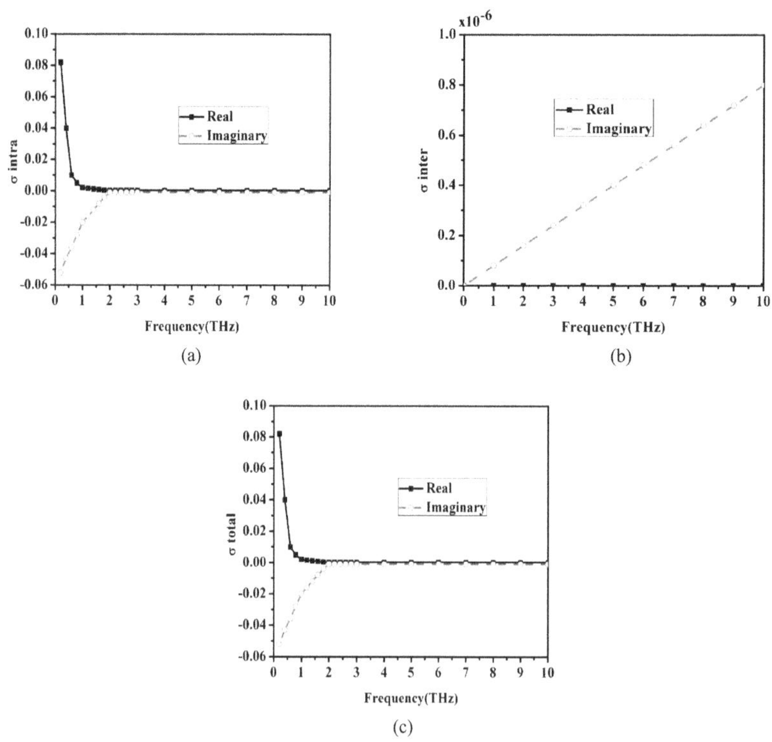

FIGURE 9.1 Graphene surface conductivity: (a) intra, (b) inter, and (c) total.

potential of graphene is a function of the applied DC-bias voltage to the graphene layer, which is given by the following equation:

$$\mu_c = he_f \sqrt{\frac{\pi c V_g}{e}} \tag{9.4}$$

where 'V_g' is the applied DC voltage. This property of tunable surface conductivity offers frequency-agile features to graphene-based antenna design.

9.3 SINGLE ANTENNA UNIT

Using the theoretical background described in the previous section, a graphene antenna is designed and analyzed in this section. Unlike other antenna design, terahertz antenna designs are first started with creating graphene material (using the properties listed in Table 9.1) in the 3D Electromagnetic (EM) simulator-Computer simulation technology (CST). Figure 9.2 shows the general flowchart of the graphene-based antenna designs. After the graphene material is created, the graphene-based antennas are designed, optimized, and analyzed. Finally, MIMO antenna is created using the single antenna unit and the performance metrics are evaluated. For example, the graphene antenna structure along with its parameters is plotted in Figure 9.3. The structure consists of three layers. At the top, a rectangular patch engraved with two L-shaped slots connected via 50 Ω feed line made up of graphene layer is shown in Figure 9.3a. For the middle layer, a silicon substrate with high relative permeability of 11.9 and conductivity σ (25×10^{-5} S/m) is used. At the bottom, the use of a graphene-based ground plane etched with circular slot is shown in Figure 9.3b. The front side view of the antenna is displayed in Figure 9.3c. The dimensions of all the three layers used in the antenna design are given in Table 9.2. The antenna design and simulations are carried out using CST software.

Figure 9.4 shows the simulated reflection coefficient ($|S_{11}|$) of the proposed antenna. The design and analysis of the suggested antenna is carried out using 3D CST. It is the dual-band antenna resonates at 1.33 THz (band-1) and 2.51 THz (band-2) with reflection coefficient (S_{11}) of −20 and −30 dB, respectively. The −10 dB impedance bandwidth of the antenna at band-1 and -2 measures 0.25 and 0.3 THz. The fractional bandwidth of the antenna is 15% and 16% at 1.3 and 2.51 THz frequency band.

TABLE 9.1
Graphene Properties

Parameters	Value
T	300 K
T	1 ps
μ_c	1 eV
Thickness	0.001 μm
Minimum frequency	0.1 THz
Maximum frequency	10 THz

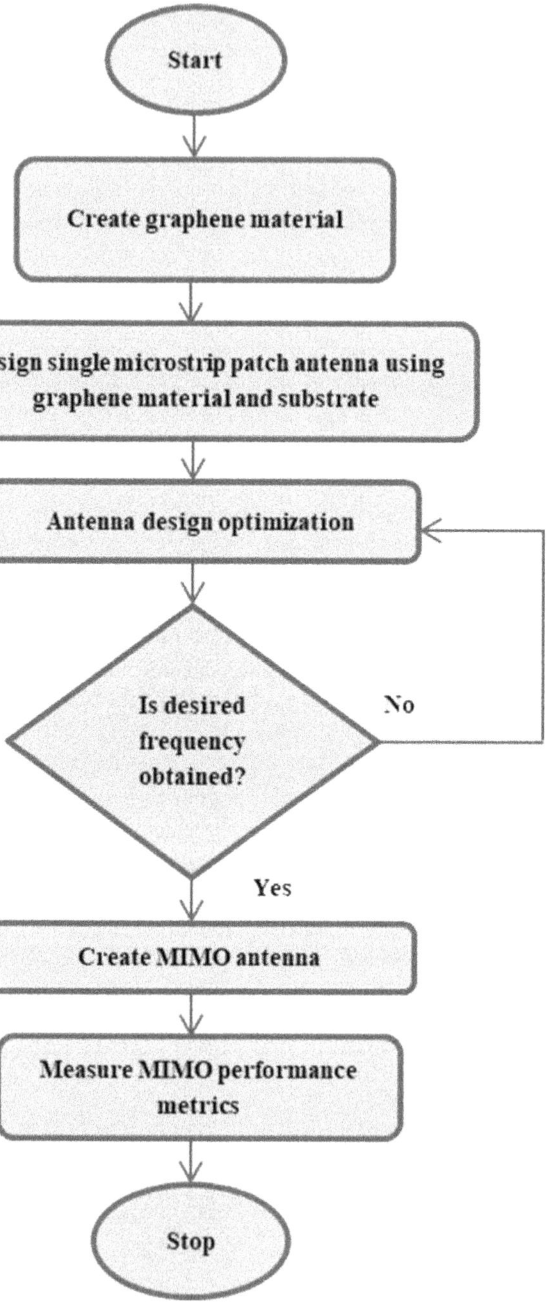

FIGURE 9.2 General flowchart of the graphene-based antenna design.

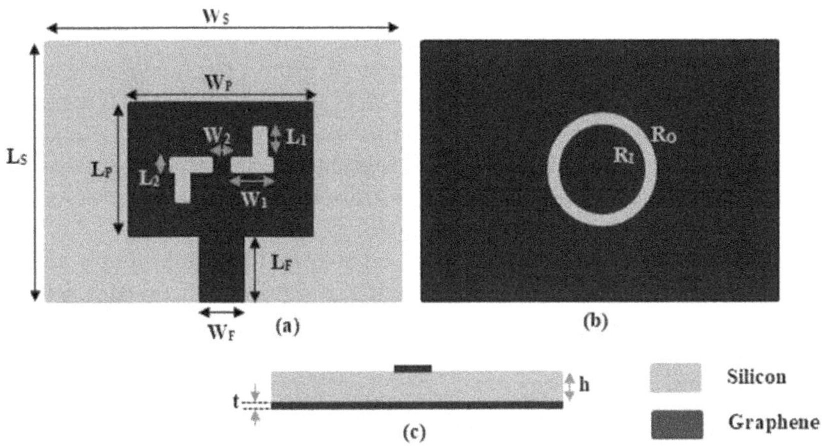

FIGURE 9.3 Layout of the single antenna unit: (a) top side, (b) bottom side, and (c) front side.

TABLE 9.2
Antenna Dimensions

Parameters	Value (μm)	Parameters	Value (μm)
W_S	30	L_S	20
W_P	15	L_P	10
W_F	4	L_F	5
W_1	3	L_1	1.82
W_2	1	L_2	1
h	2.5	R_I	4
t	0.1	R_0	3

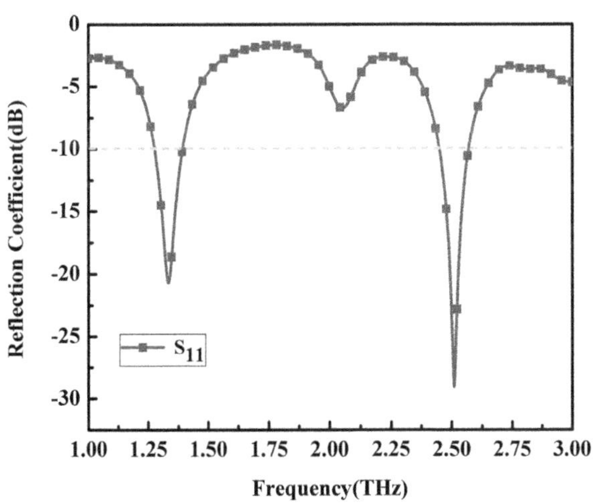

FIGURE 9.4 Single antenna-unit reflection coefficient.

9.3.1 Characteristic Mode Analysis (CMA)

Characteristic Mode Analysis (CMA) is one of the popular electromagnetic numerical techniques that use Method of Moments (MoM) for analyzing the electromagnetic behavior of the antenna by solving MoM impedance matrix, resulting in a collection of 'characteristic currents', or the so-called fundamental modes of the given antenna structure. This characteristics current gives valuable insights into antenna radiation pattern and polarization, as well as helps in design optimization.

CMA is calculated using the CST multi-layer solver treating conductor and substrate as lossless materials without input excitation signal. CMA explains how each mode contributes to electromagnetic radiation at particular frequency.

The set of characteristic currents is evaluated using the following mathematical equations [29,30]:

$$\vec{J} = J_m c_m \tag{9.5}$$

$$\text{MS}_m = \frac{1}{|1 + j\lambda_m|} \tag{9.6}$$

$$\beta_m = 180° - \tan^{-1}(\lambda_m) \tag{9.7}$$

where J_m, c_m, and λ_m are characteristic current, modal weighting coefficients, and eigenvalue. The modal significance (MS_m), characteristic angle (β_m), and eigenvalue are the important entities of the characteristic mode theory, which are used to identify the desired modes. The characteristic current is said to be dominant, if modal significance is greater than 0.7 (ideally zero), characteristic angle is 180°, and eigenvalue is close to zero at the given frequency. Otherwise, current is subservient at that frequency. Figure 9.5a–c shows the modal significance, characteristic angle, and eigenvalue of the graphene antenna designed. At 1.3 THz frequency, the mode-1 and mode-2 have $\text{MS}_m > 0.7$, $\beta_m = 180°$, and zero eigenvalue shows the dominant modes of resonance. Alternatively, the MS_m of mode-3 at 1.3 THz is less than 0.7, which makes it a non-dominant mode. Similarly, mode-1, mode-2, and mode-3 are called dominant characteristic currents of the second band (2.5 THz). In this simulation, the number of characteristic modes is limited to three modes; however, it can be increased to m modes, which increase the accuracy at cost of computation complexity and simulation time. The radiation pattern of all the characteristic modes corresponding to 1.3 and 2.5 THz is shown in Figure 9.6.

9.3.2 Parametric Analysis

To investigate the effect of various parameters on bandwidth, frequency shift, and S-parameters, parametric investigation has been conducted using selected parameters. During this investigation, only one parameter is varied at a time while keeping remaining parameters constant. Figure 9.7 shows the parametric investigation of the designed graphene antenna for W_2, μ_c, and R_0 parameters.

9.3.2.1 Effect of Varying W_2

The distance between two L-shaped slots in a graphene patch is denoted as 'W_2'. From the perusal of Figure 9.7a, it is observed that the variation of the gap from 0.5

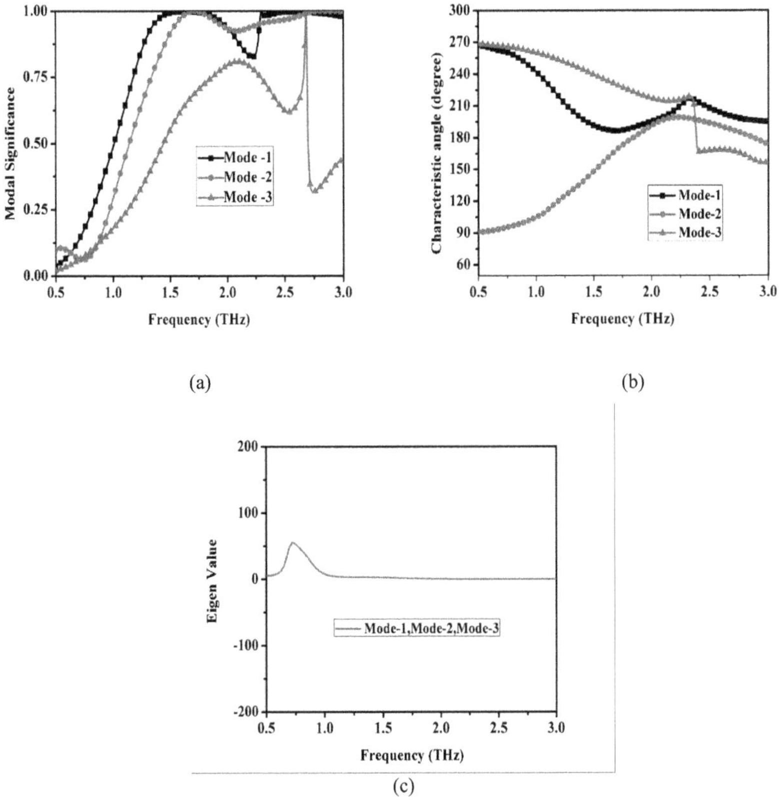

FIGURE 9.5 Characteristics mode analysis: (a) modal significance (MS_m), (b) characteristic angle (β_m), and (c) eigenvalue.

to 2 μm affects the return loss of the antenna. At 0.5 μm gap, the return loss of −18 dB (band-1) and −15 dB is obtained, while for 1 and 2 μm gap, the return loss of −21 dB for band-1 and −28 dB for band-2, and −22 dB for band-1 and −32 dB for band-2, is obtained, with a minor frequency shift in band-2 from 2.4 to 2.5 THz.

9.3.2.2 Effect of Varying μ_c

Figure 9.7b depicts the frequency-reconfigurable nature of the graphene antenna when the chemical potential applied of the graphene material is varied between 0.8 and 1.2 eV with a step size of 0.2 eV. This plot reveals that the resonance of the graphene antenna increases as the chemical potential increases and vice versa. The resonant frequency shifted from 1.3 to 1.4 THz (band-1) and 2.4 to 2.6 THz (band-2) when the chemical potential is varied from 1 to 1.2 eV. Similarly, the frequency shifts from 1.3 to 1.2 THz (band-1) and 2.4 to 2.2 THz (band-2) when the chemical potential is decreased from 1 to 0.8 eV. Therefore, it is proved that this design has a frequency reconfigurability feature, which can be achieved by varying the applied DC-bias voltage to the graphene material.

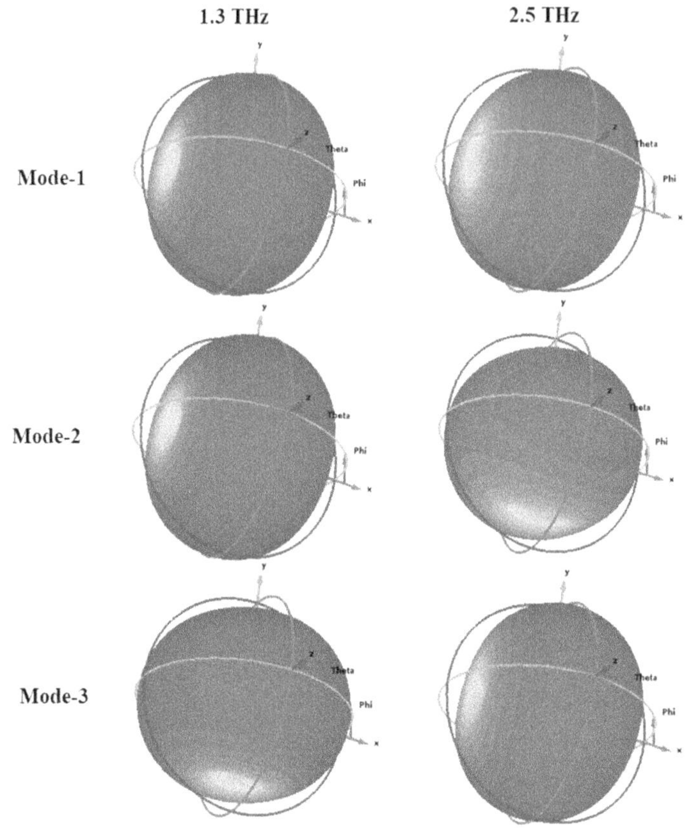

FIGURE 9.6 Mode-1, mode-2, and mode-3 radiation pattern at 1.3 and 2.5 THz.

9.3.2.3 Effect of Varying R_0

The ground plane of the proposed graphene antenna consists of a circular slot of inner radius R_I and outer radius R_0. The outer radius is varied from 4 to 5.5 µm with incremental step of 0.5 µm while keeping R_I unchanged. Figure 9.7c shows the effect of varying R_0. It is clearly visible from the graph that the outer radius has significant control over the second band. The return loss is below −10 dB for the second band for the lowest value of the outer radius and achieves −32 dB of return loss for 5.5 µm radius. In other words, the antenna produces a single band when $R_I = R_0$ and a dual band when $R_I > R_0$.

9.4 MIMO DESIGN

MIMO antennas are capable of increasing communication speed and link reliability by combining multiple transmit-receive antennas. The antenna diversity can be achieved through the replication of single antenna unit in a parallel configuration. To ensure continuous and uniform power level across different ports, the ground planes of all the MIMO elements are interconnected. In the MIMO antenna, the antenna

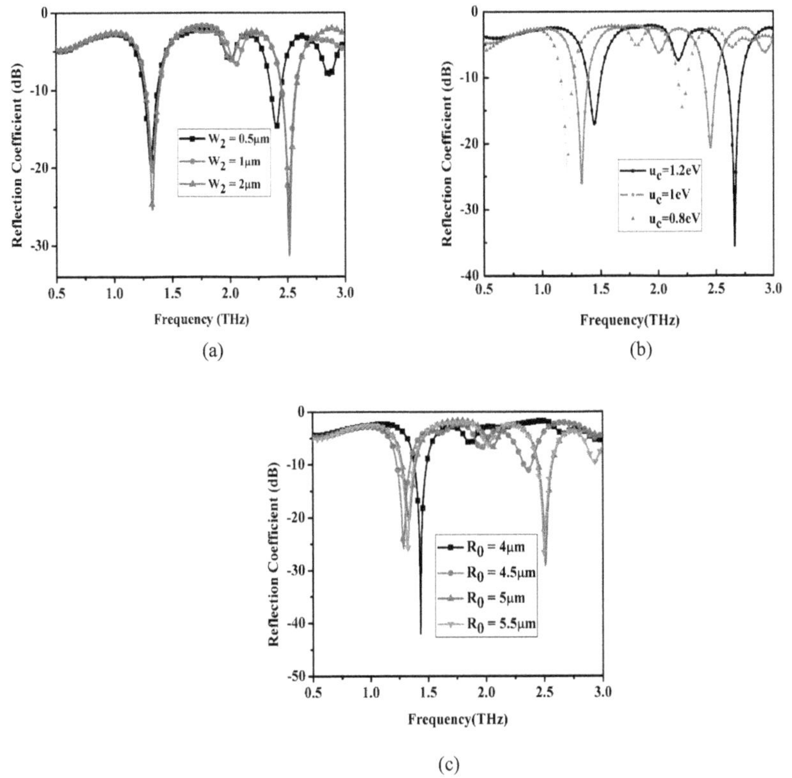

FIGURE 9.7 Parametric study: (a) variation of W_2, (b) variation of μ_c, and (c) variation of R_0.

elements that are placed close to each other cause mutual coupling. Mutual coupling is the energy absorbed by an adjacent element when other element of the MIMO in its vicinity is radiating. Mutual coupling between *i* and *j* ports is defined by the following equation [31]:

$$\mathrm{MC}_{ij} = e^{\frac{-2d(a+3)}{\lambda}} \qquad i, j = 1, 2, 3, \ldots, n \tag{9.8}$$

where '*d*' is the distance between *i* and *j* port and '*a*' is the coupling controlling parameter. Mutual coupling affects radiation pattern, reflection coefficient, and impedance of the MIMO antenna. From equation (9.8), it is evident that increasing the distance between the two ports decreases mutual coupling, but the overall size of the antenna increases, which results in space constraints for portable devices. Therefore, utmost care is essential while designing MIMO antennas. In this section, the single antenna unit created in Section 9.3 is used to construct two-port and four-port MIMO antennas.

9.4.1 Two-Port MIMO

The two-port MIMO configuration is designed via the replication of the single antenna unit cell in a parallel configuration along the *x*-axis. As mentioned earlier, the ground planes of the individual elements were interconnected to maintain constant power levels across the entire MIMO antenna. Figure 9.8a and b shows the

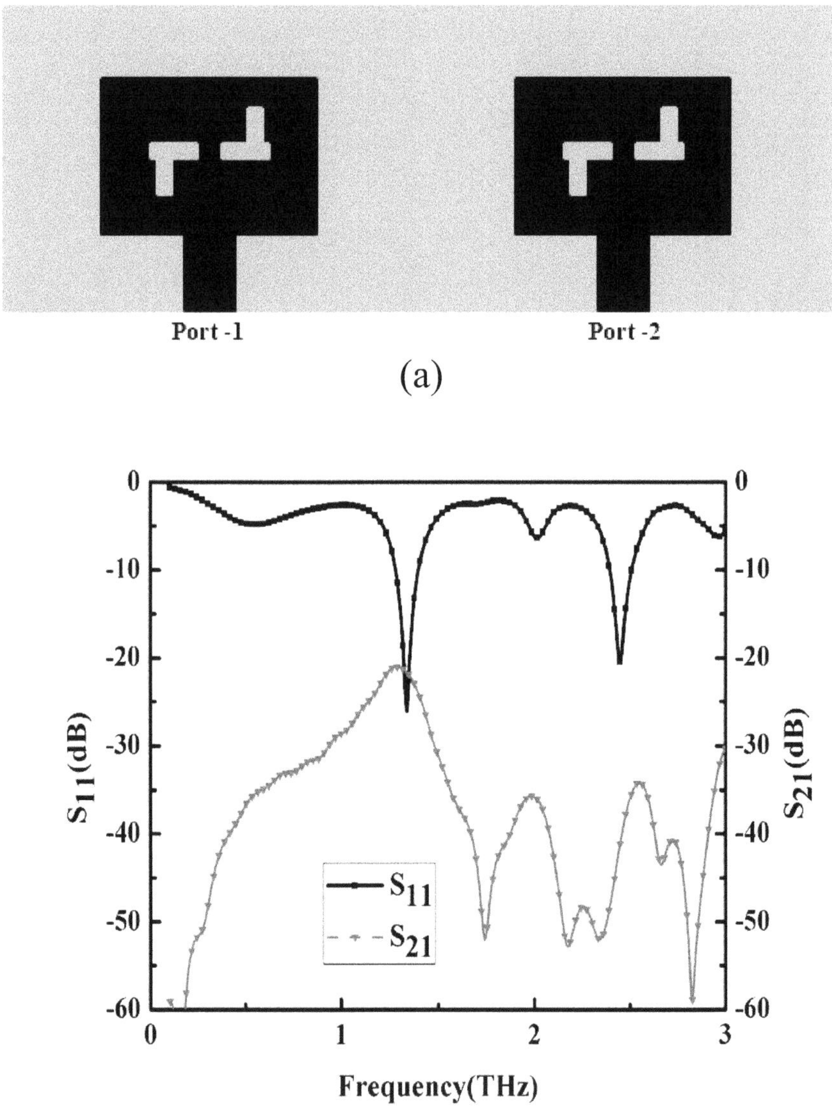

FIGURE 9.8 (a) Two-port MIMO antenna and (b) *S*-parameters.

two-port MIMO antenna with reflection coefficient (S_{11} and S_{22}) and transmission coefficient (S_{12} and S_{21}). Like single antenna unit, two-port MIMO antenna also produces dual band with reflection coefficients of −25 and −20dB at 1.3 and 2.5 THz. The transmission coefficient of designed graphene two-port MIMO is <−20dB at 1.3 THz and <−35dB at 2.5 THz, which is far better than the standard acceptable value (−15dB). The transmission coefficients are used to calculate isolation between the ports, which is given by the following equation [31]:

$$\text{Isolation} = -20\log\left(|S_{21}|\right)\text{ dB} \tag{9.9}$$

9.4.2 Four-Port MIMO

To further improve the data augmentation and channel capacity of terahertz communication, the four-port MIMO antenna is created. The four-port MIMO is constructed by placing four graphene single-elements in parallel at all the four corners of the substrate top layer, as depicted in Figure 9.9a. The simulated scattering parameters of the graphene-based four-port MIMO are shown in Figure 9.9b. Good coherency is observed between S_{11}, S_{22}, S_{33}, and S_{44} parameters. The isolation between adjacent ports (port-1 and port-2, port-3 and port-4) and diagonal ports (port-1 and port-3, port-2 and port-4) is also found to be satisfactory.

9.5 SIMULATED RESULTS

9.5.1 Radiation Pattern

The 1D and 3D radiation pattern of the suggested quad-port MIMO antenna is displayed in Figure 9.10a and b. The far-field radiation pattern is plotted for 1.3 and 2.5 THz frequency. It exhibits dipolar radiation pattern along the *x-o-z* plane and omnidirectional pattern along the *y-o-z* plane at 1.3 THz. Similarly, omnidirectional and dipolar radiation pattern is recorded along the *x-o-z* and *y-o-z* planes for 2.5 THz frequency, respectively. The two frequency bands possess different radiation patterns at their respective planes. Hence, the MIMO antenna acquires pattern diversity naturally, which yields good isolation between the elements.

9.5.2 Field Distribution

To further explore the mutual coupling behavior of the proposed MIMO antenna, the surface current density at 1.3 and 2.5 THz has been plotted in Figure 9.11. The magnitude of the field distribution is calculated by exciting port-1 while terminating other ports with a matched load. When port-1 is excited, a strong field is produced around radiator edges and the feed center, and almost negligibly small amount of field is visible on other radiators, as shown in Figure 9.11. Therefore, it is evident that the quad ports of the graphene MIMO antenna are independent of each other resulting in good isolation. One of the primary reasons for the low mutual coupling is pattern diversity behavior of the antenna (discussed in Section 9.5.1).

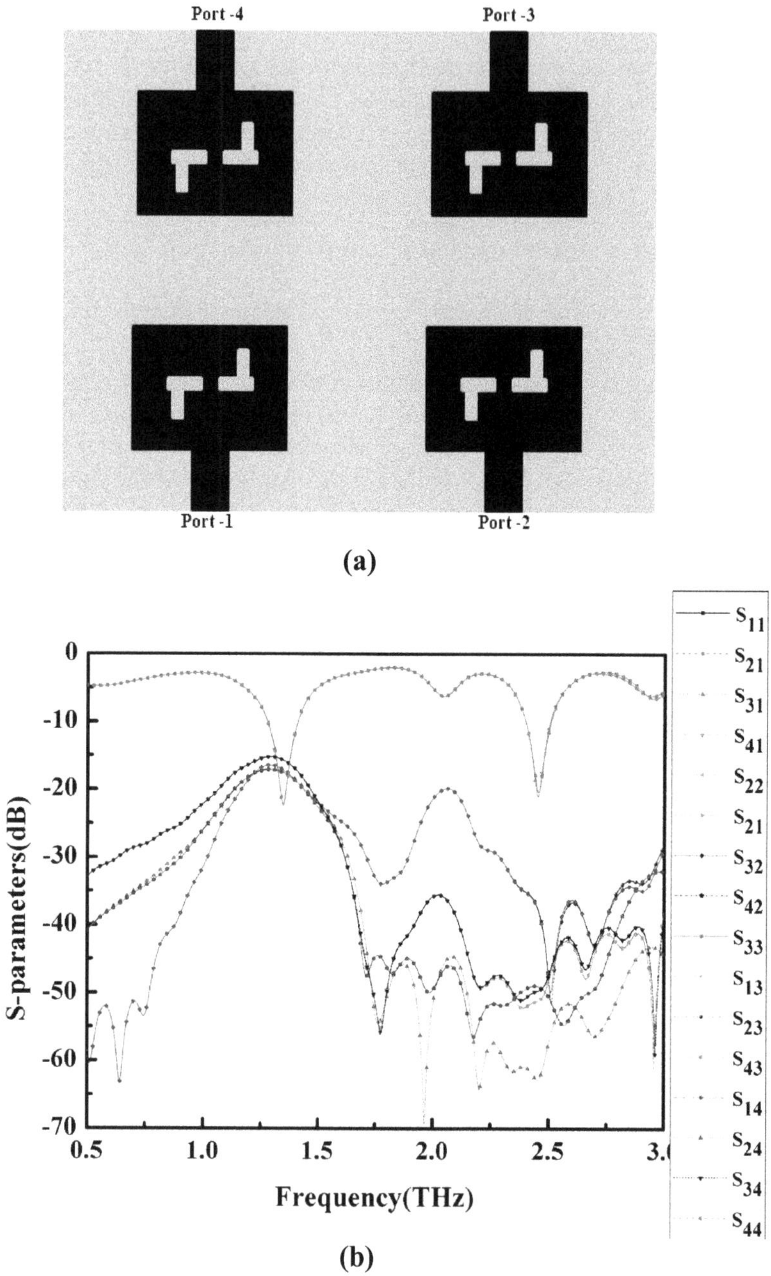

FIGURE 9.9 (a) Four-port MIMO and (b) *S*-parameters.

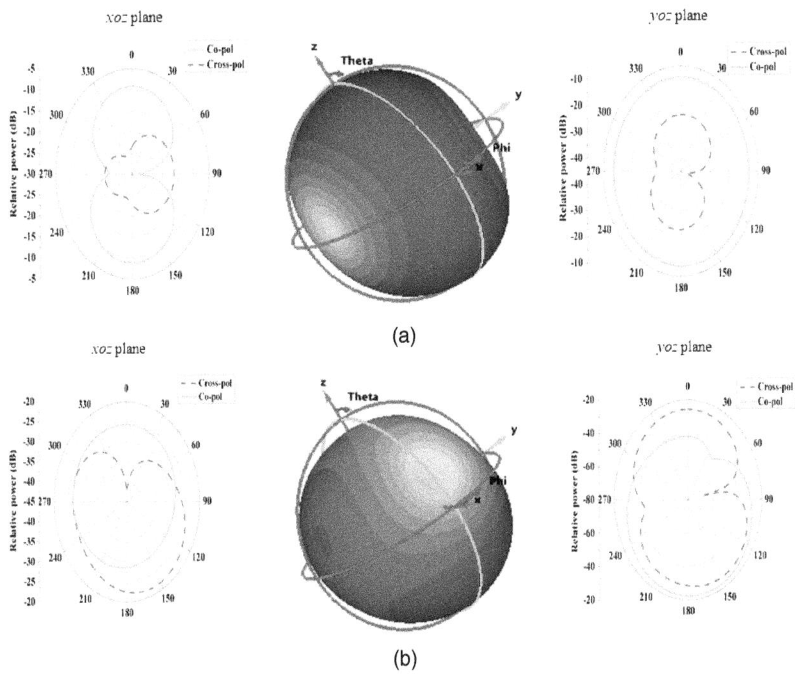

FIGURE 9.10 Radiation pattern at (a) 1.3 THz and (b) 2.5 THz.

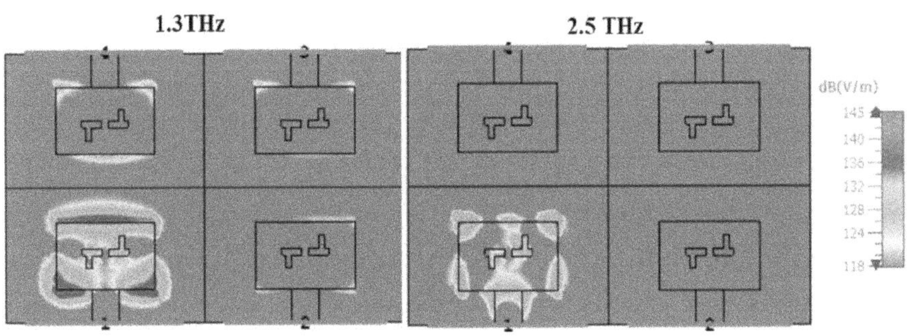

FIGURE 9.11 E-field distribution at 1.3 and 2.5 THz.

9.5.3 MIMO Performance Metrics

The diversity performance parameters like envelope correlation coefficient (ECC), mean effective gain (MEG), total active reflection coefficient (TARC), diversity gain (DG), and channel capacity loss (CCL) are calculated to understand the port independency and multi-channel propagation performance of the dual-band MIMO antenna.

The performance metric that estimates the correlation between MIMO elements is called as the envelope correlation coefficient. There are two ways for estimating ECC: (1) far-field radiation and (2) *S*-parameters method. The ECC calculated by

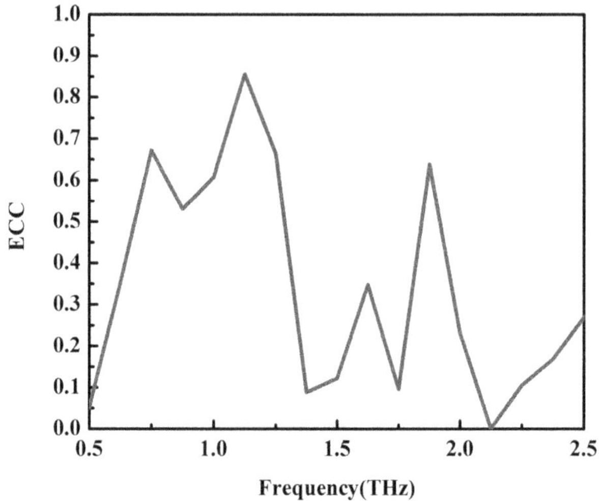

FIGURE 9.12 Envelope correlation coefficient.

utilizing the far-field radiation pattern is highly recommended. This is because the ECC calculation based on radiation pattern is accurate. In some cases, both the methods are recommended when the radiation efficiency of the antenna is >95%. ECC value less than 0.5 indicates good isolation between the ports. Equations (9.10) and (9.11) are used to calculate ECC. Figure 9.12 shows the ECC value of the proposed MIMO antenna, which is 0.2 at band-1 and 0.05 at band-2, respectively, which is quite below than the accepted value [32].

$$\mathrm{ECC} = \frac{\left|\iint_{4\pi} d\Omega U_1(\theta,\phi)^* \cdot U_2(\theta,\phi)\right|^2}{\iint_{4\pi} d\Omega \left|U_1(\theta,\phi)\right|^2 \cdot \iint_{4\pi} d\Omega \left|U_2(\theta,\phi)\right|^2 \cdot} \tag{9.10}$$

$$\mathrm{ECC} = \frac{\left|\sum_{n=1}^{M} S_{i,n}^* S_{n,j}\right|}{\sqrt{\left|\prod_{k=i,j}\left[1-\sum_{n=1}^{M} S_{i,n}^* S_{n,k}\right]\right|}} \tag{9.11}$$

Next metrics, which determines the dependability and quality of the MIMO antenna in 6G wireless communication, is diversity gain. The relation between ECC and DG is given in equation (9.12). When ECC is zero, DG is maximum (10 dB) and minimum when ECC is high. Figure 9.13 shows the simulated DG of the dual-band quad-port MIMO antenna. Diversity gains of 8 and 9 dB are achieved at 1.3 and 2.5 THz [33].

$$\mathrm{DG} = 10 \times \sqrt{1-\left|\mathrm{ECC}\right|^2} \tag{9.12}$$

Another important diversity antenna performance metric is the total active reflection coefficient, which is calculated using equation (9.13). TARC is calculated using

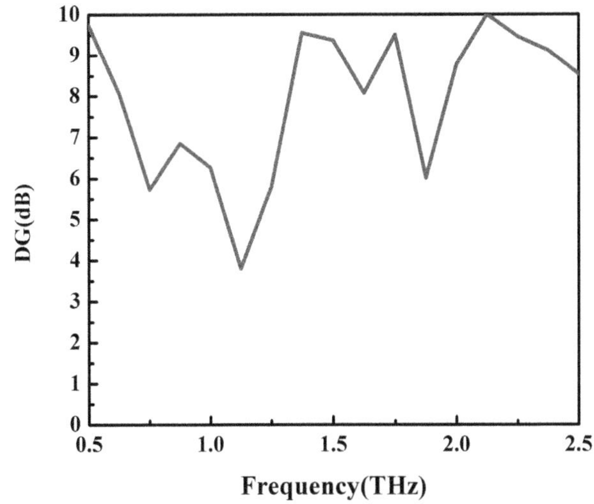

FIGURE 9.13 Diversity gain.

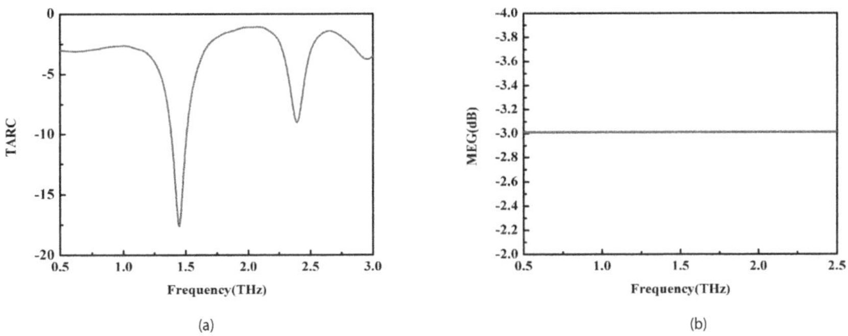

FIGURE 9.14 (a) Total active reflection coefficient and (b) mean effective gain.

S-parameters. The quad-port MIMO antenna TARC is below −10 dB for both the resonance at zero phase angle of the input signals. Figure 9.14a displays the calculated TARC value of the graphene MIMO antenna. The mean effective gain of the port-1 is evaluated using equation (9.14) [33]. The calculated value is well within the accepted range, as shown in Figure 9.14b.

$$\text{TARC} = \frac{\sqrt{((|S_{11} + S_{12}|^2) + (|S_{22} + S_{21}|^2))}}{\sqrt{2}} \tag{9.13}$$

$$\text{MEG}_1 = 0.5\left[1 - |S_{11}|^2 - |S_{12}|^2\right] \tag{9.14}$$

Finally, the maximum allowable information that can be transferred in a given communication channel without loss is defined as channel capacity loss (CCL). The CCL for the MIMO systems without information loss is defined to be <0.4 bps/Hz [33]. Equations that utilize scattering parameters as input to calculate CCL are as follows:

$$\mathrm{CCL} = -\log_2 \det\left(x^k\right) \tag{9.15}$$

$$x^k = \begin{pmatrix} x_{11} & x_{12} \\ x_{21} & x_{22} \end{pmatrix} \tag{9.16}$$

$$x_{11} = 1 - \left(|S_{11}|^2 + |S_{12}|^2\right) \tag{9.17}$$

$$x_{12} = -\left(S_{11}^* S_{12} + S_{21}^* S_{12}\right) \tag{9.18}$$

$$x_{22} = 1 - \left(|S_{22}|^2 + |S_{21}|^2\right) \tag{9.19}$$

$$x_{21} = -\left(S_{22}^* S_{21} + S_{12}^* S_{21}\right) \tag{9.20}$$

9.6 CONCLUSION

In this chapter, reconfigurable terahertz MIMO based on graphene material has been discussed. Using the graphene, a single antenna-unit operating at 1.3 and 2.5 THz was designed and optimized first. Then two-port and four-port MIMO antennas are created by the replication of single element along the x-axis and y-axis. Fundamental modes of resonance have been estimated using characteristic mode analysis. Isolation of −15 dB at 1.3 THz and −30 dB at 2.5 THz has been achieved with the four-port MIMO. MIMO performance metrics (ECC, DG, MEG, TARC, and CCL) are estimated and compared with the acceptable values. With these findings, it is ensured that the graphene-based antennas are highly suitable for 6G communication, biomedical imaging and sensing, medical treatments, and security scanning.

REFERENCES

1. I. F. Akyildiz, J. M. Jornet, C. Han, Terahertz band: Next frontier for wireless communications, *Communications Physics*, 12, 16–32, 2014.
2. D. Kumar, V. Kumar, Y. Subhash, P. Giri, G. Varshney, THz UWB antenna with tunable band-notch creation using resonant/non-resonant graphene loop, *Optical Engineering*, 61(10), 107106, 2022.
3. J. M. Jornet, I. F. Akyildiz, Channel modeling and capacity analysis for electromagnetic wireless nanonetworks in the terahertz band, *IEEE Transactions on Wireless Communications*, 10, 3211–3221, 2011.
4. M. Alsabah, M. A. Naser, B. M. Mahmmod, S. H. Abdulhussain, M. R. Eissa, A. Al-Baidhani, 6G wireless communications networks: A comprehensive survey, *IEEE Access*, 9, 148191–148243, 2021.
5. K. R. Mahmoud, A. M. Montaser, Design of multi-resonance flexible antenna array applicator for breast cancer hyperthermia treatment, *IEEE Access*, 32034, 31, 2022.
6. I. Malhotra, K. R. Jha, G. Singh, Terahertz antenna technology for imaging applications: A technical review, *International Journal of Microwave and Wireless Technologies*, 10, 271–290, 2018.

7. J. M. Jornet, I. F. Akyildiz, Channel modeling and capacity analysis for electromagnetic wireless nanonetworks in the terahertzband, *IEEE Transactions on Wireless Communications*, 10, 3211–3221, 2011.
8. S. Dash, A. Patnaik, Performance of graphene plasmonic antenna in comparison with their counterparts for low-terahertz applications, *Plasmonics*, 13, 2353–2360, 2018.
9. R. G. Vaughan, J. B. Andersen, Antenna diversity in mobile communications, *IEEE Transactions on Vehicular Technology*, 36(4), 149–172, 1987.
10. M. Shalini, M. G. Madhan, Design and analysis of a dual-polarized graphene based microstrip patch antenna for terahertz applications, *Optik*, 194, 163050, 2019.
11. U. Raj, M. K. Sharma, V. Singh, S. Javed, A. Sharma, Easily extendable four port MIMO antenna with improved isolation and wide bandwidth for THz applications, *Optik*, 247, 167910, 2021.
12. W. Wang, Y. Wu, W. Wang, Y. Yang, Isolation enhancement in dual-band monopole antenna for 5G applications, *IEEE Transactions on Circuits and Systems-II: Express Briefs*, 68(6), 1867–1871, 2021.
13. P. Garg, P. Jain, Isolation improvement of MIMO antenna using a novel flower shaped metamaterial absorber at 5.5 GHz WiMAX band, *IEEE Transactions on Circuits and Systems-II: Express Briefs*, 67(4), 675–679, 2020.
14. M. Alibakhshikenari, B. S. Virdee, P. Shukla, C. H. See, R. Abd-Alhameed, M. Khalily, Interaction between closely packed array antenna elements using meta-surface for applications such as MIMO systems and synthetic aperture radars, *Radio Science*, 53(11), 1368–1381, 2018.
15. R. Mark, H. V. Singh, K. Mandal, S. Das, Mutual coupling reduction using near-zero ε and μ metamaterial-based superstrate for an MIMO application, *IET Microwaves, Antennas & Propagation*, 8(6), 479–484, 2020.
16. D. Pandurangan, N. Mishra, Design and analysis of two-port self-decoupled highly efficient MIMO antenna for 5G new radio mid-band (n1/n2/n3) wireless communication, *IEEE Access*, 13, 7110–7122, 2025.
17. M. Ameen, R. K. Chaudhary, Isolation enhancement of metamaterial-inspired two-port MIMO antenna using hybrid techniques, *IEEE Transactions on Circuits and Systems-II: Express Briefs*, 70(6), 1966–1970, 2023.
18. M. Li, L. Jiang, K. L. Yeung, A general and systematic method to design neutralization lines for isolation enhancement in MIMO antenna arrays, *IEEE Transactions on Vehicular Technology*, 69(6), 6242–6253, 2020.
19. D. K. Choudhary, N. Mishra, I. Singh, N. O. Parchin, G. Singh, Various mutual coupling reduction techniques for 5G MIMO antenna, in D. K. Choudhary (Ed.), *Microwave Devices and Circuits for Advanced Wireless Communication Design and Analysis*, 3rd edition, CRC Press, 2024, ch.3, pp. 37–63.
20. G. Moreno, H. Mehrpour Bernety, A. B. Yakovlev, Reduction of mutual coupling between strip dipole antennas at terahertz frequencies with an elliptically shaped graphene monolayer, *IEEE Antennas and Wireless Propagation Letters*, 15, 1533–1536, 2016.
21. S. Dash, C. Psomas, A. Patnaik, I Krikidis, An ultra-wideband orthogonal-beam directional graphene-based antenna for THz wireless systems, *Scientific Reports*, 12, 22145, 2022.
22. S. A. Khaleel, H. Eki, N. O. Parchin, M. B. Saleh, MTM-inspired graphene-based THz MIMO antenna configurations using characteristic mode analysis for 6G/IoT applications, *Electronics*, 11, 2152, 2022.
23. N. K. Maurya, S. Kumari, P. Pareek, L. Singh, Graphene-based frequency agile isolation enhancement mechanism for MIMO antenna in terahertz regime, *Nano Communication Networks*, 35, 100436, 2023. ISSN 1878–7789.

24. M. F. Ali, R. Bhattacharya, G. Varshney, Graphene-based tunable terahertz self-diplexing/MIMO-STAR antenna with pattern diversity, *Nano Communication Networks*, 30, 100378, 2021. ISSN 1878–7789.
25. M. Alibakhshikenari, C. H. See, B. Virdee, R. A. Abd-Alhameed, Meta-surface wall suppression of mutual coupling between microstrip patch antenna arrays for THz-band applications, *Progress in Electromagnetics Research Letters*, 75, 105–111, 2018.
26. B. Smari, M. Labidi, F. Choubani, Mutual coupling reduction in metamaterial antenna for terahertz application, *Applied Physics A*, 125, 1–7, 2019.
27. G. W. Hanson, Dyadic Green's functions and guided surface waves for a surface conductivity model of graphene, *Journal Applied Physics*, 103(6), 064302, 2008.
28. C. S. Kaipa, A. B. Yakovlev, G. W. Hanson, Y. R. Padooru, F. Medina, F. Mesa, Enhanced transmission with a graphene-dielectric microstructure at low-terahertz frequencies, *Physical Review B*, 85(24), 245407, 2012.
29. R. Harrington, J. Mautz, Theory of characteristic modes for con-ducting bodies, *IEEE Transactions on Antennas and Propagation*, AP-19(5), 622–628, Sep. 1971.
30. B. B. Q. Elias, P. J. Soh, A. A. Al-Hadi, P. Akkaraekthalin, G. A. E. Vandenbosch, A review of antenna analysis using characteristic modes, *IEEE Access*, 9, 98833–98862, 2021, https://doi.org/10.1109/ACCESS.2021.3095422
31. P. Das, A. K. Singh, K. Mandal, Metamaterial loaded highly isolated tunable polarisation diversity MIMO antennas for THz applications, *IEEE Transactions on Antennas and Propagation*, 54, 250, 2022. https://doi.org/10.1007/s11082-022-03641-8
32. N. K. Maurya, R. Bhattacharya, CPW-fed dual-band compact Yagi-type pattern diversity antenna for LTE and WiFi, *Progress in Electromagnetics Research C*, 107, 183–201, 2021.
33. K. V. Babu, S. Das, G. Varshney, G. N. J. Sree, B. T. P. Madhav, A micro-scaled graphene-based tree-shaped wideband printed MIMO antenna for terahertz applications, *Journal of Computational Electronics*, 21, 289–303, 2022.

10 A Compact Graphene Antenna for Advanced 6G Communication and Sensing Technologies

Jayant Kumar Rai, Poonam Tiwari, Praveena A, Rachit Jain, Pinku Ranjan, and Rakesh Chowdhury

10.1 INTRODUCTION

Due to the significant developments in wireless communication systems over the last few decades, wireless communication services require fast data rates, excellent spectral efficiency, and improved fading mitigation over large frequency bands. While THz is lower than the infrared spectrum, it is more significant than microwave frequencies. Managing 100 Gbps of data flow, which is necessary for next-generation communication systems, is one advantage of this frequency range [1]. Microstrip Patch Antennas (MPA) can meet the needs of future modern wireless communication systems because of their small size, easy manufacturing process, appropriate impedance matching, low cost, and planar design [2–6]. The antenna's performance is influenced by the patch's size, feeding method, substrate height, and dielectric constant [7]. In antenna design, copper has historically been the most often used metal since it performs best at lower frequencies. Micrometer-sized copper antennas' radiation efficiency and propagation losses are significantly reduced in THz bands due to copper's reduced conductivity and skin depth. The THz frequency range allows the next-generation high-speed wireless communication systems to meet user demands for higher data rates and more capacity. In the THz frequency band, conventional antennas can be replaced with graphene-based antennas that are micrometers in size [8,9]. Although copper is more accessible and less expensive than graphene, it is unable to provide better gain and efficiency at higher frequencies due to its higher attenuation and poorer conductivity [10].

In [11], a photonic band gap (PBG) substrate is used in a THz MPA, which operates between 0.5 and 0.8 THz. Simulated rectangular patch antennas are designed to optimize antenna performance for communication and sensing applications at 0.65 THz on a variety of substrates, such as periodic and aperiodic photonic crystal substrates. The second antenna construction is particularly noteworthy among them due to its wide bandwidth beyond 230 GHz, gain of 9.19 dB, low return loss of −83.73 dB, and radiation efficiency of 90.84%. In [12], at 300 GHz, a dipole-reflector

DOI: 10.1201/9781003599111-10

single-element antenna first achieves a gain of 5.14 dBi and a broad impedance bandwidth (IBW) of 38.6%. After that, the gain is kept constant, rising to 8.01 and 9.6 dBi in two further variations with more directors. At last, a 1×4 element array exhibits remarkable potential for wireless networks with a peak gain of 13.6 dBi and 89% efficiency. In [13], a THz antenna is constructed for breast cancer detection using metamaterial ideas. The antenna is a straightforward rectangular patch design with a Complementary Split Ring Resonator (CSRR) integrated into it. The antenna runs at a frequency of 1 THz with a gain of 20dB. In [14], using traditional CMOS technology, an on-chip antenna is created by sandwiching 500 µm thick polyimide substrates between a 3 µm aluminum ground plane. The design consists of 3×3 matrix radiation patches on top of the ground plane with carved concentric dielectric and concentric metal rings. With an average radiation gain of 8.15 dBi, a wide bandwidth of 0.350–0.385 THz, and on-chip integration at 6×6×1 mm^3 dimensions, the structure's effective aperture area improvement is based on a bottom-layer microstrip feedline. Additionally, an efficiency of 65.71% is attained. In [15], conventional CMOS technology results in a high-performance 0.4-THz on-chip antenna (OCA) with improved gain. The back chamber of the OCA suppresses surface waves and focuses radiation on the low-resistance substrate by using substrate-integrated waveguides (SIWs) to link the bottom and top metal layers. On the upper metal layer, a slotted radiator enhances the performance of broadside radiation. In [16], the performance of an OCA utilizing metasurface technology is examined in the THz band. On a thin gallium arsenide substrate, the antenna is made up of 11×11 arrays of circular patches with slots. The leaky-wave feeding slot lines are attached through the perforations in the center patches. Using a waveguide port excitation, the antenna achieves over 60% radiation efficiency and over 10 dBi gain between 0.41 and 0.47 THz. In [17], the THz frequencies (700–850 GHz) are the focus of an MPA design. At such high frequencies, size limitations provide difficulties for traditional MPA. This is addressed by the antenna's multi-layer nature, which makes modeling easier. The antenna continuously has a reflection coefficient below −10 dB and a broad bandwidth. Over the frequency range, its realized gain varies between 6.4 and 9.7 dBi. In [18], an antenna array was created utilizing SiGe BiCMOS technology, which removed the requirement for post-processing. Several shorted pins, an annular ring patch, and a ground that forms a hollow to increase gain and reduce surface waves are all features of the OSAR design. A 1×2 OSAR antenna array is constructed with a die size of 550×1,100 μm^2. In [19], at 0.72 THz, a graphene-based MPA was developed for wireless communication. Between 0.53 and 0.84 THz, the antenna impedance bandwidth is 37.50%. In [20], a square patch surface slot incorporates a tiny, micro-sized, wideband THz antenna with a rhombus-shaped element made of gold metal. The antenna operates on a 300×300 μm^2 polyimide substrate with a $\varepsilon_r = 4.3$ and a thickness of 45 µm. In [21], applications in the THz frequency range are designed for high-gain antennas to meet the needs of ultra-high-speed communication systems and modern wearable smart devices. The usage of a polyimide substrate with a PBG crystal allows the antenna to achieve remarkable performance at 0.63 THz in terms of directivity (9.99 dBi), gain (9.45 dB), and improved VSWR (<1). In [22], prototype 2×3 arrays with graphene-coated metallization were created on a polyimide dielectric substrate that operated between 0.19 and 0.20 THz. Enhancing antenna array performance is essential for sub-THz integrated circuits when power from sub-THz sources is limited. Conventional arrays increase radiation gain, but mutual coupling issues affect tiny arrays. These impacts are successfully suppressed by using SIW technology. Comparisons between the metamaterial properties with sub-wavelength slots and reference arrays revealed gains of 6.3 dB, 28 dB, and 34% in efficiency isolation, respectively. In [23], enhancements to an OCA for integrated circuits using millimeter and terahertz wavelengths are meant to increase impedance bandwidth, radiation gain, and efficiency while reducing substrate loss. With a circular radiation patch with crossed-shaped gaps and an open-ended microstrip feedline, the antenna integrates SIW and metamaterial technology. With an average efficiency of 5.3% and a radiation gain of 6.9 dBi over the operating frequency range of 0.285–0.325 THz, this $1.35 \times 1 \times 0.06\,mm^3$ device is suitable for integrated-circuit applications. In [24], there is a study on an optimized wideband 1×2 MIMO fractal antenna, such as imaging and high-speed communication. The fractal design involves partial ground plane components and was developed from a simple rectangular patch antenna. A 45 μm polyimide substrate with $\varepsilon_r = 3.5$ is used for particle swarm optimization, which improves the antenna's performance. Simulations using computers evaluate variables such as radiation patterns and diversity features. This fractal MIMO antenna offers a potential alternative for the next THz technologies due to its excellent isolation (≤ -25 dB) and wide 400 GHz bandwidth (0.35–0.75 THz). Referencei [25] describes a tree-shaped microstrip MIMO antenna for graphene-based terahertz applications. An excellent 88.14% impedance bandwidth is obtained. Variations in frequency and S-parameters in relation to the chemical potential of graphene are examined. Sensing, imaging, biological applications, high-speed transmission, and security scanning are all made possible by the THz spectrum. In [26], THz applications may need the creation of a small MPA. The conducting substance is graphene, and the antenna's Rogers RT/Duroid 6010 substrate gives it incredible properties. Its $130 \times 100\ \mu m^2$ optimized dimensions allow it to achieve a broad 450 GHz impedance bandwidth (0.57–1.02 THz), notable improvements in bandwidth characteristics, and robust performance measures, including gain, radiation efficiency, VSWR, and reflection coefficient. Reference [27] describes a wideband antenna inspired by the compact planar coronavirus. The radiating portion of the antenna is supplied via a 50 Ω coplanar waveguide using the 3D model of the virus. The antenna is composed of gold on a polyamide substrate (45 μm thick, $\varepsilon_r = 3.5$) and measures $300 \times 300\ \mu m^2$. The antenna provides 47.98% fractional bandwidth spanning 0.3627–0.5918 THz, making it appropriate for various uses such as wireless communication and the Internet of Things. The far-field findings show omnidirectional radiation with 93% efficiency and an average gain of 4 dBi over the frequency range. In [28], a triple band hexagonal-shaped THz antenna is presented. A 5 dBi gain is achieved in the suggested antenna.

The suggested 6G THz antenna has a high gain, is compact, and is easy to configure. Additionally, various substrate and conducting material combinations are used to analyze and apply the 6G antenna construction. The following are the characteristics of the suggested 6G THz antenna:

i. The suggested 6G THz antenna operates from 2.29 to 2.88, 3.06 to 7.02, and 7.91 to 9.04 THz in Band-1, Band-2, and Band-3, respectively.
ii. The performance of the suggested 6G THz antenna has been investigated, and the impacts of several conducting materials like gold, aluminum, copper, and graphene and dielectric materials like polyamide, Rogers RT/duroid 5880, SiO, and quartz are explored.
iii. The size of the substrate is $20 \times 20\,\mu m^2$.
iv. The peak gain of the suggested 6G THz antenna is 6 dBi.

10.2 ANTENNA CONFIGURATION AND ANALYSIS

10.2.1 Antenna Configuration

The proposed 6G THz antenna is designed on the polyamide substrate. The width and length of the substrate are 20 μm. The thickness of the substrate is 10 μm. The full ground plane is created below the substrate. The ground plane's width and length are the same as the substrate. The antenna radiator is created using graphene. The other dimensions of the suggested 6G THz antenna are illustrated in Figure 10.1.

10.2.2 Mathematical Modelling of Graphene

One of the many significant features of graphene is that its one layer may be considered a meta surface due to its very thin thickness. Additionally, using the Kubo Formulation, the surface conductivity of graphene can potentially be determined in scalar form. The following describes graphene's conductivity [28,29]:

$$\sigma_{\text{Graphene}} = \sigma_{\text{Intraband}} + \sigma_{\text{Interband}} \tag{10.1}$$

where

$$\sigma_{\text{Intraband}} = \frac{2 \times q_e^2 \times k_B \times T}{\pi \times h^2 \times \left(\tau^{-1} + j\omega\right)} \times \ln\left\{2 \times \cosh\left(\frac{\mu_c}{2 \times k_B \times T}\right)\right\} \tag{10.2}$$

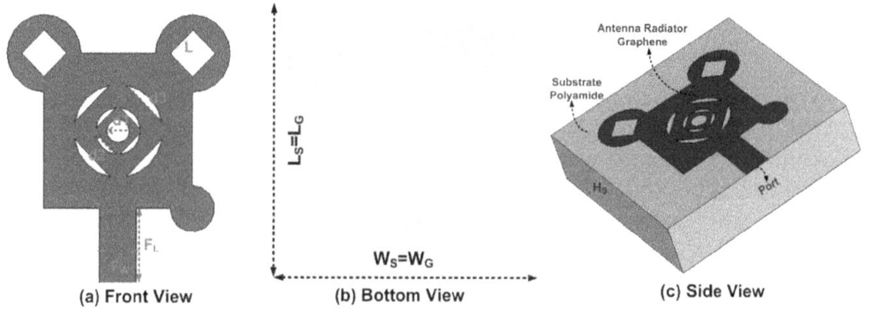

FIGURE 10.1 The layout of the proposed 6G THz (Dimensions (μm): $L_S = L_G = 20$, $W_S = W_G = 20, H_S = 10, F_W = 5, F_L = 11, d_1 = 3, d_2 = 6, d_3 = 12, L = 3$).

$$\sigma_{\text{Interband}} = -j\times\left(\frac{q_e^2}{4\times\pi\times h}\right)\times\ln\left\{\frac{2\times|\mu_c|-(\omega-j\times\tau^{-1}\times h)}{2\times|\mu_c|-(\omega+j\times\tau^{-1}\times h)}\right\} \tag{10.3}$$

$$k_B\times T \ll |\mu_c|,\quad h\omega$$

wherek_B is the Boltzmann constant, q_e is the electron charge, ω is the angular frequency, τ is the relaxation time, μ_c is the chemical potential, and h is the Planck constant.

$$\mu_c \propto E_F \tag{10.4}$$

where E_F is the Fermi level

From equation (10.4), it is clear that the chemical potential of graphene is related to the Fermi level.

The change in μ_c changes σ_{Graphene} by injecting the carriers. Graphene's chemical potential and bias voltage can be connected through the following equation:

$$V_b = \frac{q_e^2\times\mu_c^2\times t}{\pi\times h^2\times v_F^2\times\varepsilon_r\times\varepsilon_0} \tag{10.5}$$

where V_b is the bias voltage, t is the thickness, v_F is the velocity of fermi, ε_r is the relative permittivity, and ε_0 is the vacuum permittivity.

10.2.3 Step Evaluation

Various design steps of the suggested 6G THz antenna are illustrated in Figure 10.2. The Frequency Range, bandwidth, and bandwidth % of various design steps are

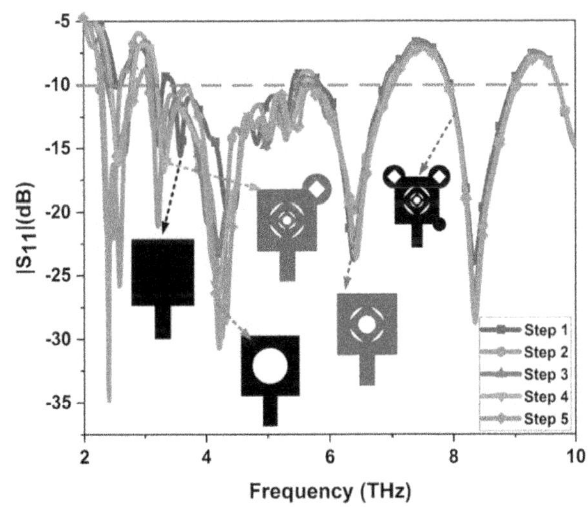

FIGURE 10.2 Effect of S_{11} due to the various design steps of the suggested antenna.

TABLE 10.1
Performance of Various Design Steps of the Proposed THz Antenna

Step	Frequency Range (THz)	Bandwidth (THz)	Impedance Bandwidth (%)
Step-1	3.47–5.41	1.94	43.69
	5.84–6.88	1.04	16.35
	7.94–8.95	1.01	11.96
Step-2	3.11–4.49	1.38	36.32
	5.79–6.90	1.11	17.51
	7.92–9.01	1.09	12.88
Step-3	3.19–5.52	2.23	50.11
	5.83–6.93	1.10	17.24
	7.92–9.01	1.09	12.88
Step-4	3.20–5.57	2.37	54.11
	5.74–6.96	1.22	19.21
	7.92–9.01	1.09	12.88
Step-5	2.29–2.88	0.59	22.86
	3.06–7.02	3.96	78.88
	7.91–9.04	1.13	13.34

given in Table 10.1. In the initial design, Step-1, the rectangular patch is created, and it is operating in three bands. The frequency ranges are 3.47–5.41, 5.84–6.88, and 7.94–8.95 THz. In the next design, Step-2, the circular slot is created inside Step-1, and its operating frequencies are 3.11–4.49, 5.79–6.90, and 7.92–9.01 THz. In the next design, Step-3, a square is created inside Step-2, and its operating frequencies are 3.19–5.52, 5.83–6.93, and 7.92–9.01 THz. In the design, Step-4, another square is created inside Step-3, and its operating frequencies are 3.20–5.57, 5.74–6.96, and 7.92–9.01 THz. The last design, Step-5, covers operating frequencies from 2.29 to 2.88, 3.06 to 7.02, and 7.91 to 9.04 THz.

10.2.4 Parametric Analysis

The antenna dimensions are optimized using parametric analysis, as shown in Figure 10.3. With a 0.5 μm step size, the circular slot's radius (R_1) ranges from 5.5 to 6.5 μm. With a 0.5 μm step size, the bottom circle's radius (R_2) ranges from 2.5 to 3.5 μm. With a 1 μm step size, the top circle's radius (R) ranges from 4 to 6 μm. With a 1 μm step size, the patch's length ranges from 9 to 11 μm. With a 2 μm step size, the patch's width varies from 8 to 12 μm. The patch's feed width ranges from 2 to 3 μm.

10.2.5 Effect of Various Materials on Substrate and Antenna Radiator

The effect of various materials on the substrate and antenna radiator is given in Table 10.2 and illustrated in Figure 10.4. The effects of various materials such as Rogers 5880, polyamide, SiO_2, and quartz are used as the substrate. The effects of various materials such as Aluminum, Copper, Gold, and graphene are used for

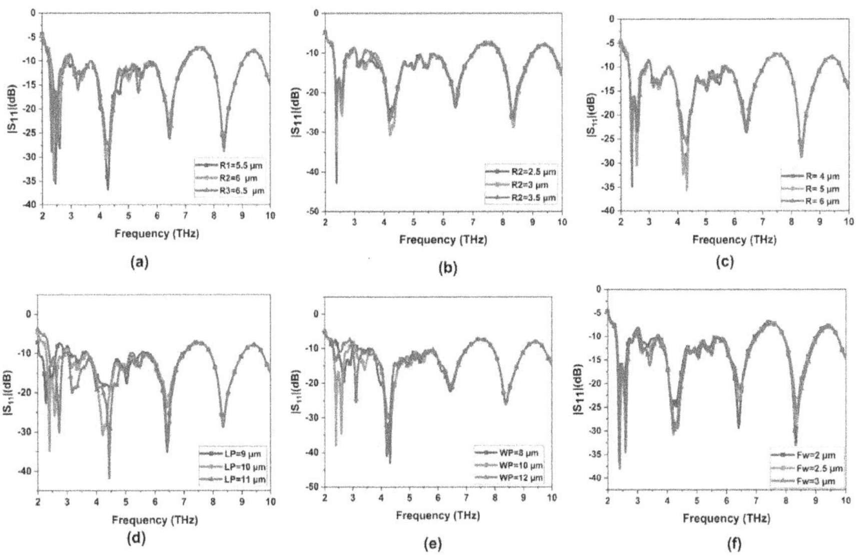

FIGURE 10.3 Parametric analysis of various parameters of 6G THz antenna.

TABLE 10.2
Various Materials for Substrates and Antenna Radiator

Substrate/ Antenna Radiator	Rogers 5880	Polyamide	SiO_2	Quartz
Aluminum	6.90–8.32	5.91–6.42	4.96–5.56	5.51–6.43
Copper	6.90–8.31	5.94–6.42	4.96–5.56	5.54–5.94
Gold	6.93–8.29	6.01–6.38	5.02–5.54	4.31–4.64
Graphene	2.73–3.51	2.29–2.88	3.37–4.88	3.62–5.43
	4.56–7.71	3.06–7.02	8.024–9.37	5.51–6.69
		7.91–9.04		

antenna radiators. From Table 10.2, we can observe that the Polyamide substrate with graphene antenna radiator provides the multiband and highest bandwidth compared to other combinations.

10.2.6 Result and Discussion

The return loss of the proposed 6G THz antenna is illustrated in Figure 10.5. The proposed 6G THz antenna provides multiband characteristics. In Band-1, it is operating from 2.29 to 2.88 THz. The bandwidth in Bnad-1 is 0.59 THz. The impedance bandwidth % in Band-1 is 22.82%. In Band-2, it is operating from 3.06 to 7.02 THz. The bandwidth in Bnad-2 is 3.96 THz. The impedance bandwidth % in Band-2 is 78.57%. In Band-3, it is operating from 7.91 to 9.04 THz. The bandwidth in Bnad-3

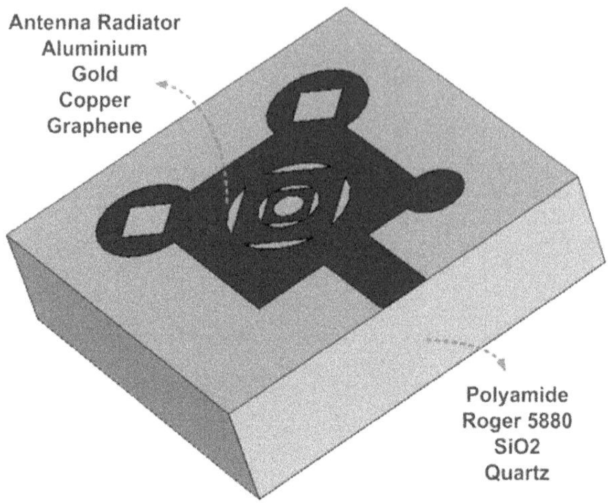

FIGURE 10.4 Effect of various materials on substrates and antenna radiator.

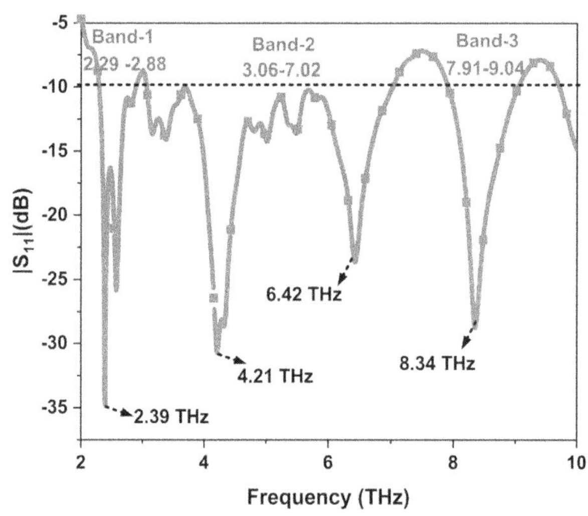

FIGURE 10.5 S_{11} of the proposed 6G THz antenna.

is 1.13 THz. The impedance bandwidth % in Band-3 is 13.34%. The surface current distribution of the suggested 6G antenna is illustrated in Figure 10.6. The gain of the proposed 6G THz antenna is illustrated in Figure 10.7. The proposed 6G THz antenna achieved the maximum peak gain of 6 dBi. The efficiency of the suggested 6G antenna is illustrated in Figure 10.8. The directivity of the suggested 6G antenna is illustrated in Figure 10.9. The 3 dB polar plot gain of the suggested 6G antenna is illustrated in Figure 10.10. The 2D radiation patterns at various resonating frequencies in the *XZ* and *YZ* planes are illustrated in Figure 10.11. The characteristics of the radiation patterns are given in Table 10.3.

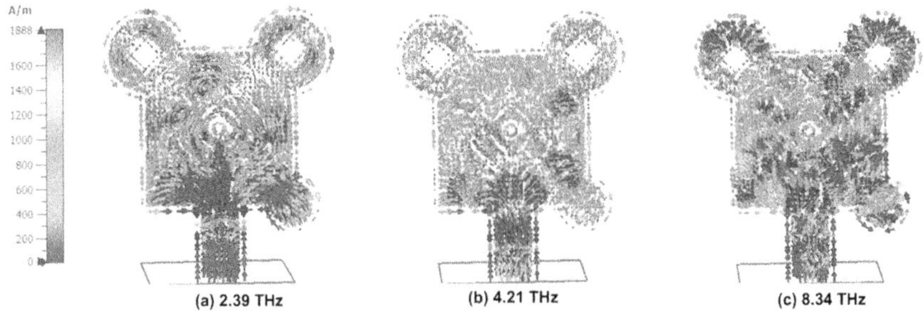

FIGURE 10.6 Surface current distribution at various frequencies of the 6G THz antenna.

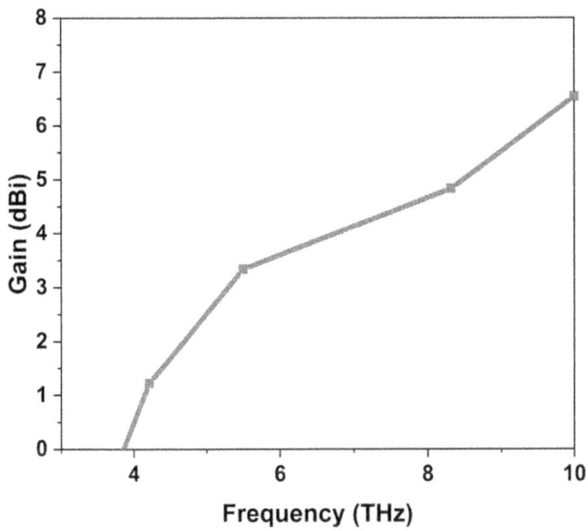

FIGURE 10.7 Gain of the proposed THz antenna.

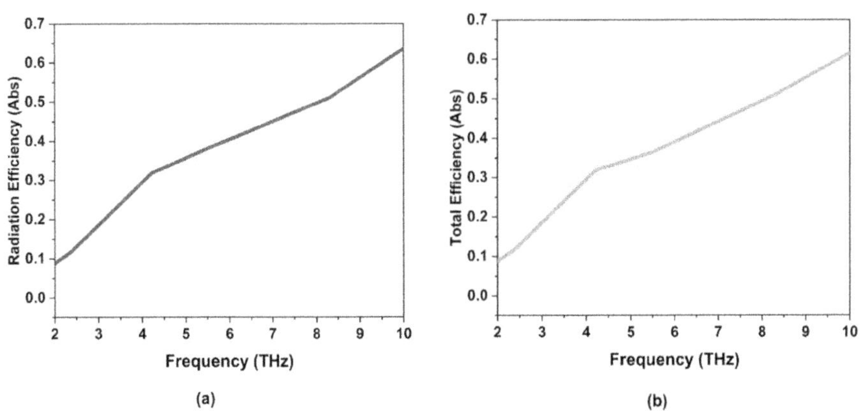

FIGURE 10.8 Efficiency of the 6G antenna.

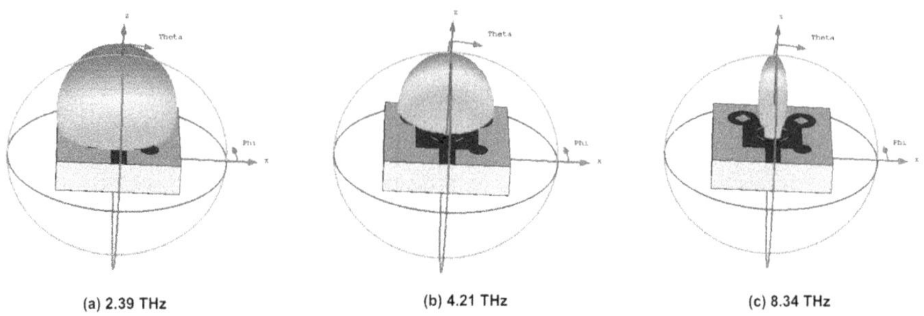

FIGURE 10.9 Directivity at various resonating frequencies of the 6G antenna.

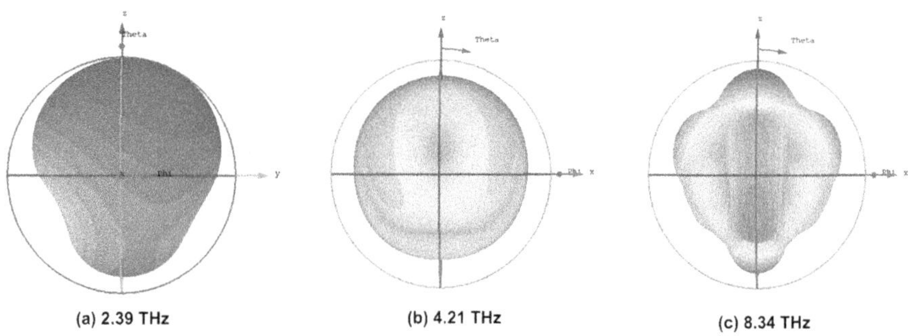

FIGURE 10.10 3 dB polar plot gain at various resonating frequencies of the 6G antenna.

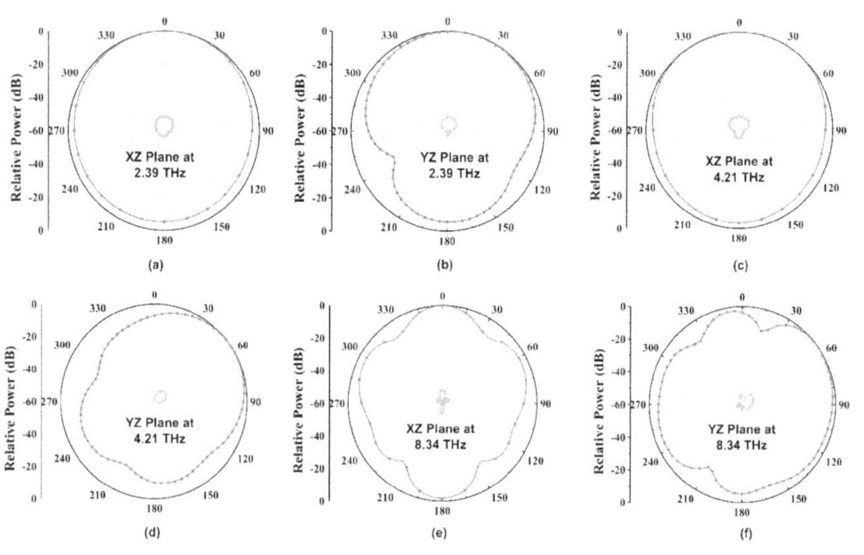

FIGURE 10.11 Radiation patterns at various frequencies.

10.3 COMPARATIVE ANALYSIS OF THE PROPOSED 6G THZ ANTENNA

The proposed 6G THz antenna is compared with the other existing work, and it is given in Table 10.4. Table 10.4 shows that the proposed 6G THz antenna is compact and provides high gain compared to the other existing work.

TABLE 10.3
Characteristics of the Radiation Patterns

Resonating Frequency (THz)	Plane	Polarization	Main Lobe Magnitude	Main Lobe Direction	Angular Width (3 dB)	Side Lobe Level
2.395	*XZ*	Co	9.36 dB (V/m)	0°	145.6°	
		Cross	−42.2 dB (A/m)			
	YZ	Co	9.69 dB (V/m)	20°	111.3°	−5.3 dB
		Cross	−41.8 (A/m)			
4.213	*XZ*	Co	9.2 dB (V/m)	25°	151.4°	−3.4 dB
		Cross	−42.3 dB (A/m)			
	YZ	Co	16 dB (V/m)	55°	73.4°	−9.1 dB
		Cross	−35.5 (A/m)			
8.314	*XZ*	Co	16 dB (V/m)	0°	31.7°	−1.7 dB
		Cross	−35.5 dB (A/m)			
	YZ	Co	19.6 dB (V/m)	50°	73°	−2.6 dB
		Cross	−31.9 (A/m)			

TABLE 10.4
Comparative Analysis of the Suggested Work with Other Existing Work

References	Substrate	Size of Substrate (μm²)	Frequency Range (THz)	Bandwidth (THz)	Impedance Bandwidth (%)	Gain (dBi)
[11]	Polyimide	600 × 600	0.5–0.8	0.3	46.15	9.19
[30]	Rogers RO3003	210 × 180	0.68–0.698	0.018	2.6	6.79
[31]	Arlon–100	500 × 500	1.04–1.1	0.07	6.54	7.99
[12]	InP and BCB	322 × 280	0.294–0.410	0.116	32.8	5.14
[13]	PEC	800 × 800	0.9–1.07	0.17	17.2	20
[14]	Polyamide	6,000 × 6,000	0.35–0.385	0.035	9.52	8.15
[15]	Silicon	490 × 450	0.38–0.42	0.04	10	6.5
[32]	Silicon and polycarbonate	20,000 × 3,500	0.29–0.316	0.026	8.6	11.71
[16]	GaAs	8,600 × 8,600	0.41–0.47	0.06	13.64	10
[17]	RT/Duroid 6006	600 × 700	0.7–0.85	0.15	19.35	6.4

(Continued)

TABLE 10.4 (*Continued*)
Comparative Analysis of the Suggested Work with Other Existing Work

References	Substrate	Size of Substrate (μm²)	Frequency Range (THz)	Bandwidth (THz)	Impedance Bandwidth (%)	Gain (dBi)
[18]	SiO_2	500×1,000	0.303–0.320	0.017	5.44	4.1
[19]	Arlon 1000	120×120	0.53–0.84	0.31	45.26	Not given
[20]	Polyimide	300×300	0.445–0.714	0.269	46.41	5.7
[21]	Polyimide	700×670	0.616–0.64	0.024	3.82	9.45
[22]	Polyimide	20,000×13,500	0.19–0.20	0.01	5.12	6.3
[23]	Silicon	1,350×1,000	0.285–0.325	0.04	13.11	6.9
[24]	Polyimide	600×300	0.35–0.75	0.4	72.72	5.49
[33]	Graphene	93×120	0.88–1.18	0.3	29.12	2.45
[25]	Polyamide	600×300	0.276–0.711	0.435	88.14	10
[26]	Rogers RT/ Duroid 6010	100×130	0.57–1.02	0.45	56.6	3.47
[27]	Polyamide	300×300	0.3627–0.5918	0.2291	48	4
[28]	Polyamide	38×46	2.51–5.05	2.54	67.19	5
			5.99–7.43	1.44	24.04	
			7.94–9.63	1.69	21.28	
Proposed work	**Polyamide**	**20×20**	**2.29–2.88**	**0.59**	**22.86**	**6**
			3.06–7.02	**3.96**	**78.88**	
			7.91–9.04	**1.13**	**13.34**	

10.4 CONCLUSION

This chapter presented a multiband graphene antenna for 6G communication and sensing technologies. The proposed antenna operates from 2.29 to 2.88, 3.06 to 7.02, and 7.91 to 9.04 THz in Band-1, Band-2, and Band-3, respectively. The peak gain of the proposed 6G antenna is 6 dBi. Hence, it is suitable for 6G communication and sensing applications.

ACKNOWLEDGMENT

This work is sponsored by the Department of Telecommunications (DoT), under the Telecom Technology Development Fund (TTDF), Ministry of Communications, Government of India (Proposal ID: TTDF/6G/382).

REFERENCES

1. Koenig, S., Lopez-Diaz, D., Antes, J., Boes, F., Henneberger, R., Leuther, A., & Kallfass, I. (2013). Wireless sub-THz communication system with high data rate. *Nature Photonics*, 7(12), 977–981.
2. Khan, M. U., Sharawi, M. S., & Mittra R. (2015). Microstrip patch antenna miniaturization techniques: A review. *IET Microwaves, Antennas & Propagation*, 9(9), 913–922.

3. Pozar, D. M. (2005). *Microwave Engineering*, John Wiley & Sons Inc., New York.
4. Rai, J. K., Ranjan, P., Chowdhury, R., & Jamaluddin, M. H. (2024). Quad Port MIMO Al_2O_3 ceramic based integrated reconfigurable wideband sensing and communication dielectric antenna for 5G cognitive radio. *IEEE Transactions on Components, Packaging and Manufacturing Technology*, 15, 94–103.
5. Rai, J. K., Ranjan, P., Chowdhury, R., & Jamaluddin, M. H. (2024). Design and optimization of dual port dielectric resonator based frequency tunable MIMO antenna with machine learning approach for 5G new radio application. *International Journal of Communication Systems*, 37, e5856. doi:10.1002/dac.5856
6. Rai, J. K., Ranjan, P., Kumar, S., Chowdhury, R., Kumar, S., & Sharma, A. Machine learning-enabled two-port wideband MIMO hybrid rectangular dielectric resonator antenna for n261 5G NR millimeter wave. *International Journal of Communication Systems*, 37, e5898.
7. Dash, S., & Patnaik, A. (2018). Material selection for THz antennas, *Microwave and Optical Technology Letters*, 60(5), 1183–1187.
8. Jornet, J. M., & Akyildiz, I. F. (2010). Graphene-based nanoantennas for electromagnetic nanocommunications in the terahertz band. *Proceedings of the Fourth European Conference on Antennas and Propagation*, Barcelona, Spain, pp. 1–5.
9. Singh, G., Sandha, K. S., & Kansal, A. (2024). Impact of sole layer duple substrates on GA-based optimised graphene antennas for THz applications. *Nano Communication Networks*, 40, 100508.
10. Ranjan, P., Mishra, N., Radovanovic, J., Potrebić Ivaniš, M., Murmu, L., & Rai, J. K. (2024). Refractive index sensing: Study and analysis for SARS-CoV-2 detection. *Optical and Quantum Electronics*, 56(7), 1–13.
11. Hocini, A., Temmar, M. N., Khedrouche, D., & Zamani, M. (2019). Novel approach for the design and analysis of a terahertz microstrip patch antenna based on photonic crystals. *Photonics and Nanostructures-Fundamentals and Applications*, 36, 100723.
12. Ullah, S., Ruan, C., Haq, T. U., & Zhang, X. (2019). High performance THz patch antenna using photonic band gap and defected ground structure. *Journal of Electromagnetic Waves and Applications*, 33(15), 1943–1954.
13. Tripathi, S. K., & Kumar, A. (2019). High gain miniaturised photonic band gap terahertz antenna for size-limited applications. *Australian Journal of Electrical and Electronics Engineering*, 16(2), 74–80.
14. Vettikalladi, H., Sethi, W. T., Abas, A. F. B., Ko, W., Alkanhal, M. A., & Himdi, M. (2019). Sub-THz antenna for high-speed wireless communication systems. *International Journal of Antennas and Propagation*, 2019, 1–9.
15. Geetharamani, G., & Aathmanesan, T. (2019). Metamaterial inspired THz antenna for breast cancer detection. *SN Applied Sciences*, 1(6), 1–9.
16. Alibakhshikenari, M., Virdee, B. S., See, C. H., Shukla, P., Salekzamankhani, S., Abd-Alhameed, R. A., ... & Limiti, E. (2020). Study on improvement of the performance parameters of a novel 0.41–0.47 THz on-chip antenna based on metasurface concept realized on 50 μm GaAs-layer. *Scientific Reports*, 10(1), 11034.
17. Kushwaha, R. K., Karuppanan, P., & Dewang, R. K. (2020). Design of a SIW on-chip antenna using 0.18-μm CMOS process technology at 0.4 THz. *Optik*, 223, 165509.
18. Alibakhshikenari, M., Virdee, B. S., Khalily, M., See, C. H., Abd-Alhameed, R., Falcone, F., & Limiti, E. (2020). High-gain on-chip antenna design on silicon layer with aperture excitation for terahertz applications. *IEEE Antennas and Wireless Propagation Letters*, 19(9), 1576–1580.
19. Alibakhshikenari, M., Virdee, B. S., See, C. H., Abd-Alhameed, R. A., Falcone, F., & Limiti, E. (2020). High-gain metasurface in polyimide on-chip antenna based on CRLH-TL for sub-terahertz integrated circuits. *Scientific Reports*, 10(1), 4298.

20. Mahmud, R. H. (2020). Terahertz microstrip patch antennas for the surveillance applications. *Kurdistan Journal of Applied Research*, 5(1), 16–27.
21. Zhu, H., Li, X., Qi, Z., & Xiao, J. (2020). A 320GHz octagonal shorted annular ring on-chip antenna array. *IEEE Access*, 8, 84282–84289.
22. Shamim, S. M., Uddin, M. S., Hasan, M. R., & Samad, M. (2021). Design and implementation of miniaturized wideband microstrip patch antenna for high-speed terahertz applications. *Journal of Computational Electronics*, 20, 604–610.
23. Krishna, C. M., Das, S., Nella, A., Lakrit, S., & Madhav, B. T. P. (2021). A micro-sized rhombus-shaped THz antenna for high-speed short-range wireless communication applications. *Plasmonics*, 16(6), 2167–2177.
24. Ahmad, I., Ullah, S., Ullah, S., Habib, U., Ahmad, S., Ghaffar, A., … & Limiti, E. (2021). Design and analysis of a photonic crystal based planar antenna for THz applications. *Electronics*, 10(16), 1941.
25. Alibakhshikenari, M. (2021). High-isolation antenna array using SIW and realized with a graphene layer for sub-terahertz wireless applications. *Scientific Reports*, 11(1), 1–14.
26. Alibakhshikenari, M., Virdee, B. S., Althuwayb, A. A., Aïssa, S., See, C. H., Abd-Alhameed, R. A., … & Limiti, E. (2021). Study on on-chip antenna design based on metamaterial-inspired and substrate-integrated waveguide properties for millimetre-wave and THz integrated-circuit applications. *Journal of Infrared, Millimeter, and Terahertz Waves*, 42, 17–28.
27. Vasu Babu, K., Das, S., Varshney, G., Sree, G. N. J., & Madhav, B. T. P. (2022). A micro-scaled graphene-based tree-shaped wideband printed MIMO antenna for terahertz applications. *Journal of Computational Electronics*, 21(1), 289–303.
28. Nickpay, M. R., Danaie, M., & Shahzadi, A. (2022). Wideband rectangular double-ring nanoribbon graphene-based antenna for terahertz communications. *IETE Journal of Research*, 68(3), 1625–1634.
29. Babu, K. V., Das, S., Sree, G. N. J., Madhav, B. T. P., Patel, S. K. K., & Parmar, J. (2022). Design and optimization of micro-sized wideband fractal MIMO antenna based on characteristic analysis of graphene for terahertz applications. *Optical and Quantum Electronics*, 54(5), 281.
30. Krishna Ch, M., Islam, T., Suguna, N., Kumari, S. V., Devi, R. D. H., & Das, S. (2023). A micro-scaled graphene-based wideband (0.57–1.02 THz) patch antenna for terahertz applications. *Results in Optics*, 12, 100501.
31. Kiani, S., Rezaei, P., & Fakhr, M. (2023). On-chip coronavirus shape antenna for wide band applications in terahertz band. *Journal of Optics*, 52(2), 860–867.
32. Vishwanath, Varshney, G., & Sahana, B. C. (2024). Design of tunable THz dielectric resonator antenna with cross-slot for circular polarization. *Optical and Quantum Electronics*, 56(6), 922.
33. Rai, J. K., Patel, U., Tiwari, P., Ranjan, P., & Chowdhury, R. (2024). Machine learning enabled compact frequency-tunable triple-band hexagonal-shaped graphene antenna for THz communication. *International Journal of Communication Systems*, 38, e6044.

11 Frequency-Tunable Graphene-Based THz Devices

Sureshkumar S, Naveen Kumar Maurya, and Vegesna S M Srinivasavarma

11.1 INTRODUCTION

The terahertz (THz) regime occupies the intermediate range of the electromagnetic spectrum, resting between the millimeter-wave and far-infrared transition band. It typically spans from 0.1 THz (100 GHz) to 10 THz, often known as THz gap [1–3,7,12,19]. The wavelengths within this range extend from 30 μm to 3 mm, effectively bridging the gap between radio waves and visible light [1,4] (Table 11.1).

Non-ionizing THz radiation can propagate along a line of sight [1,9,7]. Similar to microwave radiation, it is able to permeate an extensive selection of non-conducting material, including ceramics, clothing, plastic, wood, masonry, and paper [2–4]. However, its penetration depth is generally shallower than that of microwave radiation [9,11,31]. While THz radiation exhibits limited penetration through clouds and fogs, it is completely obstructed by liquids (such as water) and metals. THz spectrum finds its applications in communications and sensors [2,3,7]. THz communication provides numerous fascinating prominences like dense system design, wide bandwidth (BW), and higher data rate, [5,12,19]. The THz regime has an ultra-wideband (UWB) frequency bandwidth, and its emission is non-ionic in nature [13,22]. THz communication is being explored for the 6G network, which aims to support data rates of up to 1 TBPS. This can support holographic calls, immersive virtual reality, and high-speed broadband [21].

Non-ionizing THz radiation is functional with medical imaging, especially for diagnosing skin cancer and other diseases at an early stage [17–20]. THz imaging is

TABLE 11.1
Terahertz Frequency Spectrum

THz Band	Frequency Range (THz)	Wavelength Range (μm)
Lower THz	0.1–1	3,000–300
Mid-THz	1–5	300–60
Upper THz	5–10	60–30

DOI: 10.1201/9781003599111-11

applicable to find deformity in materials or layers within complex structures, which is valuable in manufacturing in smart factories [3,6,7]. THz communication can support data exchange between thousands of sensors and devices in real time [8,9,18].

Graphene is a solitary layer arrangement of carbon atoms structured in a hexagonal lattice, derived from mass graphite [2–7]. Often regarded as a wonder material, it exhibits exceptional properties [9–11]. Graphene is highly ductile, capable of undergoing significant tensile strain without breaking. It has developed as one of the most prominent materials for designing antennas with frequency-reconfigurability and circuit elements in THz spectrum [13,22]. Its remarkable attributes include tunable complex surface conductance, high thermal conductivity, superior mechanical strength, and optical transparency [26–31]. Frequency tunability in graphene is a vital asset that enables devices to carry out an operation across a wide range of spectrums in the THz regime [1–3,13,14]. This tunability arises from graphene's unique band structure, which allows for manipulation of its electronic properties through external factors such as electric fields or DC bias voltage [16–18]. Through the use of an external gate voltage to graphene [2,21], the Fermi level (the energy level at which the likelihood of locating an electron is 50%) can be modulated [10,13].

Gate voltage control in Graphene Field-Effect Transistors (GFETs) enables high-speed switching and tunable RF applications [2,10–12]. GFETs are capable of driving at extremely high frequencies, providing them suitable for applications in telecommunications and signal processing [1,2,5]. For example, this technique is used in GFETs to create tunable radiofrequency (RF) and THz devices [4,5,8]. With gate voltage control, these devices can dynamically shift their operating frequency. Equations (11.1)–(11.3) present Kubo's formula for calculating the conductivity of graphene [13,31]. A comprehensive understanding of graphene's conductivity can be achieved by analyzing its two key components: intra- and inter-band conductivity. As expressed in Equation (11.1), the graphene's total conductivity is defined by the sum of intra-band (σ_{intra}) and inter-band (σ_{inter}) conductivity [1–3,31]. Equation (11.2) provides the formula for calculating the σ_{intra} of graphene, and equation (11.3) gives the formula for calculating σ_{inter}.

In equation (11.2), ω is the angular frequency, T is the room temperature, k_B is the Boltzmann constant, $\hbar$ is the decreased Planck's constant, and Γ is the scattering rate, all of which are mutually connected to the relaxation time [1,13,31]. In this THz frequency spectrum, σ_{inter} is imperceptible [3,19–21]. Therefore, the overall conductivity of graphene is contingent on σ_{intra}. Equation (11.2) indicates that σ_{intra} is the purpose of chemical potential (μ_c), which is capable of being simply varied by altering a DC bias potential, thereby modifying σ_{intra}. By effectively tuning μ_c, we can manipulate graphene's conductivity and, consequently its resonant frequency, enabling the realization of tunable THz devices [9,20,23]. This alteration in the intricate surface conductivity grants frequency-adaptive characteristics to graphene-integrated designs [2,3,22]. As τ cannot be further tuned after fabrication, μ_c is the preferred parameter for tuning σ_{intra}.

Equation (11.4) gives the association in mid of applied DC voltage (V_{dc}) along with the chemical potential (μ_c) [25–27].

$$\sigma_{\text{graphene}} = \sigma_{\text{total}} = \sigma_{\text{intra}} + \sigma_{\text{inter}} \tag{11.1}$$

$$\sigma_{\text{intra}} = -J\frac{e^2 k_B T}{\pi\hbar^2(\omega - j\Gamma)}\left[\frac{\mu_c}{k_B T} + 2\ln\left(e^{-\mu_c/k_B T} + 1\right)\right] \tag{11.2}$$

$$\sigma_{\text{inter}} = \frac{-je^2}{4\pi\hbar}\ln\left(\frac{2|\mu_c| - (\omega - j\Gamma)\hbar}{2|\mu_c| + (\omega - j\Gamma)\hbar}\right) \tag{11.3}$$

$$V_{\text{dc}} = V_o + \frac{e t_s \mu_c^2}{\pi \mathcal{E}_0 \mathcal{E}_r \hbar^2 v_f^2} \tag{11.4}$$

$$\mu_c = \hbar v_f \sqrt{\frac{\pi \mathcal{E}_0 \mathcal{E}_r v_{\text{dc}}}{e t_s}} \tag{11.5}$$

Assuming $V_o = 0$ and rearranging equation (11.5), where V_o is the voltage compensation, v_f is the Fermi pace in graphene, ε_r is mentioned as dielectric constant of SiO_2, and t_s is the substrate thickness, it is able to tune the chemical potential (μ_c) of graphene by varying V_{dc} [1–3]. This, consequently, modifies the overall conductivity of graphene (σ_{graphene}) and enables frequency adjustability in graphene-based devices [9,13].

Remarkably, graphene exhibits resonance behavior via the stimulation of surface plasmon polaritons (SPPs) in THz spectrum. To ensure uniform propagation of V_{dc} and achieve tunability, the graphene pattern must interconnect all unit cells [2]. In this chapter, we will delve into the following tunable THz devices in detail: (1) antennas, (2) absorbers, (3) biosensors, and (4) filters.

11.2 GRAPHENE-BASED TUNABLE ANTENNAS

Graphene's electrical conductivity can be flexibly modified through gate voltage or chemical doping. This enables the regulation of antenna's frequency resonance and impedance, allowing for adaptable antennas [2–4]. Graphene's high carrier mobility contributes to efficient radiation and reception of THz waves. Graphene's flexibility and transparency make it suitable for applications requiring flexible or transparent antennas. Graphene can be patterned into dipole structures for THz radiation [2,3]. Graphene loop antennas can be designed for specific THz frequency. Graphene can be used in array configurations for beam steering and focusing of THz radiation. Developing scalable and cost-effective fabrication methods is crucial for widespread adoption. Graphene-based THz antennas with existing technologies, such as silicon-based electronics, are important for practical applications [2–4]. In this section, we thoroughly examine the advanced graphene-based antennas outlined in recent publications.

The key finding of this work [3] is the proposed graphene-based, frequency-adjustable isolation refinement technique for THz MIMO antennas. This mechanism significantly improves isolation, achieving approximately 30.41 dB at the operating frequency. It also provides tunability in the range of 1 THz in the transmission function by varying the chemical potential [3]. The introduced decoupling structure was simple, compact, and versatile, making it applicable to other THz devices.

It effectively suppresses mutual coupling produced by the emission modes of MIMO elements without colliding with the resonant frequency of individual elements [1,2]. When concerned with a slot-ring-based, dual-polarized MIMO/diversity antenna, this technique achieves a frequency spectrum of 0.83 THz (ranging between 5.68 and 6.51 THz) with high isolation and gain.

The perimeter of the slot ring plays a crucial role in establishing the basis of the antenna's resonance, which is primarily magnetic. The arrangement of the distribution along the edges of the lower ground plane is essential and holds significant importance. This specific arrangement influences the overall performance, behavior, and characteristics of the system, making it a key factor in determining its efficiency and functionality. Extending this design to a MIMO antenna with a shared ground plane would likely result in substantial mutual coupling [3]. The coupling mechanism between MIMO elements primarily involves propagation along the antenna's longitudinal axis and across the substrate width. To mitigate this mutual coupling, decoupling structures can be integrated into the central region of the common ground plane.

Figure 11.1 illustrates the schematic of the microstrip line, showing its length, width, and spacing. The antenna achieves optimal matching at the frequency range of 6 THz, providing a 10 dB return-loss bandwidth (RLBW) in the range of 0.97 THz. It operates effectively within the frequency spectrum from 5.69 to 6.4 THz frequency band, with isolation exceeding 17.15 dB at 6 THz. The electric-field (*E*) dispersion of the MIMO antenna at the 6 THz frequency spectrum was simulated with one port active, while the remaining ports were terminated. The coupling

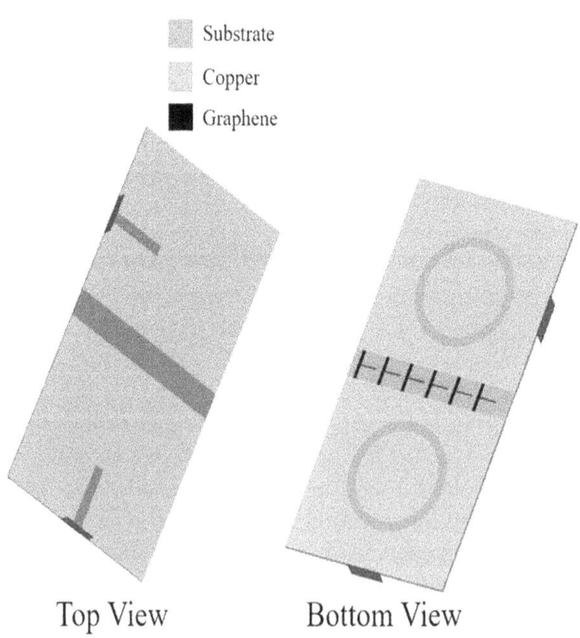

FIGURE 11.1 Schematic of microstrip-line-guided slot ring-based antenna.

factor along the previously mentioned coupling paths can be mitigated by placing decoupling structures at the midpoint of the recurring ground plane, isolating the couple of MIMO elements.

The design of a frequency-agile decoupling mechanism with a compact footprint significantly enhances the performance of THz antennas and devices. Beyond its ability to tune the transmission coefficient over a 1 THz range, the proposed decoupling structure effectively suppresses coupling leading to some radiation mode of the radiating elements. This innovative design allows for adaptable isolation enhancement while preserving the resonant spectrum of the distinct MIMO elements.

In article [7], the authors describe the characterization and development of a tunable antenna based on graphene performing at the microwave frequency spectrum that resides at 15 GHz. The presented antenna design contains a graphene sheet as the emitting element and a copper gate electrode placed behind the antenna. The simulation results show that as the graphene material resistance varies from 0.00028 to 2.8 Ω^2, the resonance frequency remains mostly unchanged, but the reflection coefficient magnitude and gain change significantly [7]. The antenna short of the gate electrode exhibited a resonance frequency shift of 1.7 GHz differentiated with the simulation, likely because of fabrication challenges such as imprecise cutting of the graphene material and misalignment with the substrate. By deriving equations relating the chemical potential to the gate voltage and dielectric interlayer thickness, the authors suggest that a gate voltage of at least 2.85 kV and a dielectric interlayer thickness of less than 0.26 μm would be required to effectively tune the chemical potential and, consequently, the antenna's performance. The analytical approach to enhance the tuning characteristics through gate voltage and dielectric spacer thickness optimization is a unique and interesting contribution that could guide future research in this field.

Article [2] provides the design and development of a tunable THz a couple of port graphene patch antenna capable of functioning in either a self-diplexing mode or a multi-input-multi-output (MIMO) antenna mode with diversity of pattern. The antenna is designed using a proximity-coupled feeding technique, which allows for independent tunability with a couple of radiating patches. The optimized antenna structure is to provide high isolation (over 40 dB) between the two ports across the tunable frequency range of 1.5–3 THz. This high isolation enables the antenna to operate in a simultaneous-transmit-and-receive (STAR) mode. The antenna dimensions are chosen such that one patch operates in the TE mode, while the other operates in the TM mode.

The tunability of the antenna is realized by modifying the chemical potential of the graphene patches, which changes their surface conductivity and resonant frequency. The analysis of a microstrip antenna with tetradecagonal ring-shaped for super broadband MIMO applications was proposed by Singhal [4]. The antenna features a tetradecagonal radiator with an altered semi-circular ground plane, resulting in an impedance bandwidth ranging from 14.8 to 15.1 THz. Two diversity MIMO antenna configurations are also inspected namely spatial diversity and pattern diversity. The spatial diversity configuration has a functional spectrum of 0.25–15.4 THz, while the pattern diversity configuration has a bandwidth of 0.2–15.3 THz. Both configurations exhibit an isolation of $S_{21} \leq -20$ dB in almost all their performing bands.

The MIMO performance parameters, including envelope correlation coefficient, loss in the channel capacity, diversity gain, are also found to be within acceptable limits for both configurations. A Gain-enhanced PEC reflector-backed slot-loop antenna was proposed by Jin [5]. The antenna they presented consists of a rectangular loop with an internally inset-fed rectangular patch (slot-loop), a couple of symmetrical L-shaped emitting branches, and a PEC reflector. With a dense dimension of 36 mm × 17.5 mm × 7.2 mm, the antenna ensures a −10 dB reflection coefficient bandwidth ranging from 7.94 to 14.45 GHz. Parametric analyses are conducted to examine the results related to the span of the driven element and reflector, as well as the vertical dimension of the radiating branches. The concise values of these parameters are intent on to attain the best impedance matching and gain level in the expected band. The simulated radiation patterns, measured gain, and antenna efficiency are presented, demonstrating its wideband and high-gain characteristics.

A quadruple port MIMO extendable antenna using a single antenna element (A4) is proposed by Raj et al. [6]. In this, a single antenna component (A4) is enlarged to a dual-port and then to a quadruple-port MIMO configuration. In the dual-port MIMO antenna, the elements were arranged orthogonally, and inverted L-shaped ground stubs are employed to create distance among the elements. The quadruple-port MIMO antenna additionally enhances the size of the antenna to $67.5 \times 67.5 \times 2.4\,\mu m^3$ to accommodate the four elements in an orthogonal arrangement. The mirror imaged and L-shaped ground segments are also enlarged to refine the bandwidth and isolation, which is provided for above −20 dB throughout the functional frequency range.

This topic also explains a precise analysis of MIMO performance parameters, including envelope correlation coefficient (ECC), channel capacity loss (CCL), diversity gain (DG), and total active reflection coefficient (TARC). The acquired results for these parameters are within the admissible range, indicating good diversity results of the designed antenna. The proportions and results of the antenna are distinguished with the available literature, and it is discussed to offer improved bandwidth and enhanced isolation, making it suitable for various THz use cases such as cancer diagnosis, mixers and multipliers, healthcare, deep space networks, aerospace, and wireless body area networks. Article [7] presents the analysis and development of a double-polarized microstrip patch antenna made of graphene material. The presented antenna contains a unitary graphene layer excited by dual orthogonally oriented microstrip feed lines, enabling the generation of vertical and horizontal polarized THz radiation. Configurable specification analysis is performed to fine-tune the antenna's efficiency. The outcome of modifying the graphene's chemical potential, substrate size, and feed line width on the return loss, isolation, and radiation characteristics is investigated. The antenna is designed to demonstrate a maximal return loss of −38 dB, isolation exceeding 45 dB, and an efficiency of approximately 90% at the resonant frequency of 3.98 THz. It also reports a peak gain of 9.6 dB and a half-power beamwidth of 45°. In article [8], the authors present the design with a band-stop function within the antenna's operating range. The authors incorporated a metasurface design that utilizes an electric inductive capacitive resonator for achieving the above-mentioned functionality. Incorporating the metasurface resonator between the two antennas provided at least 10 dB improvement in mutual coupling reduction. The metasurface exhibited a band-stop functionality in the defined range of the antenna and hence the electromagnetic interaction into the radiating elements got controlled significantly [32, 33].

11.3 GRAPHENE-BASED THZ ABSORBERS

At THz frequencies, the interaction of light with graphene primarily involves intra-band transitions, where electrons within the same energy band absorb photons and transition to higher energy levels within that band. The electrical conductivity of graphene can be regulated through the use of an extrinsic electric field or via chemical doping.

The tunability enables for dynamic supremacy over the absorption properties of absorbers in the THZ range graphene-based devices. Integrated graphene into structures that exhibit coherent perfect absorption, where incident THz radiation is completely absorbed due to destructive interference. In this section, we discuss numerous THz range graphene-based absorbers detailed in the latest literature.

In article [10], the authors introduce the design and development of a graphene-based metasurface absorber (MPA) that efficiently absorbs THz radiation across a wide bandwidth. The proposed MPA features a periodic array of graphene patches with two distinct dimensions, separated from a gold bottom reflector by a SiO_2 spacer layer. Numerical and analytical results demonstrate that the presented MPA achieves a broadband absorption response spanning nearly 3 THz, from 3.64 to 6.55 THz. This broadband absorption is attributed to generating surface plasmon ecstasy within the graphene patches. Furthermore, the authors investigate the dependence of the absorption spectrum on the density of the SiO_2 spacer layer and the chemical potential of graphene, revealing that the absorption bandwidth and spectral location can be actively tuned. In article [11], the authors introduce a novel polarization-insensitive graphene-based metamaterial absorber (MA) in the THz range with a broad bandwidth. The presented MA comprises three concentric annular strips and a single octagonal graphene unit. By carefully modifying the positions and measurements of these components, the hybridization and stacking effects are enhanced, leading to strong and broad absorption results. The developed MA results in an absorptivity exceeding 90% within the frequency spectrum of 1.027–1.958 THz, with most of the absorptivity surpassing 95%. Notably, the structure demonstrates insensitivity to both polarization and angle of incidence. An analysis of the electric field intensity distribution reveals that the absorption in the broadband is achieved through the resonance of electric dipoles at various frequencies.

Lv et al. [12] present the design and development of a broadband MA in THz regime based on a hybrid graphene–gold pattern. The proposed MA structure comprises a top layer combining a patterned graphene and a gold disk, and polyethylene cyclic olefin copolymer (TOPAS) was a middle spacer layer along with the bottom gold ground plate.

The authors demonstrate that by varying the number of slits cut into the graphene (from zero to four), the absorption effect can be tuned. Notably, broadband absorption is achieved by introducing four symmetrically distributed slits. This phenomenon is attributed to increased field confinement surrounding the nano-slits and field hybridization into different structural components. By adjusting the chemical potential of graphene from 0 to 0.5 (eV), the proposed MA structure is able to operate as a modulator, exhibiting a low and high modulation depth of 69.5% and 92.7%, respectively. To further enhance performance, the authors introduced a modern integration of the graphene pattern on the gold disk with the top layer, resulting in an even wider absorption bandwidth exceeding 4.0 THz. This recently innovated MA structure maintains perfect modulation characteristics, with a range of high and low modulation depth of 91.5% and 70.57%, respectively. In further, the presented MA structures exhibit dynamic tunability, polarization insensitivity, and wide-angle incidence, making them suitable for numerous features in the THz frequency range, including absorbers, detection devices, modulators, and optoelectronics. In article [13], the authors present the design of a graphene-based tunable UWB metasurface absorbers designed for THz range applications. The proposed metasurface consists of a silicon dioxide (SiO_2) substrate layered into a top graphene patterned layer and a continuous bottom layer of graphene, each with a thickness of 1 nm. The design incorporates a separate DC bias configuration for separate biasing a pair of graphene layers. Ultrathin profile in the metasurface is exhibited with a thickness of only 5.1 µm (λ_0/588.235) and a periodicity of 13.5 µm (λ_0/222.22).

The absorber results in an absorption bandwidth of 13.9 THz (FBW = 197.16%), encompassing the electromagnetic frequency ranges from 0.1 to 14 THz with 90% absorptivity. The structure exhibits polarization insensitivity to normal incidence of plane waves due to its structural symmetry. Furthermore, it maintains an absorptivity exceeding 80% among the frequency spectrum of 0.1–14 THz for incidence angles up to 60° for both Transverse Electric (TE) and Transverse Magnetic (TM) polarizations. The variation in the chemical potential (μ_c) of the top patterned graphene layer provides tunable absorption characteristics, with an increase in μ_c resulting in a blue switch in the absorptivity response. In article [14], the authors discuss the design and performance analysis of a novel polarization-insensitive, wideband, and tunable absorber in the THz regime utilizing a slotted-square graphene meta-ring.

The presented absorber applications use a single-layer device configuration, with a top layer mounted on graphene over a gold-backed polyimide lossy dielectric substrate.

The absorber in Figure 11.2a exhibits efficient absorption, exceeding 95%, across a wide THz band spanning from 2.2 to 4.6 THz. It maintains an absorption level exceeding 80% even under various obliquely incident operating conditions for the TE and TM wave excitation modes. Notably, the design results in polarization-insensitive characteristics, with the extent of absorption being unchanged by changes

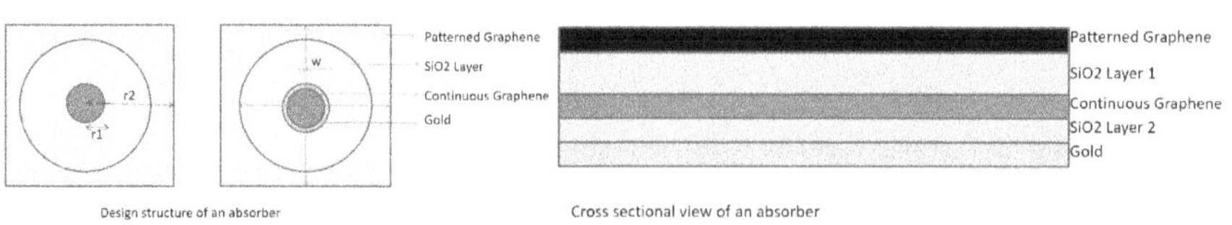

FIGURE 11.2 (a) Design structure of an absorber and (b) absorber cross-sectional view.

in the polarization angle of the incident THz waves. Furthermore, the absorption characteristics in the proposed absorber can be dynamically altered from 95% to 15% by varying the chemical potential of graphene from 1 to 0.1 (eV).

This feature enables the absorber tunability to function as a fast, controllable switch, with potential applications in the implementation of high frequency switching circuits and Boolean logic. Pan et al. [15] analyze a metamaterial structure consisting of a single layer of graphene on top, a silica layer in the middle, and a gold layer at the bottom. The graphene layer features a frame and five square graphene structure. The authors employ the chemical vapor deposition (CVD) procedure to grow and transfer the monolayer graphene onto the SiO_2 substrate. The metamaterial exhibits an absorption rate exceeding 0.9 within the 3–6.7 THz band, with a peak absorption of 0.998 at 5.2 THz. The absorption spectra remain identical for both the TE and TM polarization modes, indicating polarization insensitivity. The authors calculate the relative impedance of the metamaterial and demonstrate that it closely matches the free space impedance when the metamaterial exhibits high absorption, which is consistent with impedance matching theory.

Furthermore, the metamaterial results in a notable degree of stability against incident angles of electromagnetic waves, maintaining an absorption rate exceeding 0.9 up to an incidence angle of 30°. A novel design for a broadband tunable THz absorber utilizing a single ultrathin layer of graphene is proposed in article [16]. The proposed absorber contains a patterned graphene metasurface, a polyethylene cycloolefin copolymer (TOPAS) structured as dielectric layer, and a gold film. The graphene metasurface incorporates a cross-shaped graphene patch with four edges cut off at the center of the unit cell, along with four arc-shaped graphene patches in Figure 11.2b at all the corners of the unit cell. Numerical simulations show that the proposed absorber is able to absorb exceeding 97% within the broadband frequency range of 1.45–2.99 THz under standard occurrence conditions, resulting in an exact frequency range of 1.54 THz. The absorber also exhibits both incident angle and polarization insensitivity, providing maximum absorption across a broad range of incidence angles. The property of tunability in the graphene was verified, revealing that the absorbance is able to dynamically tune from 11% to 100% by regulating the chemical potential of the graphene through external bias voltage.

In article [17], the design and analysis of a broadband, polarization-insensitive, and graphene-supported tunable THz metamaterial absorber (GMMA) is presented. The designed GMMA employs a monolayer component topology incorporating a composition of square, circular, and cross-shaped graphene-based meta-resonators positioned on a lossy dielectric substrate backed by a gold reflector layer. The GMMA exhibits unprecedented absorption exceeding 90% across a wide frequency range from 2.3 to 6.4 THz. This broad absorption bandwidth is attained through cautious design and optimization of the multi-shaped meta-resonators, which induce multiple plasmonic resonances that collectively couple to provide the broadband absorption characteristics. The GMMA also demonstrates a polarization-insensitive response, maintaining absorption higher than 70% for obliquely incident angles across the extensive operating frequency range. The spectra amplitude of the GMMA is able to dynamically be tuned from 19% to more than 90% by modulating the chemical potential of graphene from 0 to 0.7 (eV), showcasing its tunability and reconfigurability. Furthermore, the

GMMA exhibits a low polarization conversion ratio (PCR) of less than 16% across the wideband, indicating its suitability for various applications requiring polarization-insensitive performance.

The simplified design structure, wideband absorption application, and GMMA tunability offer a promising pathway for the development of high-speed optical switches, detectors in THz range, and other optoelectronic modules. A graphene-based metasurface absorber having tunability designed for wideband absorption in the lower mid-infrared (MIR) spectrum is proposed in article [18]. The proposed structure contains a graphene layer deposited on top of an amorphous silicon dioxide (SiO_2) substrate, which is backed by a thin gold layer. The patterned graphene layer is a square fractal-like slot structure to achieve broadband absorption. The designed absorber exhibits over 90% absorptivity across a wide frequency spectrum of 3.69–9.77 THz, covering a fractional bandwidth of 90.3% [8]. This wideband absorption is attained by producing a square fractal slot layout in the graphene layer. Additionally, graphene with a square pattern is deposited into the slotted area to enhance coupling and absorption bandwidth. The absorber is polarization-independent under normal occurrence because of the fourfold symmetry in the graphene structure. The absorber maintains its wideband absorption performance under oblique incidence angles up to 40 degrees for both TE and TM polarizations. The size and thickness of the designed absorber is mentioned around $\lambda/11$ in accordance with the lower frequency, and the periodicity of the unit cell is around $\lambda/13$, supporting the effective homogeneity state in the sub-wavelength region.

Detailed analysis of the absorption mechanism includes the calculation of effective permittivity and permeability, electric field distributions, and surface current patterns at the three definite absorption peaks of 4.21, 6.78, and 9.27 THz. The authors also distinguish the performance of the proposed absorber with the previously reported graphene-based metasurface absorbers, highlighting the advantages of the presented design configured with broadband absorption, compactness, and the ease of fabrication [18].

11.4 GRAPHENE-BASED TUNABLE BIOSENSORS

THz radiation interacts with biomolecules, triggering various mechanisms in graphene, including conductivity changes, plasmon resonance, and the photo-thermoelectric effect. These interactions enable the detection of disease biomarkers, viruses, bacteria, contaminants, and pathogens in biological samples and food products. They also facilitate the study of drug–biomolecule interactions at the molecular level and the sensing of pollutants or toxins in water and air. Integrating graphene sensors with microfluidic platforms automates and increases the throughput of biological sample analysis. Furthermore, developing sophisticated algorithms is crucial for analyzing and interpreting the complex THz spectra obtained from biosensors. THz waves are appropriate for biosensing because of the non-ionizing nature, allowing for the analysis of materials without disruption. Metasurfaces have emerged as one of the most effective and cost-efficient methods for biosensing due to their unique properties.

MPAs with sharp absorption peaks enable narrowband biosensing by interacting with analytes such as body fluids. In one study, a metasurface absorber was investigated for its potential in detecting glycosuria. The metasurface was exposed to an analyte (urine sample) with varying refractive indices corresponding to different glucose concentrations. A 0.002 change in the refractive index resulted in a 0.0002 THz shift in the metasurface's resonance frequency, demonstrating a high sensitivity of 100 GHz/RIU. This research highlights the potential of the proposed polarization-insensitive, gold-based metasurface absorber for THz-based bio-sensing applications, particularly for glycosuria detection. The metasurface absorber resonates at 11.912 THz with a fractional bandwidth of 0.353% (bandwidth of 0.042 THz) and exhibits 99.1% absorptivity.

The symmetrical design of the metasurface's top pattern guarantees insensitivity to polarization, yielding identical absorption properties for both TE and TM polarizations. Modal analysis indicates that the absorption is due to the concentration of the electric field at the outer edge-points of the metasurface pattern and the corresponding conductor and dielectric losses. The defining characteristic of this work resides in the design of a simple, easy-to-fabricate, and polarization-insensitive metasurface absorber that can effectively detect changes in the refractive index associated with glucose concentration in urine samples. The high sensitivity and narrowband absorption characteristics of the metasurface make it a promising candidate for future advancements in THz-based biosensing applications.

Article [19] presents a novel THz biosensor made of metamaterials for the early detection of NMSC (Non-Melanoma Skin Cancer). The absorption characteristics of the proposed biosensor are at 0.503 and 1.076 THz, i.e., multi-band, which are achieved through a fractal-shaped curved aluminum resonator on a polyimide substrate. The absorber is designed to have near-perfect absorption (over 99%) at the two resonant frequencies. Parametric analyses are conducted to examine in detail the results of numerous design parameters, such as the absorber's length, dielectric substrate's thickness, and gap sizes, on the absorption characteristics.

The resonance mechanism of the absorber is explained in detail, highlighting the role of plasmonic resonance and the distribution of the surface current at the two absorption frequency points. The performance of the developed biosensor is estimated with regard to sensitivity, figure of merit, and quality factor. The sensor measures a sensitivity of 95.76 GHz/RIU (Radio Interface Unit) and 100 GHz/RIU at the initial pair of absorption peaks, respectively, indicating its potential for finding alternates in the surrounding medium's refractive index. The sensor proposed also demonstrates a high figure of merit (4.85 and 3.29) and a quality factor (25.65 and 34.40) at the respective resonant frequencies, highlighting its suitability for biosensing applications. In article [20], the design and development of MPA operating in frequency spectrum spanning from 15 to 35 THz (THz spectrum) is detailed. The proposed MPA contains a thin ZnSe (Zinc Selenide) substrate structure with copper placed inside it. The ZnSe is placed alongside a metal backplate to attenuate electromagnetic wave emission. The MPA exhibited significant absorption of 98.44% and 99.28%, respectively, at frequencies 22.46 and 28.95 THz. For the TE and TM polarizations, the MPA achieved insensitivity to an impact angle from 0° to 60°. This polarization independence is attributed to the symmetric resonator design. The authors also investigated the result of different resonator materials, substrate types, and resonator thicknesses on the MPA's absorption ability. Given the dual-band absorption of the proposed MPA, it holds significant potential for applications in the harvesting of thermal energy, stealth materials technology, and imaging.

A novel plasmonic absorber based on graphene, which demonstrates two narrowband perfect absorption peaks in the mid-infrared region, is introduced in article [21]. The absorber consists of a periodically arranged H-shaped graphene array, which is separated from a metallic ground plate by a broad dielectric spacer. The analysis results indicate that the proposed structure displays two distinct absorption peaks at wavelengths of 6.3 and 8.6 μm, with absorption coefficients of 99.65% and 99.80%, respectively. The absorption bandwidths, characterized by the full width at half-maximum (FWHM), are measured at 90 and 188 nm for the respective peaks. The dipole resonance mode is enhanced by the presence of the graphene ribbon at the 6.3 μm wavelength, while the absorption peak at 8.6 μm is associated with a novel hybridized mode that is separate from the dipole resonance.

The potential to actively control the absorption characteristics through external parameters, such as Fermi energy and dielectric refractive index, unaltering the physical structure, represents a unique and intriguing feature with potential applications in various mid-infrared optical devices, including filters, detectors (sensing), and energy storage systems. The absorber proposed in [22] consists of a plus-shaped graphene strip and four dual concentric graphene rings, designed using a Silicon Oxide (SiO_2) substrate over a silver ground plane. The graphene strip shaped in the plus form facilitated absorption in the range of 5.0–5.32 THz, whereas the concentric rings achieve over 90% absorption from 2.36 to 3.24 THz. The variation in the chemical potential of graphene allows for reconfigurability within the two absorption bands. Consistent with the fourfold symmetry of the design, the absorptivity remains stable across multiple polarizations and a wide angle of incidence exceeding 60°. The surface current distribution analysis confirms that the peak absorption frequencies are independently controlled by different resonating structures within the absorber. The proposed graphene-based absorber exhibits several unique features that make it suitable for THz sensing and imaging applications. The key highlight of this work is that, by varying the chemical potential of the graphene, the following two key properties of the absorber can be reconfigured. They are: (1) controlling the absorption bands independently and (2) the polarization insensitivity over a wide incident angle. These characteristics demonstrate the potential of the designed absorber for versatile applications in the THz frequency regime.

Figure 11.3 shows the THz absorber layout structure. It is designed over the SiO_2 substrate with the dielectric constant 3.8. The thickness (HS) of the substrate is 1.6 μm. On the top side of the substrate, four dual concentric rings and a plus-shaped strip based on a graphene material are designed. Its thickness is taken as 0.34 nm. Graphene is widely used to design the absorber due to its high absorption rate in a lower THz frequency band, i.e., 0.1–5.0 THz. On the bottom side of the substrate, a silver material is used. Different resonant sections of the absorber independently control the peak absorption frequencies. This allows for more precise tuning and optimization of the absorption bands. The absorber achieves maximum absorption of 99.81%, 97.34%, and 99.74% at 2.56, 3.08, and 5.16 THz, respectively. These high absorption rates make it highly efficient.

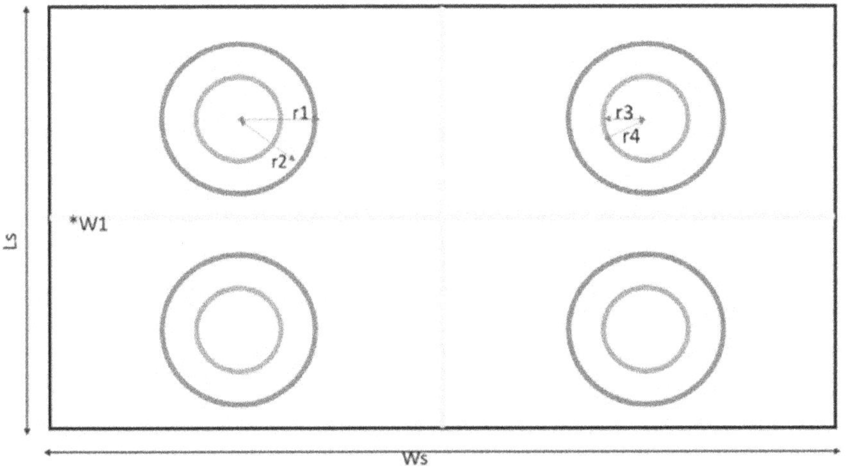

FIGURE 11.3 Proposed layout structure for the graphene-based dual-band absorber.

Sahu et al. [23] introduced a comprehensive design and analysis of a flexible metasurface (MS) absorber operating in the THz frequency range. The research begins with the presentation of a microwave multiband metasurface absorber that demonstrates significant absorption capabilities, highlighting six distinct peaks that surpass 98% within the 2–18 GHz frequency range. The paper then details the role of a copper-based MS absorber in the THz frequency spectrum, which achieves impressive performance metrics, including 99% absorptivity at 3.48 THz and a narrow absorption bandwidth of 0.087 THz, all while maintaining an ultrathin profile of 2.4 μm. This copper-based MS absorber serves as a foundational element for the innovation of a THz biosensor aimed at glucose detection. The biosensor indicates high sensitivity to the refractive index (RI) of the encasing environment, thereby improving precision of analysis at a consistent analyte thickness of 6.0 μm. In order to enhance the sensing capabilities, a novel integrated sensor architecture that integrates graphene and copper is investigated for the detection of glucose, malaria, and cancer cells. The observations point to the significant potential of the proposed MS framework for a wide array of THz sensing applications, including medical diagnostics and disease detection, utilizing MS technology to connect electromagnetic sensing with biosensing.

In article [24], the absorber demonstrates dual-band absorption capabilities, covering frequency ranges from 0.6 to 1.75 and 3.25 to 4.08 THz, with fractional bandwidths of 98.29% and 22.67%, respectively. The design incorporates a minimum thickness of 0.06λ, aligned with the lower resonant frequency. The graphene-based dual circular ring metamaterial absorber (GDCRMA) exhibits polarization insensitivity up to 90° and angle insensitivity up to 45°. The metamaterial characteristics of the absorber are characterized by negative values in its effective permittivity and permeability. Furthermore, the intensity of absorption is capable of optimizing by varying the chemical potential of the graphene material. An equivalent circuit for the

GDCRMA has been developed, demonstrating a decent performance through simulated results. The proposed absorber's capacity to maintain a wide bandwidth with dual-band absorption, along with its tunable absorption amplitude and polarization insensitivity, underscores its unique advantage of other absorbers.

These attributes make the designed absorber a suitable one for applications in high-frequency switching systems, Boolean logic devices, imaging, and wireless communication. The key insights uncovered from the document include the design of a highly efficient, dual-band, and tunable THz absorber using graphene-based metamaterials, the demonstration of polarization and angle insensitivity, and the development of an equivalent circuit model that accurately captures the absorber's behavior. These advancements in THz absorber design highlight the potential of graphene-based metamaterials for various practical features in the THz domain.

11.5 GRAPHENE-BASED TUNABLE FILTER

Graphene-based THz filters offer dynamic control over THz waves due to graphene's unique properties. By adjusting its conductivity through methods like applying a gate voltage or chemical doping, the interaction of graphene with THz radiation can be precisely tuned. Graphene supports plasmons, which are collective electron oscillations that can be tuned to resonate at specific THz frequencies. This allows for selective filtering of THz waves. Furthermore, the filtering characteristics of these devices can be dynamically adjusted by modifying graphene's conductivity. Achieving the desired filtering characteristics requires careful patterning of graphene into specific shapes and structures. The choice of substrate also significantly impacts the filter's performance.

Finally, graphene-based THz filters can be integrated with other THz devices, such as detectors and sources, to create complete and functional THz systems. This section summarizes various graphene-based THz filters proposed in the literature. Also, the key advantages of graphene-based THz filters, such as their tunability, plasmonic properties, and potential for integration into larger systems, are emphasized. A study on a tunable graphene-based THz bandpass filter designed for space applications is discussed in [25]. The filter utilizes a dual-mode open-loop (DMOL) resonator and operates at a resonant frequency of 6.35 THz. The design leverages graphene's unique properties, particularly it enables to support surface plasmon polaritons (SPPs) in the THz range, which enables dynamic tunability. The filter's structure consists of a double substrate with quartz and silicon layers, with gold material edged on the top and bottom layers.

A graphene layer is placed into the top metal layer and the quartz substrate. The design permits modifying both bandwidth and center frequency. The tunability of the bandwidth is achieved by altering the geometry of the resonator, specifically the width (W_4) of the open loop. Center frequency tuning will be done by adjusting the chemical potential of the graphene layer.

The three-dimensional representation of the represented band-pass filter (BPF) with a DMOL resonator, fabricated using a dual-substrate structure consisting of 50 nm thick layers of quartz and silicon, is illustrated in Figure 11.4. The top and bottom layers are made of 0.2 μm thick gold material with a conductivity of 4.561×10^7 S/m.

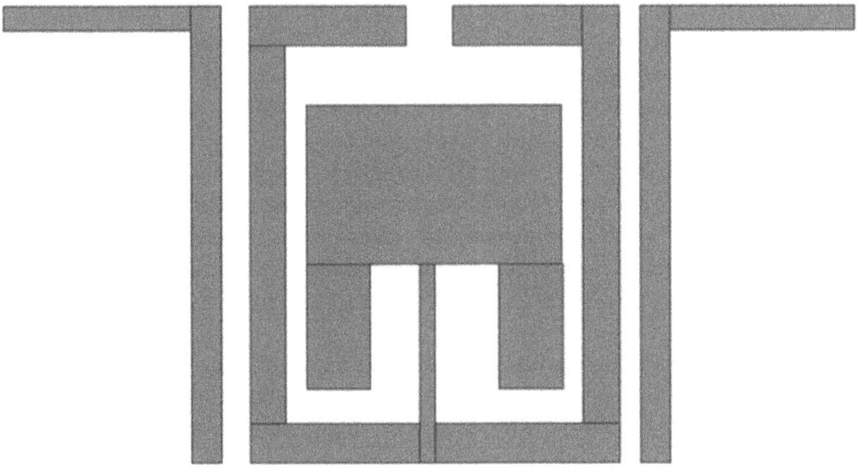

FIGURE 11.4 Designed layout of BPF with DMOL resonator.

A dual-layer substrate, composed of silicon and quartz with dielectric constants of 11.9 and 3.75, respectively, is positioned atop of the ground plane. The study reveals that the filter can achieve an adjustable span of around 0.16 THz by adjusting the chemical potential from 0.95 to 0.99 eV. The filter demonstrates two distinct resonant peaks corresponding to odd and even modes, with the even mode resonating at 6.31 THz and the odd mode at 6.39 THz. Additionally, the passband attenuation is approximately −20 dB, with a bandwidth of 0.20 THz. One of the most interesting aspects of this study is the filter's potential for space applications.

The authors highlight that graphene-based coatings have been verified with highly resistant extreme temperature shifts typical in space environments, ranging from −180°C to +130°C. This resilience makes the proposed filter particularly suitable for outer space conditions. Additionally, the dual-tuning capability of both bandwidth and center frequency sets this design apart from previous works, which primarily focused on center frequency tuning alone. This increased flexibility renders the filter highly versatile for various space applications, where accurate regulating bandwidth and frequency is crucial for improving communication networks and signal processing in challenging conditions.

In article [26], the advantages of THz technologies, such as wide bandwidth, material penetration, and device miniaturization, were discussed. It highlights the challenges of using ordinary metals at THz frequencies and the need for noble metals that can support surface plasmon propagation. Graphene is introduced as a promising metamaterial for THz devices due to its distinctive features, including the ability to support surface plasmons and the tunability of its conductivity through applied bias voltage [8]. It describes the design of the graphene-based band stop filter using a transmission line model with open-circuited shunt stubs connected through modular elements. The design process involves transforming a low-pass prototype filter to the desired band stop filter characteristics using frequency transformation, Richard's transformation, and Kuroda's identity. The key design parameters, such as the characteristic impedances of the shunt stubs and connecting unit elements, are derived from the low-pass prototype [8]. It represents the simulation performance for the designed graphene-based band stop filter, showing the desired frequency response.

It demonstrates the tunability of the filter by modifying the chemical potential of the graphene layer, which can shift the resonant frequency over a range of 0.1 THz while sustaining absolute bandwidth [8]. The unique insight from work described in article [27] is the exploration of graphene as a promising material for tunable THz band stop filters. The ability to dynamically adjust the filter's resonant frequency by controlling the graphene's chemical potential presents an interesting and versatile solution for various THz applications, such as sensors, material analysis, and DNA characterization.

This study [27] explores a tunable dual-band bandpass filter utilizing graphene for THz spectroscopy of polymers. The filter is engineered to function at resonant frequencies of 1 and 1.5 THz. The envisioned configuration comprises a series of TEM-mode transmission-line resonators with alternating open and short-circuited ends. Based on a low-pass Chebyshev prototype, the filter integrates a graphene layer between the dielectric and conductor layers to enable plasmonic wave propagation. The paper details the theory and models behind graphene's electrical properties and its application in filter design.

It explains how the graphene's chemical potential, which can be altered by changing the applied bias voltage, affects its conductivity and enables tunability of the filter. The design equations for the interdigital bandpass filter are provided, along with the geometrical dimensions and equivalent circuit element values. By altering the width and length of the central resonator, a tuning range of 70 and 40 GHz, respectively, is achieved. Notably, by modifying the chemical potential of graphene from 0.09 to 0.11 eV, the filter can be electrically tuned across a frequency spectrum of 0.14 THz while preserving a fixed absolute bandwidth of 50 GHz. This tunable graphene-based dual-band bandpass filter holds significant potential for spectroscopy applications, particularly in detecting structural defects within the polymer industry. One of the most compelling aspects of this study is the groundbreaking use of graphene to achieve tunability in THz filters.

The ability to electrically tune the filter's response by simply adjusting the bias voltage across the graphene layer offers a significant advantage over traditional filters that rely on physical alterations for tuning. This approach not only provides a wider tuning range but also allows for more precise and dynamic control of the filter characteristics. Furthermore, the dual-band nature of the filter, with passbands at 1 and 1.5 THz, makes it particularly suitable for polymer spectroscopy applications, as many plastic materials are permeable to THz radiation. This combination of tunability and dual-band operation in the THz regime represents a notable advancement in filter design for non-destructive testing and quality control in the polymer processing industry.

The filter proposed in article [28] consists of graphene nanoribbons deposited on a two-layer dielectric substrate of silica (SiO_2) and silicon (Si). The filter operates by blocking specific frequencies while allowing others to pass, utilizing the propagation of SPP waves on the graphene surface. The researchers analyzed the filter's performance by varying parameters such as the distance between nanoribbons (L_2), the width of nanoribbons (w_2), and the chemical potential of graphene (μ_c). They found that changing w_2 had a significant impact on resonance frequencies and bandwidth, while varying L_2 only slightly shifted resonance frequencies. The chemical potential, controllable through applied voltage, allowed for dynamic tuning of the operational bandwidth of the device. The filter demonstrated excellent performance characteristics, including insertion losses reaching around −55 dB and a bandwidth of 7.3 THz.

The authors observed two primary resonant frequencies at 6.67 and 8.16 THz. By varying the chemical potential from 0.3 to 0.7 eV, they were able to shift the operating range and improve the S_{21} parameter to below −50 dB. One of the most interesting insights from this research is the device's potential for dynamic control through electrostatic doping of graphene. By applying an external voltage, researchers can alter the chemical potential of graphene, effectively changing the filter's operating range on demand. This feature, combined with the filter's compact size and high performance, makes it a promising component for future integrated circuits in optical communication systems operating in the THz band. The ability to tune the filter's characteristics electrically opens the possibilities for adaptive and reconfigurable THz systems, which could be crucial for advancing technologies in areas such as high-speed wireless communications, security scanning, and medical imaging.

In article [29], the authors introduce the concept of graphene electromagnetic band gap (GEBG) structures and analyze their characteristics using equivalent circuit models. The study focuses on two main types of GEBG structures, those based on parallel planar waveguides and those based on coplanar waveguides (CPW). For parallel planar waveguide-based GEBG structures, the authors examine two configurations: ribbon arrays and concentric rings. In the ribbon array GEBG structure, they illustrate that the cutoff frequency is actively adjusted by modifying the chemical potential of graphene ribbons. Meanwhile, the concentric ring GEBG structure reveals that the resonant frequency can be modified by altering the sequence of chemical potentials across different rings. Both structures exhibit tunable electromagnetic wave propagation properties within the THz range. Additionally, the study investigates a spiral-shaped GEBG structure utilizing a coplanar waveguide, demonstrating the capability to switch between single-band and dual-band operation by repositioning the graphene material.

The authors also investigate the effect of cascading multiple spiral GEBG-CPW units, showing improved roll-off characteristics and broader bandwidth with an increase in the number of units. One of the most interesting insights from this study is the versatility and tunability of GEBG structures. By manipulating the chemical potential of graphene through electrostatic gating or chemical doping, these structures can achieve dynamic control over electromagnetic wave propagation. This feature opens possibilities for creating reconfigurable THz devices with applications in communications, nanoelectronics, and MMIC/RFIC technologies. The ability to suppress spurious passbands and switch between different operational modes (such as single-band to dual-band) demonstrates the potential for GEBG structures to enhance the performance and flexibility of various electromagnetic devices in the THz range.

A plasmonic band-stop reflecting filter [34] derived from a periodic graphene patch is presented and quantitatively analyzed using the Finite-Difference Time-Domain (FDTD) approach. The designed filter features a simple three-layer configuration consisting of graphene, a dielectric layer, and metal. Simulation outputs indicate that a prominent absorption peak is achievable, and by adjusting the chemical potential, the absorption spectrum can be flexibly tuned over an extensive frequency band. Unlike conventional active filters, this design eliminates the need for mechanical movement, making it tailored for different high-functionality implementations where stability is crucial. Additionally, its ultrathin structure minimizes both magnitude and mass. The filter is compatible with both TE and TM polarizations and can serve as an extremely sensitive sensor.

11.6 APPLICATION

Graphene as a material can be utilized to produce and emit THz radiation, potentially resulting in compact and tunable THz sources. Graphene-based detectors can detect and measure THz radiation, making them useful in imaging, spectroscopy, and sensing applications. Because of its tunable conductivity, graphene can modulate the amplitude or phase of THz waves, making it useful in communication and signal processing applications. Graphene may be utilized to make antennas that transmit and receive THz radiation, which has advantages in terms of tunability and flexibility. Graphene-based filters may selectively transmit or block specific THz frequencies, allowing for applications in spectral analysis and filtering. Graphene-based THz modulators and detectors allow for high-speed data transfer across small distances. THz communication systems can provide prominent levels of security since THz transmissions are difficult to intercept. This feature is quite appealing in defense applications where information confidentiality is of paramount importance. Graphene could play a crucial part in the development of next-generation wireless networks that operate in the THz range, allowing for substantially larger data rates and bandwidths. Graphene-based THz antennas have the adequate potential for deployment in 6G networks with data speeds of up to 1 Tbps. Graphene adds the advantage of tunability to the antennas, filters, and absorbers. This tunability improves the versatility of the devices. Skin cancer and other disorders detection can be done by using graphene-based sensors. Graphene-based sensors are well suited for the next-generation Internet of Things (IoT) technology named as Internet of Nano Things given their extremely smaller circuit size requirement. The graphene-based sensors can be used for sensing various body parameters. Researchers demonstrated their ability to detect non-melanoma skin cancer, glycosuria, and other biological samples as well. They can also be used for detecting faults in materials or layers of complicated structures. They find applications in non-destructive testing for manufacturing and smart factories. THz waves may penetrate through clothing, paper, cardboard, and other materials, which is ideal for security scanning in strategic places.

11.7 CONCLUSION

Graphene-based tunable THz devices are a huge step forward in solving the THz gap. This study lays the foundation for a new generation of highly efficient, adaptive, and multifunctional devices by leveraging graphene's exceptional properties. Continued improvements in material science, fabrication, and integration methods will expedite the adoption of graphene-based technologies, allowing them to reach their full potential across a wide range of sectors and applications. This chapter gives a thorough description of frequency-tunable graphene-based THz devices. The THz range, which extends from 0.1 to 10 THz, is between the microwave and infrared regimes. Graphene, a material derived from graphite, has adjustable characteristics making it suitable for THz applications. The chapter covers a variety of graphene-based THz devices, such as biosensors, antennas, absorbers, and filters. This chapter goes into graphene's unique features, which enable its usage in THz devices. Graphene's conductivity can be dynamically modulated through gate voltage or chemical doping. This adjustability enables the tuning of resonance frequency and impedance in graphene-based antennas, rendering them programmable. The chapter also investigates graphene-based absorbers, which can achieve high absorption rates across a wide frequency range in the THz domain. These absorbers frequently use metasurfaces and can be tailored for specialized applications such as biosensing. Graphene-based THz filters are also explored, emphasizing its potential to selectively transmit or block specific THz frequencies. These filters may be dynamically modified by altering the chemical potential of graphene, enabling flexible spectral analysis and filtering applications. One of the most intriguing aspects of this chapter is graphene's multifarious potential for tackling the issues of the so-called 'terahertz gap'. The ability to dynamically modify the characteristics of graphene-based devices through electrical or chemical methods offers exceptional versatility in THz applications. This tunability, coupled with graphene's unique electrical and optical properties, opens up the possibility of creating adaptive and reconfigurable THz devices. Such devices have the potential to revolutionize industries including high-speed wireless communications, security scanning, and medical imaging. This shows that graphene-based THz technology has the potential to significantly advance space communication and sensor capabilities.

REFERENCES

1. N. K. Maurya, S. Kumari, P. Pareek, G. Varshney, Highly-efficient tunable dipole-driven Yagi–Uda antenna with end-fire radiation for terahertz application, *Nano Communication Networks*, 38, 100480, 2023.
2. M. F. Ali, R. Bhattacharya, G. Varshney, Graphene-based tunable terahertz self-diplexing/MIMO-STAR antenna with pattern diversity, *Nano Communication Networks*, 30, 100378, 2021.
3. S. N. H. Sa'don, M. H. Jamaluddin, M. R. Kamarudin, F. Ahmad, Y. Yamada, K. Kamardin, I. H. Idris, N. Seman, Characterization of tunable graphene antenna, *AEU - International Journal of Electronics and Communications*, 118, 153170, 2020.
4. S. Singhal, Tetradecagonal ring shaped terahertz super wideband MIMO antenna, *Optik - International Journal for Light and Electron Optics*, 208, 164066, 2020.
5. Y. Jin, J. Choi, Gain-enhanced PEC reflector backed slot-loop antenna, *Journal of Electromagnetic Waves and Applications*, 34, 468–482, 2020.
6. U. Raj, M. K. Sharma, V. Singh, S. Javed, A. Sharma, Easily extendable four port MIMO antenna with improved isolation and wide bandwidth for THz applications, *Optik - International Journal for Light and Electron Optics*, 247, 167910, 2021.
7. M. Shalini, M. G. Madhan, Design and analysis of a dual-polarized graphene based microstrip patch antenna for terahertz applications, *Optik*, 194, 163050, 2019.

8. B. Smari, M. Labidi, F. Choubani, Mutual coupling reduction in metamaterial antenna for terahertz application, *Applied Physics A*, 125, 695, 2019.
9. N. K. Maurya, S. Kumari, P. Pareek, L. Singh, Graphene-based frequency agile isolation enhancement mechanism for MIMO antenna in terahertz regime, *Nano Communication Networks*, 35, 100436, 2023.
10. G. Yao, F. Ling, J. Yue, C. Luo, J. Ji, J. Yao, Dual-band tunable perfect metamaterial absorber in the THz range, *Optics Express*, 24, 1518–1527, 2016.
11. H. Yu, P. Xiang, Y. Zhu, S. Huang, X. Luo, Polarization insensitive broadband concentric-annular-strip octagonal terahertz graphene metamaterial absorber, *IEEE Photonics Journal*, 14, 4600906, 2022.
12. Y. Lv, W. Liu, J. Tian, R. Yang, Broadband terahertz metamaterial absorber and modulator based on hybrid graphene-gold pattern, *Physica E: Low-dimensional Systems and Nanostructures*, 140, 115142, 2022.
13. N. K. Maurya, J. Ghosh, Ultrathin tunable UWB metasurface absorber design by exciting plasmonic modes in bilayer graphene stack, *Diamond & Related Materials*, 145, 111065, 2024.
14. S. Zakir, R. M. H. Bilal, M. A. Naveed, M. A. Baqir, M. U. A. Khan, M. M. Ali, M. A. Saeed, M. Q. Mehmood, Y. Massoud, Polarization-insensitive, broadband, and tunable terahertz absorber using slotted-square graphene meta-rings, *IEEE Photonics Journal*, 15, 4600108, 2022.
15. M. Pan, J. Weng, B. Zhou, J. Su, H. Tang, B. Fan, L. Jiang, L. Lin, Y. Qiu, Terahertz bandwidth absorption metamaterial based on composite structure of single layer graphene, *Diamond and Related Materials*, 140, 110535, 2023.
16. N. Yi, R. Zong, J. Gong, R. Qian, Dynamically tunable broadband absorber with a single ultra-thin layer of graphene in the terahertz regime, *Materials Science in Semiconductor Processing*, 136, 106161, 2021.
17. M. A. Jabbar, M. A. Naveed, M. Zubair, M. Q. Mehmood, Y. Massoud, Wideband polarization insensitive tunable graphene-supported terahertz metamaterial absorber, *IEEE Photonics Journal*, 15, 2201908, 2023.
18. S. K. Ghosh, V. S. Yadav, S. Das, S. Bhattacharyya, Tunable graphene-based metasurface for polarization-independent broadband absorption in lower mid-infrared (MIR) range, *IEEE Transactions on Electromagnetic Compatibility*, 62, 346–354, 2020.
19. S. K. Tripathy, A. Gorai, T. M. Behera, R. Ghatak, A miniaturized dual band terahertz metamaterial-based absorber as a biosensor for non-melanoma skin cancer diagnostic, *Optik*, 316, 172048, 2024.
20. Y. I. Abdulkarim, F. Ö. Alkurt, H. N. Awl, F. F. Muhammad Sharif, M. Bakır, S. Dalgac, M. Karaaslan, H. Luo, An ultrathin and dual band metamaterial perfect absorber based on ZnSen for the polarization-independent in terahertz range, *Results in Physics*, 26, 104344, 2021.
21. H. Meng, L. Wang, G. Liu, X. Xue, Q. Lin, X. Zhai, Tunable graphene-based plasmonic multispectral and narrowband perfect metamaterial absorbers at the mid-infrared region, *Applied Optics*, 56, 6022–6027, 2017.
22. A. Verma, O. P. Meena, Dual-band polarization insensitive graphene-based reconfigurable absorber in a terahertz regime, *Applied Optics*, 63, 1934–1938, 2024.
23. D. Sahu, E. Panwar, R. Panwar, Graphene and copper constituted hybrid metasurface enhanced terahertz sensing for biomedical applications, *IEEE Sensors Letters*, 7, 4504004, 2023.
24. S. Jorwal, A. Dubey, R. Gupta, S. Agrawal, A tunable, dual band graphene terahertz absorber with equivalent circuit modeling, *Optical and Quantum Electronics*, 56, 1377, 2024.

25. G. C. Ram, N. K. Maurya, S. Yuvaraj, K. B. Sindhuri, D. R. Varma, D. G. Kumar, K. P. Vasavi, S. H. Rao, Graphene based THz filter design using open-loop resonator for space applications, *Diamond and Related Materials*, 151, 111820, 2025.
26. G. C. Ram, P. Sambaiah, S. Yuvaraj, M. V. Kartikeyan, Tunable band stop filter using graphene in terahertz frequency band, *AEU - International Journal of Electronics and Communications*, 144, 154047, 2022.
27. G. C. Rama, P. Samiyah, S. Yuvaraj, M. V. Kartikeyan, Graphene based tunable bandpass filter for terahertz spectroscopy of polymers, *Optik*, 268, 169792, 2022.
28. G. Melo, W. Castro, C. Oliveira, Tunable plasmonic band stop filter based in graphene nanoribbon, *Optik*, 261, 169100, 2022.
29. Y. Dong, P. Liu, W. Y. Yin, G. Li, B. Yi, Planar electromagnetic band-gap structure based on graphene, *Physica E*, 70, 176–182, 2015.
30. N. Kumar, S. K. Ghosh, S. Bhattacharyya, Thermally switchable metasurface for controlling transmission in the THz-gap, *Plasmonic*, 19, 1799–1807, 2024.
31. N. K. Maurya, P. Pareek, L. Singh, S. Kumari, J. Ghosh, Polarization-insensitive metasurface-biosensor for glycosuria sensing in terahertz regime, *IEEE Wireless Antenna and Microwave Symposium (WAMS)*, Visakhapatnam, India, 2024.
32. D. Samantaray, S. K. Ghosh, S. Bhattacharyya, Modified slotted patch antenna with metasurface as superstrate for dual-band applications, *IEEE Antennas and Wireless Propagation Letters*, 22, 109–113, 2023.
33. D. Samanta Ray, S. Bhattacharyya, A gain-enhanced slotted patch antenna using metasurface as superstrate configuration, *IEEE Transactions on Antennas and Propagation*, 68, 6548–6556, 2020.
34. J.-Y. Liu, T.-J. Huang, P.-K. Liu, Tunable polarization-independent terahertz band-stop filter based on graphene metasurface, *43rd International Conference on Infrared, Millimeter, and Terahertz Waves*, Nagoya, Japan, 2018.

12 Synergizing Terahertz Radiation and Surface Plasmons

Innovations in Filters and Sensing Applications

G. Challa Ram, S. Yuvaraj, and Naveen Kumar Maurya

12.1 INTRODUCTION

Over the last 15 years, millimetre-wave technology has undergone significant advancements, especially in its integration with 5G networks. This has led to substantial improvements in data transfer rates, reduced latency, and increased network capacity. Beyond telecommunications, millimetre-wave technology has also made inroads into diverse applications such as advanced imaging systems, radar technology, and emerging fields like autonomous vehicles and smart cities. Despite the benefits it offers, millimetre-wave technology has limitations in the context of future wireless communications. Concerns arise due to its limited range, vulnerability to obstacles, and the requirement for a denser infrastructure. The struggle of mmWave signals to penetrate obstacles results in reduced coverage and reliability. The growing need for increased bandwidth and faster data rates in modern communication system has become increasingly apparent in recent years. The demand for increasing wireless data rates is growing exponentially, leading to the anticipation that multi-gigabit-per-second links will become a reality in the next 5 years [1–3]. Within a communication system, the Shannon–Hartley theorem establishes a connection between the channel capacity (C), signal-to-noise ratio (SNR $= S/N$), and available bandwidth (BW), as expressed in equation (12.1). This theorem sets forth the theoretical maximum bit-rate achievable with an arbitrarily small bit-error rate (BER), given a specific average signal power (S) over a channel with bandwidth (BW) affected by additive white Gaussian noise (N):

$$C = \mathrm{BW}\log_2\left(1+\frac{S}{N}\right) \tag{12.1}$$

As indicated by equation (12.1), it is clear that channel capacity and, consequently, data rates can be increased by either expanding the bandwidth while maintaining

DOI: 10.1201/9781003599111-12

the same signal power (S) or increasing the given SNR with the same bandwidth. However, since there are limitations on increasing the signal power beyond a certain point due to interference and health safety concerns, researchers are actively investigating the high-frequency spectrum to meet the growing demand. At microwave frequencies, the available bandwidth for communication systems is restricted. Even in the millimetre band at 60 GHz, only 7 GHz of bandwidth is accessible, which is insufficient to meet the ever-growing demand for high data rates [4]. This serves as a motivation for researchers and scientists to explore the terahertz (THz) frequency bands for communication, as they are expected to address the need for high data rates owing to the significantly larger available bandwidth.

The THz frequency region, situated between millimetre waves and infrared wavelengths, covers a broad spectrum ranging from 100 GHz to 10 THz, showcasing immense potential for various applications [4]. In terms of wavelength, this range spans from 3 to 0.03 mm, as depicted in Figure 12.1. Often referred to as the THz gap, this frequency range signifies a technological frontier that is currently in the early stages of development [5]. The THz frequency range offers opportunities for applications in imaging, communication, and sensing. Due to its characteristics, THz radiation holds the capability to penetrate various materials, making it valuable for non-invasive imaging in medical diagnostics and security screening. Moreover, THz communication holds promise for high-speed and high-capacity wireless data transfer [6].

In comparison to traditional millimetre-wave links, THz communication links offer a suite of advantageous features. The primary advantage lies in the inherently broader usable bandwidth attributed to the higher carrier frequency of THz links. This characteristic enables these links to accommodate significantly more data, potentially meeting the exponentially growing demands for data intensive services such as augmented reality, uninterrupted video streaming with high quality, and other bandwidth-hungry applications [5].

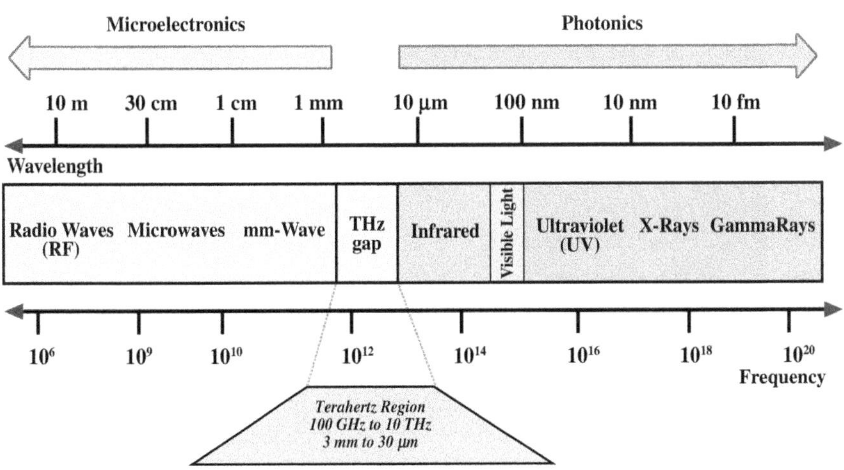

FIGURE 12.1 Electromagnetic spectrum.

Moreover, THz communication links exhibit promising prospects for ensuring secure communications. THz waves are able to interact with various molecular structures, and their ability to penetrate certain materials makes them potentially suitable for secure communication applications, adding an extra layer of confidentiality and security to wireless transmissions. These collective advantages position THz communication as a potentially transformative technology in the wireless communication landscape. It holds the promise of catering to the ever-increasing demands for higher data rates, ensuring more reliable and secure communication, and opening avenues for innovative applications across diverse sectors. Over the past two decades, the landscape of wireless networking has witnessed a remarkable surge in point-to-point and collective bandwidth needs.

Meeting these escalating demands necessitates a paradigm shift towards pioneering solutions. Advanced modulation techniques have emerged as a beacon of hope in this pursuit. These techniques not only promise higher point-to-point data rates but also offer the tantalizing prospect of improved frequency sharing within designated bands and enhanced frequency reuse within defined spatial parameters. However, there is an inherent recognition that these sophisticated methods, while powerful, have their limits. THz communication stands as an uncharted frontier, offering potential liberation from the confines of present-day strategies.

Ongoing research is dedicated to unlocking the complete capabilities of THz technology, aiming to enable transformative applications across different fields. Exploring the THz frequency range is a frontier of scientific and technological inquiry, offering the possibility of breakthroughs that could revolutionize industries and open up new opportunities in communication, imaging, and sensing. THz filters emerge as crucial components, playing a pivotal role in investigating and harnessing the capabilities of the THz frequency range. These filters hold profound importance across various domains due to their unique ability to manipulate and control electromagnetic waves within this frequency band. Their significance spans a multitude of applications, driving the relentless pursuit of innovation and advancement in scientific research, technology, and industry. At the forefront of the importance of THz filters lies their role in imaging technology. THz waves possess non-ionizing properties, allowing them to penetrate various materials without causing harm. The realm of security and surveillance also benefits from THz filters. Their utilization enables the development of highly advanced security systems capable of precise and non-invasive detection of concealed objects. The THz frequency band remains relatively unexplored but holds immense potential for high-speed wireless communication and data transfer. THz filters serve as the key components in efficiently utilizing this spectrum, addressing the ever-increasing demands for faster and more reliable wireless networks. By enabling the exploration of this underutilized frequency range, THz filters contribute to the development of a new era of communication systems, paving the way for faster, more efficient wireless connectivity. In addition to their impact on medical diagnostics, security, and communication, THz sensors find extensive applications in material analysis and quality control across various industries. Their precise characterization and analysis capabilities aid in quality assurance processes, particularly in pharmaceuticals, manufacturing, and food processing. THz sensors facilitate defect detection, composition assessment, and structural integrity evaluation, ensuring the quality and integrity of products. In spectroscopy and nanotechnology, these sensors serve as invaluable tools for studying molecular properties, interactions, and nanostructures. They enable researchers to delve deeper into molecular structures, driving advancements in materials science and nanotechnology and aiding the development of innovative materials and devices. Beyond immediate applications, the exploration and advancement of THz filters and sensors hold immense promise for future technological innovations. The development and understanding of these filters and sensors lay the foundation for potential breakthroughs in computing, imaging technologies, sensing devices, and beyond. Researchers and innovators aim to fully explore the capabilities of THz filters, foreseeing their pivotal role in shaping future technologies and discoveries.

12.2 TERAHERTZ APPLICATIONS

Figure 12.2 portrays a schematic overview encompassing a wide spectrum of applications enabled within the THz frequency range. Owing to their shorter wavelengths, expanded bandwidth, and heightened interaction with atmospheric elements, THz integrated circuits (ICs) have found extensive utility across diverse engineering and scientific domains. These applications span various fields including biomedical advancements [7,8], non-destructive technologies [9], cellular communications such as 5G and beyond [10], sub-millimetre wave astronomy [11], and military and defence applications for security, imaging, and sensing [12].

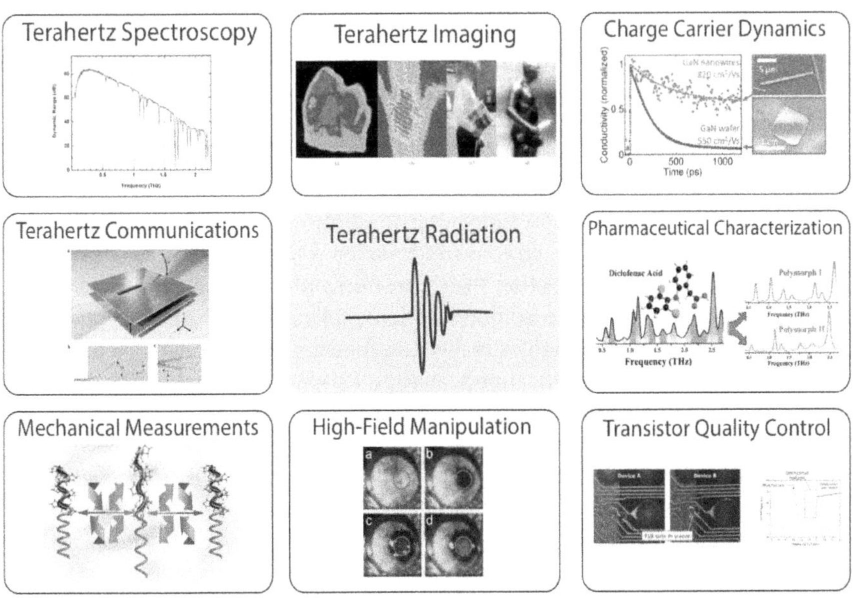

FIGURE 12.2 Applications of THz waves [13].

12.2.1 Medical and Security Applications

The medical utilization of THz technology primarily capitalizes on the human body's composition, which is predominantly composed of water molecules exhibiting resonances within the THz band. Notably, THz waves offer a safe alternative to ionizing diagnostic imaging, such as X-rays, for detecting conditions like cancer and tooth decay. Unlike X-rays, which pose risks due to their ionizing nature, THz waves are non-ionizing and harmless. The substantial absorption of THz waves by water is harnessed for identifying early-stage cancers, such as breast and skin cancer. Increased water content in diseased tissues absorbs THz radiation, facilitating the identification of abnormal tissues. Multiple studies highlight the harmful effects of ionizing radiation, like X-rays, attributing 1% of diagnosed cancers to such medical imaging. Transitioning to THz medical diagnosis could potentially mitigate this risk. THz technology also plays a pivotal role in military and civilian anti-terrorism endeavours. The oscillation frequency of THz waves aligns with the oscillation frequencies of atomic-level entities like molecules, atoms, and DNA, enabling their detection through THz imaging. This imaging method finds extensive use in security applications, facilitating the detection of concealed weapons and metals, as well as enhancing visibility through fog for identifying potential threats and adversaries.

12.2.2 Wireless Communications

THz technology has emerged as a promising frontier in wireless communications, offering unparallelled capabilities that could revolutionize the way data is transmitted and accessed wirelessly. The unique attributes of THz waves present opportunities and challenges in various facets of communication systems. One of the most significant advantages of THz technology lies in its capability to enable exceptionally high-speed data transmission. With its extraordinarily wide bandwidth, THz waves hold the potential to achieve data rates several times faster than current wireless systems. This high data transfer capability positions THz technology as a key player in future wireless communication standard, including the evolution towards 6G and beyond.

However, THz waves exhibit limitations in terms of their range and penetration capabilities. They are highly prone to attenuation and absorption by atmospheric gases and water vapour, restricting their transmission range and making them suitable primarily for short-range communication applications. Their susceptibility to obstacles and high absorption rates in the atmosphere pose challenges for long-range and non-line-of-sight communication scenarios. Despite these challenges, THz technology presents opportunities in specific applications such as wireless personal area networks (WPANs), where high-speed communication between devices in close proximity is essential. The high directionality of THz waves is advantageous for secure point-to-point communications, although it requires an unobstructed line of sight. Additionally, advancements in chip-to-chip wireless communication benefit from the establishment of low-loss and high-speed interconnections within the sub-THz frequency range. These developments enable data transfer rates reaching tens or hundreds of gigabits per second or even higher.

12.2.3 THz Sensing

Sensors play a vital role in identifying elements in diverse environments, be it solid, liquid, or gas, making them valuable in agriculture, environmental monitoring, and industrial quality control. In agriculture, they optimize plant growth by promptly conveying soil moisture levels. In environmental contexts, sensors detect contaminants in water for assessing water quality and treatment. In industries, temperature and level sensing are critical for efficient production and safety. In the medical domain, sensors are crucial for precise drug dosage and monitoring patient's physiological parameters like respiration rate, heartbeat, and blood pressure. With the rise of chronic illnesses globally, including diabetes, hypertension, and various cancers, there's a need for real-time and cost-effective diagnostic tools. Sensors are essential in various industries to detect electrochemicals and gases. Efficient sensing of analytes proves significant across nearly every domain of daily life.

With the emergence of communication standards like 5G and beyond, enabling machine-to-machine (M2M) communication and the proliferation of the Internet of Things (IoT), it's predicted that within a decade, a staggering 41.6 billion devices will be interconnected [14]. Sensors will hold a pivotal role in various applications such as automation and point-of-care scenarios. This broad spectrum of sensor applications has led to a significant surge in research across multiple domains—mechanical, electrical, optical, ultraviolet, microwave, and THz. The aim is to realize and enhance sensor capabilities for diverse applications. The deployment of billions of sensors necessitates consideration of crucial factors: design and installation costs, overall performance, and their impact on human health. Consequently, researchers across domains are driven to develop diverse sensor topologies that meet these requirements while maintaining or surpassing the performance standards set by the existing sensors.

In recent decades, microwave and THz sensors have garnered significant attention from researchers. These waves, falling within the microwave and THz spectra, possess non-ionizing properties and low power, ensuring their safety and harmlessness to biological organisms [15]. Additionally, the size of many molecules aligns with the wavelengths within the THz spectrum, making them ideal for developing biomarkers. These frequency ranges offer wireless sensing capabilities, delivering precise, non-intrusive, and contactless features. Moreover, the implementation of sensing devices is facilitated by matured and cost-effective fabrication technologies. Microwave sensors boast simple design structures, lightweight prototypes, straightforward fabrication processes, and instantaneous measurement capabilities. Their contactless nature renders them particularly suitable for harsh environments where direct human intervention isn't feasible.

In the current trend of interdisciplinary research, there's a pronounced emphasis on leveraging integrated electronic circuits and system platforms within non-electrical domains, particularly in biomedical and biochemical applications. Microwave and THz sensors, with their remarkable integration capabilities, hold significant value in this context. Filters serve as crucial components in numerous designs of microwave and THz sensors. As a result, their profile and characteristics directly influence the overall performance of sensing devices, impacting factors like miniaturization, sensitivity, and the detection threshold.

12.3 FUNDAMENTALS OF PLASMONICS

Plasmonics represents a novel domain within physics and stands as an actively investigated field within optics. It delves into the optical interactions occurring with the free conduction electrons in materials, resulting in confinement at sub-wavelength levels and an amplified intensity of the electric field on a nanoscale dimension [16]. When incident light, in the form of photons, encounters a plasmonic material, the free electrons within the material undergo oscillation, forming plasmons, quantum units of plasma oscillations. These plasmons firmly adhere to the material's surface upon interaction with light, constituting surface plasmons (SPs). SPs exhibit significant properties like notably amplified electric field strength, sensitivity to local surroundings, and the confinement of light within a restricted volume [17].

Under specific conditions, the energy carried by incident photons triggers the excitation of free electrons within metals, culminating in the formation of SPs. When the momentum of these plasmons aligns with that of light, there occurs a precise transfer of energy at a particular resonance wavelength of light. This phenomenon results in exceptional light scattering and robust light absorption, as depicted in Figure 12.3, known as surface plasmon resonance (SPR) [18]. Theoretical examination of photon

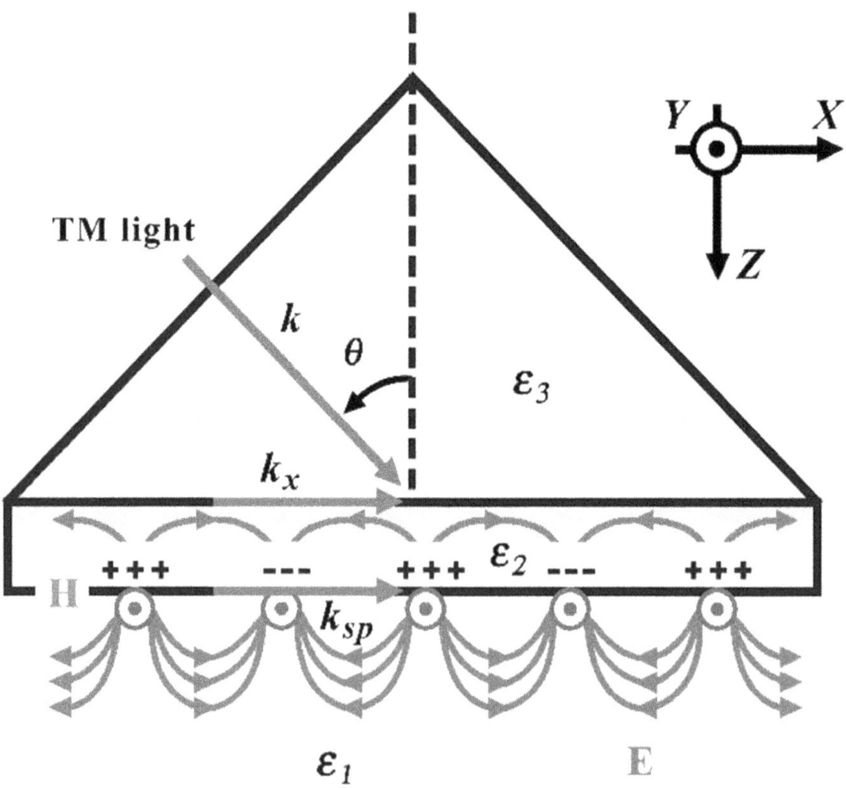

FIGURE 12.3 Schematic representation of surface plasmon resonance [18].

interaction with a metal substrate is elucidated through the utilization of Maxwell's equations [19]. This theoretical framework yields a solution represented by equation (12.2), describing the resonance condition.

$$\sqrt{\varepsilon_3}\sin\theta = \frac{\omega}{c}\sqrt{\frac{\varepsilon_1\varepsilon_2}{\varepsilon_1+\varepsilon_2}} \tag{12.2}$$

where c represents the speed of light in a vacuum, ω signifies the angular frequency of the incident photon, and ε_1, ε_2, and ε_3 denote the permittivity of the plasmonic material, the adjacent dielectric, and the prism, respectively, as illustrated in Figure 12.3.

12.3.1 Surface Plasmon Polaritons

SP polaritons are EM waves propagate along the interface between a conductor (such as a metal) and a dielectric material (such as air or glass). They are collective oscillations of free electrons excited by incident light, leading to a coupling between photons and surface charges. These waves are confined to the interface and can exhibit properties different from those of bulk electromagnetic waves. SP polaritons play a crucial role in nanophotonics, enabling the manipulation of light at the nanoscale.

Their ability to confine light beyond the diffraction limit makes them promising for applications in sensing, imaging, information processing, and enhancing light–matter interactions in various devices. Understanding and harnessing the properties of SP polaritons have opened doors to innovative technologies, allowing for the development of miniaturized photonic devices, and enhanced light-emitting devices. The intricate interplay between light and matter at the nanoscale continues to fuel research in this fascinating field.

SP polaritons represent synchronized oscillations involving the metal's conduction electrons and the electromagnetic wave. These phenomena occur naturally at optical frequencies, manifesting on the surface or at the junction between two materials where there's a transition in the signs of dielectric permittivity. The utilization of these SP polaritons at flat metal-dielectric interfaces has been instrumental in creating sub-wavelength plasmonic waveguides and components. Therefore, exploring this area further is pertinent, and it is extensively discussed below.

12.3.2 Surface Plasmon Polaritons at Metal-Dielectric Interface

SP polaritons emerge within metal-dielectric configurations and are defined by specific longitudinal and transverse electric and magnetic field components. The electromagnetic wave's magnitude diminishes exponentially in the direction perpendicular to the interface, as depicted in Figure 12.4.

Consequently, SPs represent a distinct type of propagating electromagnetic surface mode confined to the surface. The principal aim of SPs revolves around the concentration and manipulation of light at scales smaller than the wavelength, surpassing the conventional diffraction limitations in optics [20]. Within this section, an effect is made to offer deeper comprehension of SP propagation along the metal-dielectric interface by presenting a classical derivation of SP based on the established literature [21]. The Drude model's simplified expression for the ideal metal is provided as follows [21]:

$$\varepsilon_a(\omega) = 1 - \frac{\omega_p^2}{\omega^2} \tag{12.3}$$

$$\omega_p = \sqrt{\frac{N_e e^2}{m\varepsilon_0}} \tag{12.4}$$

Here, ε_a (ω) represents the frequency-dependent dielectric function of the metal, ω_p stands for the metal's plasma frequency, N_e is the electronic charge density, m is the effective mass, and ω denotes the operating frequency of the electromagnetic wave. Figure 12.4 illustrates a schematic of a flat metal-dielectric interface. The model includes two semi-infinite media, a metal and a dielectric, separated by a planar interface in the XY plane. The SP wave will be propagating along the metal-dielectric interface. This arrangement serves as the basis for deriving mathematical expressions for the supported electric (E) and magnetic (H) field components.

Figure 12.5 shows the schematic of propagation of surface wave at metal-dielectric interface. The two media are shown in the YZ plane, and the SP wave is propagating in the Y direction. Subsequently, it is necessary to define the circumstances enabling the propagation of an EM wave along the surface, specifically at $Z=0$, with field components attenuated in both the positive and negative z-directions. For an SPP to propagate across the interface, there must be an electric field component normal to the surface due to the presence of surface polarization charges. By selecting the

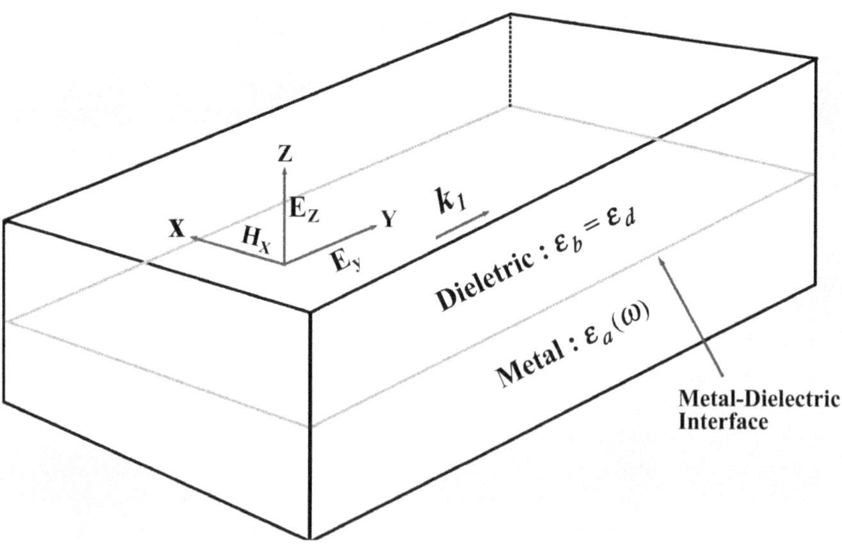

FIGURE 12.4 Surface plasmon polariton modes in two semi-infinite media [20].

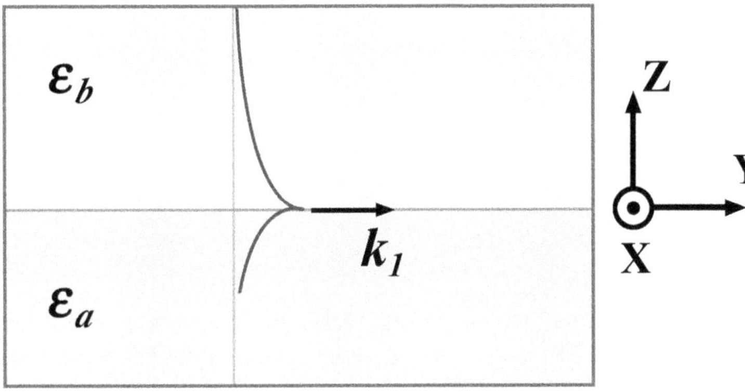

FIGURE 12.5 Surface wave propagation at the metal-dielectric interface [21].

Y-axis as the propagation direction, the electric and magnetic fields can be expressed as follows:

$$\begin{aligned} \vec{E}_1(y,z,t) &= \left(0, \overrightarrow{E_y}, \overrightarrow{E_{1z}}\right) e^{j(k_1 y - \omega t)} e^{+\alpha_a z} \quad z<0 \\ \vec{H}_1(y,z,t) &= \left(\overrightarrow{H_x}, 0, 0\right) e^{j(k_1 y - \omega t)} e^{+\alpha_a z} \quad z<0 \\ \vec{E}_2(y,z,t) &= \left(0, \overrightarrow{E_y}, \overrightarrow{E_{2z}}\right) e^{j(k_1 y - \omega t)} e^{-\alpha_b z} \quad z<0 \end{aligned} \tag{12.5}$$

Here, subscripts 1 and 2 denote the two distinct media, with k_1 representing the real propagation wave number. α_a and α_b represent positive real values, signifying the exponential decrease of the fields as they move away from the interface at $Z=0$. In this boundary-value problem, equation (12.5) must adhere to Maxwell's equations and satisfy the boundary conditions established at the interface between the two materials.

When there are no external sources, and when dealing with non-magnetic materials, Maxwell's equations are represented by the following forms:

$$\nabla \cdot \left(\varepsilon_0 \varepsilon_i \overrightarrow{E_i}\right) = 0 \tag{12.6}$$

$$\nabla \cdot \left(\mu_o \overrightarrow{H_i}\right) = 0 \tag{12.7}$$

$$\nabla \times \left(\overrightarrow{E_i}\right) = -\mu_i \frac{\partial \overrightarrow{H_i}}{\partial t} \tag{12.8}$$

$$\nabla \times \left(\overrightarrow{H_i}\right) = \varepsilon_0 \varepsilon_i \frac{\partial \overrightarrow{E_i}}{\partial t} \tag{12.9}$$

The continuity of the displacement field's normal component yields:

$$\varepsilon_a E_{1z} = \varepsilon_b E_{2z} \tag{12.10}$$

When applying the first Maxwell equation to the electric fields in (12.11), one derives the expressions for the two distinct media:

$$\begin{aligned} jk_1E_y + \alpha_a E_{1z} &= 0 \quad z<0 \\ jk_1E_y - \alpha_b E_{2z} &= 0 \quad z>0 \end{aligned} \tag{12.11}$$

This establishes the connection

$$\alpha_a E_{1z} + \alpha_b E_{2z} = 0 \tag{12.12}$$

The fusion of equations (12.10) and (12.12) yields the compatibility condition determining the presence of SPs at the interface, represented by the following equation:

$$\frac{\varepsilon_a}{\alpha_a} + \frac{\varepsilon_b}{\alpha_b} = 0 \tag{12.13}$$

Having previously addressed the boundary conditions and the requirement of field divergence nullification, we now turn our attention to the consideration of Maxwell's equations in conjunction with the curl operation. By applying the curl operation to equation (12.9) and substituting using (12.8) to eliminate the magnetic field (H)

$$\begin{aligned} \nabla\times\nabla\times\left(\vec{E_i}\right) &= \frac{\varepsilon_i}{c^2}\frac{\partial^2 \vec{E_i}}{\partial t^2} \\ \nabla^2\left(\vec{E_i}\right) &= \frac{\varepsilon_i}{c^2}\frac{\partial^2 \vec{E_i}}{\partial t^2} \end{aligned} \tag{12.14}$$

Utilizing the electric fields described in equations (12.5), along with equation (12.14), leads us to derive equation (12.15), which is presented as follows:

$$k_1^2 = \alpha_a^2 + \frac{\varepsilon_a}{c^2}\omega^2 \quad z<0 \tag{12.15}$$

$$k_1^2 = \alpha_b^2 + \frac{\varepsilon_b}{c^2}\omega^2 \quad z>0 \tag{12.16}$$

By multiplying equation (12.15) by $1/\varepsilon_a^2$ and equation (12.16) by $1/\varepsilon_b^2$, then subtracting the results while considering (12.13), we establish the compatibility relation for SP.

$$k_1^2 = \frac{\omega^2}{c^2}\frac{\varepsilon_a\varepsilon_b}{\varepsilon_a+\varepsilon_b} \tag{12.17}$$

For the metal-dielectric interface, real values of $k_1(\omega)$ are attainable for frequencies within the range where $0<\omega<\omega_{sp}$ (where both the numerator and denominator are negative) and within the range where $\omega_p>\omega$ (where both the numerator and denominator are positive).

ω_{sp} is given by

$$\omega_{sp} = \frac{\omega_p}{\sqrt{\varepsilon_d+1}} \tag{12.18}$$

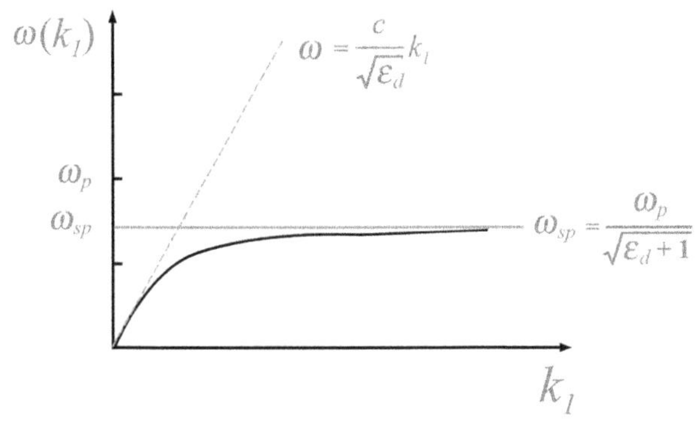

FIGURE 12.6 Dispersion curve of the SP polaritons observed at the metal dielectric interface [21].

In this specific scenario involving the interface between a metal and a dielectric with relative permittivity ε_d, the dielectric functions described in (12.17) transform into the following equation:

$$k_1^2 = \frac{\omega^2}{c^2}\varepsilon_d \frac{\omega^2 - \omega_p^2}{(\varepsilon_d + 1)\omega^2 - \omega_p^2} \tag{12.19}$$

When solving the identical equation in terms of k_1^2, it provides the expression for the dispersion curve of the interface-localized SP.

$$\omega^2 = \frac{1}{2\varepsilon_d}\left[\varepsilon_d\omega_p^2 + (\varepsilon_d + 1)c^2k_1^2 - \sqrt{\varepsilon_d^2\omega_p^4 + 2\varepsilon_d(\varepsilon_d - 1)\omega_p^2c^2k_1^2 + (\varepsilon_d + 1)^2 c^4k_1^4}\right] \tag{12.20}$$

The dispersion curve for the SP is depicted in Figure 12.6. The dashed line denotes the light line $\omega = ck_1 / \sqrt{\varepsilon_d}$. It's evident that the curve lies below the asymptotic line defined by ω_{sp}. Consequently, confined electromagnetic wave propagation is feasible for $\omega < \omega_{sp}$. The field remains highly confined and localized in close proximity to the interface. This SP wave moves with a wavelength (λ_{sp}) smaller than the wavelength in free space (λ_0), allowing for sub-wavelength confinement. There is no propagation possible within $\omega_{sp} \leq \omega \leq \omega_p$ and no localization occurs around the surface for $\omega > \omega_P$ [21].

12.3.3 Surface Plasmon Polaritons in Graphene

The THz frequency range stands out as an ideal choice for a wide range of applications owing to its compact nature and expansive bandwidth capabilities. A critical aspect for successful transmission in the THz spectrum is achieving negative permittivity. Graphene, among various materials, possesses the ability to exhibit negative permittivity, enabling the propagation of THz waves. At higher frequencies, conventional metals lose their efficacy due to their larger dimensions, making graphene a superior alternative, specifically within the THz frequency band.

FIGURE 12.7 Hexagonal honeycomb lattice structure of graphene.

Unlike ordinary metals, noble metals aid in the propagation of SPs. The negative permittivity of the material facilitates the transmission of surface plasmonic waves, which are essentially free electron oscillations occurring at the interface between a dielectric and a metal. These SP polaritons find utility in high-performance optical circuits by manipulating electromagnetic waves at an atomic scale, offering precision and efficiency. Graphene, a two-dimensional (2D) sheet as shown in Figure 12.7, presents numerous advantages over traditional metals. Its atomic-scale thickness sets it apart, being thinner than any metallic sheet available. Graphene's unique properties, including high conductivity, electrical characteristics, and flexibility, have made it a material of choice across various applications.

Serving as an interface between the metal and dielectric layers, the graphene sheet's negative permittivity greatly improves the propagation feasibility of SPs. This attribute enables potentially low-loss SPs, unlocking a myriad of possibilities in THz electronics. Furthermore, graphene's exceptional conductivity, stemming from its gapless electrical band structure, enables its conductivity to be influenced by alterations in the input voltage or chemical potential. This unique property further enhances the adaptability and controllability of graphene-based electronic systems, making it a cornerstone in futuristic applications.

The conductivity of graphene (σ) can be given as [22]

$$\sigma = \begin{bmatrix} \sigma_g & \sigma_h \\ \sigma_h & \sigma_g \end{bmatrix} \tag{12.21}$$

σ_g and σ_h in (12.21) are given by Kubo's theorem, where σ_g is given by [22]

$$\sigma_g = \sigma_{g,\text{inter}} + \sigma_{g,\text{intra}}$$

$$\sigma_{\text{g,intra}} = -j\frac{e^2 k_B T}{\pi\hbar^2\left(\omega + j\tau^{-1}\right)}\left(\frac{\mu_c}{k_B T} + 2\ln\left(e^{\frac{-\mu_c}{k_B T}} + 1\right)\right) \tag{12.22}$$

$$\sigma_{g,\text{inter}} = -j\frac{e^2}{4\pi\hbar}\ln\left(\frac{2\mu_c - (\omega - 2j\tau)\hbar}{2\mu_c + (\omega - 2j\tau)\hbar}\right)$$

where e is the electron charge, k_B is the Boltzmann constant, T is the temperature, ω is the angular frequency, τ is the relaxation time, and μ_c is chemical potential. The graphene conductivity at room temperature

$$\sigma = \frac{je^2\mu_c\tau}{\pi\hbar^2\left(\omega + j\tau^{-1}\right)} \tag{12.23}$$

The permittivity value can be given by [22]

$$\varepsilon_g = \left(\frac{\sigma_{g,i}}{\omega\Delta} + \epsilon_0\right) - j\left(\frac{\sigma_{g,r}}{\omega\Delta}\right) \tag{12.24}$$

The chemical potential (μ_c) is expressed as follows:

$$\mu_c = \hbar\upsilon\sqrt{\frac{\pi\varepsilon_r\varepsilon_0 V_b}{et}} \tag{12.25}$$

where υ is the Fermi velocity, t is the substrate thickness, and V_b is the applied voltage.

12.3.4 Electromagnetic Surface Modes at THz and Microwave Frequency

The previous discussion on SPs and their attributes positions them as potential candidates for developing sub-wavelength filters and various chemical and biosensors operating within visible and optical frequencies. However, these features are confined solely to the optical realm and aren't achievable at lower frequencies. This limitation arises because, at these lower frequencies, the dielectric function of metals is predominantly dictated by their electrical conductivity (due to considerably high dielectric permittivity values). Consequently, the penetration depth into the metal is significantly smaller compared to the wavelength, causing it to behave similar to a perfect electrical conductor (PEC). As a result, a planar metal-dielectric interface fails to support SP modes within this frequency regime [23–25]. Nonetheless, researchers persisted in exploring this domain, leading to the discovery that SP-like modes can be achieved at lower frequency ranges, such as THz/microwave, through periodic corrugation of the metal surface using 1D grooves or 2D holes or dimples [26]. In 2004, Pendry et al. analytically demonstrated that metallic surfaces perforated under specific conditions (with groove/hole width and period considerably smaller than the operating wavelength) can uphold bound electromagnetic surface modes even under perfect electric conduction [26]. As this sustained mode mimics the behaviour of SPs in optical realms, it was termed spoof surface plasmon (SSP) mode. Subsequent experimental validations at THz and microwave frequencies employing different indentation geometries supported this notion [27]. These bounded EM modes' guiding properties are contingent on the structural design parameters, hence earning the designation of SSP polariton modes. The following subsection provides a mathematical analysis delving into the origin of SSP modes and offers an analytical derivation of their dispersion relation.

12.3.5 SSPs in One-Dimensional Periodic Corrugated Metallic Structures

In this section, we delve into the SSP mode sustained by the 1D periodically corrugated metallic structure, accompanied by mathematical formulations. As illustrated in Figure 12.8, the groove dimensions are characterized by height h, width d, and period p. The periodicity extends along the y-direction. Moreover, we derive the dispersion relation of the SSP mode contingent on the geometry. In the ideal scenario of perfect conduction, the following equation [28] provides an approximate solution to the dispersion relation for the SSPs:

$$k_y = \frac{\omega}{c}\left[1+\left(\frac{d}{p}\right)^2 \tan^2\left(\frac{\omega h}{c}\right)\right]^{\frac{1}{2}} \tag{12.26}$$

In this context, k_y represents the propagation constant for the SSP mode, where d stands for the width of the groove, p denotes the periodicity, and h signifies the height of the grooves. Equation (12.26) illustrates that the dispersion curve of the SSP relies on these specific parameters of the grooved configuration.

Notably, when either d or h equals zero, it aligns with a flat surface carrying a propagation wave vector of $k_{o}=\omega/c$. However, for non-zero values of d and h, k_y surpasses k_o, causing the dispersion diagram to diverge from the light line, k_o. This deviation results in a non-radiative zone featuring reduced phase velocity and group

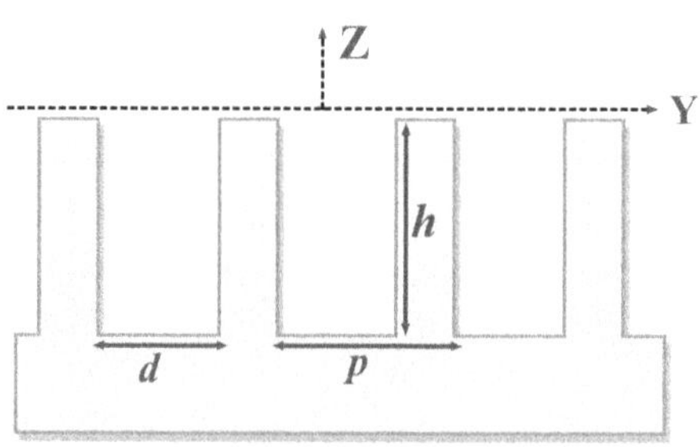

FIGURE 12.8 1D corrugated metallic surface.

velocity, confining the wave to the surface. This observation highlights the diminished group velocity of the SSP wave within the guiding structure, as delineated by the following equation:

$$v_g = \frac{d\omega}{dk_y} = \frac{c}{Q + \dfrac{2d^2 h\omega}{p^2 cQ}\tan\left(\dfrac{\omega h}{c}\right)\sec^2\left(\dfrac{\omega h}{c}\right)} \tag{12.27}$$

where Q is given by

$$Q = \sqrt{\frac{d^2}{p^2}\tan^2\left(\frac{\omega h}{c}\right) + 1} \tag{12.28}$$

Therefore, altering the geometrical parameters of the SSP enables the customization of its dispersion characteristics, consequently enhancing its ability to confine electromagnetic waves as needed. This concept empowers designers to confine EM waves within the guiding structure, thereby reducing instances of crosstalk and mutual coupling.

Additionally, Hibbins et al. [27] conducted experimental characterization on a design involving 2D holes perforated in a metal surface, confirming the support of SP-like surface modes by the corrugated structured metal surface. Expanding on this, a configuration of domino arrays using 3D non-planar metal blocks was presented in [29]. Subsequently, Shen et al. applied this principle to a flexible ultrathin dielectric film employing a comb-shaped structure [30]. They observed the support of SPs even with curved SSP cells. Moreover, [31] reported the tight confinement of the EM field on an ultrathin metal film at a sub-wavelength scale. Following these discoveries, numerous studies have emerged, focusing on SSP-based transmission lines and their relevance to THz/microwave circuits.

12.4 LITERATURE SURVEY

THz filters, encompassing both bandpass and bandstop filters, are pivotal within the THz frequency range, wielding immense significance across scientific, technological, and practical domains. It is evident that the effectiveness of the wireless communication system heavily relies on the performance of the filters [32]. These filters serve to isolate a specific range of RF/microwave signals within designated spectral boundaries while excluding undesired and extraneous frequency signals beyond that scope [33]. Their pivotal function in shaping the wireless communication system's efficiency has generated substantial interest among numerous researchers. Ongoing investigations delve into various aspects of BPFs and BSFs designed using the SPs concept.

12.4.1 Surface Plasmon-Based Filters

SP polariton-based filters are essential components in the field of photonics and plasmonics, offering unprecedented capabilities for manipulating light at nanometre scales. One key significance of SP-based filters lies in their ability to integrate optical functionalities on a sub-wavelength level [34]. By exploiting the interaction between light and SPs, these filters enable the development of compact and efficient photonic devices that can be seamlessly integrated into nanophotonic circuits. This miniaturization is crucial for advancing technologies where space and size constraints are critical factors.

Moreover, SP-based filters facilitate enhanced light–matter interactions by concentrating electromagnetic fields at the metal-dielectric interface. This concentration enhances the interaction between light and materials, enabling precise control over optical properties. The ability to confine light at the nanoscale allows for the manipulation of light in ways that were previously challenging or impossible with conventional optics, opening doors to new opportunities in optoelectronics and sensing [35].

Tuneable and selective filtering capabilities are another key advantage of SP-based filters. These filters can be engineered to exhibit tuneable spectral responses by adjusting parameters such as material compositions, geometries, or incident angles. This tunability allows for tailoring the filter's spectral characteristics to specific applications, including wavelength-selective filtering and spectral shaping. The ability to tune these filters offers flexibility and adaptability, making them suitable for optical filtering needs across different industries.

12.4.2 Graphene-Based Filters

SP polaritons denote EM waves travelling along a conductor's surface, manifested as collective electron oscillations resonating with an incident electromagnetic wave [34]. This phenomenon facilitates light coupling within sub-wavelength dimensions, enabling the confinement of light in extremely compact devices [35]. Noble metal-based plasmonic devices have provided effective solutions across scientific and commercial domains in recent decades [36]. However, the optical characteristics of noble metals lack easy adjustability, hindering dynamic control of device properties [37]. Achieving efficient, dynamically controllable plasmonic devices is pivotal as it could significantly broaden the scope of plasmonic applications.

Graphene, a sheet of carbon atoms tightly arranged in a two-dimensional honeycomb lattice [38], has showcased exceptional properties in light–matter interactions, high optical damage thresholds, and plasmons characterized by their ultra-high confinement [39]. Its linear dispersion and zero band gap nature, along with minimal free electrons available under zero doping conditions, position it as an ideal candidate for crafting dynamic and highly adaptable photonic and plasmonic devices. The unique capabilities of SPs in graphene, including tight field localization, extensive propagation lengths, and adaptable characteristics, have spurred the exploration of various excitation methods [40].

Numerous graphene-based plasmonic devices such as detectors [41], modulators [42], switches [43], encoders [44], logic gates [45], analogue to digital converters [46], optical polarizers [47], filters, and demultiplexers [48] have gained attention from various research groups. These devices serve as foundational elements for the emerging era of integrated optical circuits. Among these advancements, THz filters hold significant importance for mid-infrared wavelengths [49] and have witnessed diverse designs reported thus far [50].

Researchers have directed their efforts towards both theoretical and experimental development of THz filters, exploring conventional structures like metallic hole arrays [51], photonic crystal fibres and waveguides [52], and frequency selective surfaces [53]. However, these structures often lack compatibility with integrated THz circuits and involve intricate fabrication processes, constraining their practical usability. In contrast, graphene plasmons exhibit strong field confinement within the THz spectral range, holding the potential for superior performance in graphene-based plasmonic filters. A distinguishing factor lies in their tunability: traditional filter parameters prove challenging to modify once fabricated; while owing to graphene's unique structure and electron flexibility, graphene-based filters offer various techniques for tuning. These methods leverage the distinctive energy band structure of graphene, involving variations in substrate, the number of graphene layers, and the application of gate voltage between the graphene layer and substrate [54]. Such approaches enable the manipulation of graphene properties to change variation in filter response.

In recent times, there has been a surge in proposals concerning filters based on graphene. Weisu et al. introduced a compact plasmonic filter using a graphene rectangular ring resonator featuring double narrow gaps. Through simulations employing the FDTD method, it was observed that the transmission spectra of the filter could be influenced by several factors like physical dimensions [55]. Mengdan Kong et al. proposed a novel THz tuneable filter using graphene-based three-section L resonators [32]. The L-shaped resonator was analyzed with Z matrix method to achieve bandstop properties. The integration of graphene extended the filter's operating frequency range into the THz spectrum. Tuning was achieved by varying chemical potential. In [33], Victor et al. proposed a THz electromagnetic filter using graphene-based elements on a dielectric substrate. The device consists of a square lattice with graphene crosses that interact and resonate with dipole plasmonic modes at different frequencies, creating a transparency window between these resonances. This filtering effect, resembling Fano resonance, operates at 2.3 THz, with reflection and transmission coefficients of −18.6 and −0.92 dB, respectively. The transmission peak's position can be tuned by varying the graphene's chemical potential.

In [56], a graphene-based high pass filter for the THz band was proposed and simulated. The design utilizes a transmission line model with short-circuit stubs and employs the optimum distributed high pass filter method to achieve 0.1 dB bandpass ripple. The simulation results illustrated the insertion loss of less than 1.5 dB and a constant group delay of 0.16 ps. In [57], a micro-machined metamaterial graphene-based THz filter was presented. It utilizes metallic rings with graphene stripes in gaps to achieve tuneable characteristics. The results demonstrated a 40% tuning range by altering graphene conductivity. Notably, the device's active area occupies only 0.1% of the total unit cell area, enhancing speed during electrostatic actuation. Amin Ghahremani et al. proposed planar low-pass filters for the THz band using open stub configurations [58]. The filters were analyzed using transmission line model, where propagation constants and characteristic impedances were determined. A fifth-degree low-pass filter designed with this method achieved less than 2 dB insertion loss, a 17 dB/THz roll-off rate, and return loss exceeding 10 dB, with a constant group delay of 0.3 ps. Full-wave simulations confirmed these results, and the paper discussed graphene's chemical potential variations for predicting filter responses.

Wei Su et al. proposed an ultra-compact four-port splitting filter based on a ring resonator using graphene [59]. By varying the chemical potential value, the resonant frequency of the design can be tuned. This simple graphene-based structure, comprising graphene nanoribbon waveguides and a resonator, offers both splitting and filtering capabilities suitable for highly integrated THz optical circuits [60]. In [61], a configurable plasmonic filter waveguide employing an indium antimonide activated through a graphene layer configuration was presented. In [62], Naghizade Saleh et al. introduced a bandstop filter operating in the mid-infrared spectrum, employing a graphene-on-insulator (GOI) configuration. In [63], a tuneable bandstop filter operating in the mid-infrared range with a periodic structure utilizing bilayer graphene nanoribbons was proposed. Leveraging graphene's distinctive tuneable Fermi level, the filter's resonance wavelength demonstrated a broad tunability range from 10,020 to 15,100 nm, modulated by an applied voltage varying between 1 and 5 V. In [64], a graphene filter featuring a tuneable flat-top band-stop response, constructed with graphene ribbon resonators embedded within a graphene waveguide in a cascaded setup, was proposed. The literature on various graphene-based filter designs is presented in Table 12.1.

TABLE 12.1
Literature Survey on Graphene-Based Filters

Filter Structure	Substrate Material	Filter Type	Resonant Frequency f_r (THz)	Tuning Range (THz)	References
Series of graphene and polysilicon grating pads	Quartz	LPF	2.3	—	2014 [65]
Cross-shaped resonator	Quartz	BSF	2.39	1.78–3	2015 [33]
Rectangular ring resonator	SiO_2	Dual-band BPF	11.35, 20.99	17.3–24.8	2017 [55]
L-shaped resonators	SiO_2	BSF	1.605	1.60–17.1	2017 [32]
Transmission line model with open-circuit stubs	SiO_2	LPF	4.1	4.1–5.8	2018 [58]
Graphene ribbon resonators	SiO_2/Si	Dual-band BSF	14, 17.2	14–18.2	2018 [64]
Single stub and multi-stub structure	InSb	BSF	1.66	1.66–1.8	2018 [61]
Diamond-shaped ring resonator	SiO_2/Si	BSF	13.4, 22.6	13.4–34.3	2018 [59]
Transmission line model with short-circuit stubs	SiO_2	HPF	5	—	2019 [56]

12.4.3 Spoof Surface Plasmon-Based Filters

SSP filters have gained significant importance in metamaterials and plasmonics research due to their versatile and impactful properties. One key advantage of SSP filters is their exceptional design flexibility enabled by metamaterial structures. Unlike conventional SP filters, which rely on natural metals, SSP filters utilize engineered metamaterials, allowing precise control over geometric parameters and material properties [66]. This flexibility empowers researchers to tailor SSP filter designs to specific applications, manipulating electromagnetic wave behaviour with unprecedented precision and customization. Another crucial aspect of SSP filters is their capability to operate across a broad range of frequencies, from microwave to THz and even optical wavelengths. This broadband operation is particularly advantageous in applications requiring wideband filtering or sensing, such as wireless communications and spectroscopic analysis. SSP filters can exhibit excellent performance across multiple frequency bands, offering enhanced versatility and utility across various technological domains.

Moreover, SSP filters are designed with low-loss materials, which contributes to their high efficiency and performance. By minimizing energy losses, SSP filters maintain strong field confinement and resonance characteristics, leading to improved signal transmission and filtering capabilities. This low-loss property is instrumental in ensuring the practical viability and effectiveness of SSP-based devices in real-world applications [67]. The miniaturization potential of SSP filters is also noteworthy. Their sub-wavelength-scale features enable the development of compact and lightweight filtering components suitable for integration into modern electronic and photonic systems. This miniaturization capability is crucial for advancing technologies that demand high-performance filters in compact form factors, supporting the development of portable and efficient devices.

Transmission lines are essential for developing electronic and photonic ICs. Analyzing and designing these transmission lines is crucial for implementing SSP-based ICs. Planar technology, known for its easy fabrication, simplified mounting of surface devices, cost-effectiveness, and low-profile nature, stands out as an attractive option for constructing high-performance ICs. Numerous studies have emerged following the successful demonstration of guiding SSP modes on planar metallic structures, some documented in [68–70]. Exploring the control of SSP modes through unit cell geometry parameters, researchers have employed various shapes and structures like rectangular [68], dumbbell-shaped [69], and trapezoidal grooves [70]. These investigations delve into understanding and designing SSP-based transmission lines within the THz frequency range. For instance, [71] proposed planar plasmonic waveguides using periodically corrugated metal films. Investigations extended to verifying the guiding abilities of SSP structures in the microwave frequency domain, observed in works reported from [72–74].

Circuit analysis of SSP unit cells in [72] led to microwave frequency structure development via circuit modelling, demonstrating lower crosstalk and reduced ohmic loss compared to conventional counterparts, thus lowering mutual coupling theoretically and experimentally. These studies have generated significant interest in developing SSP-based transmission lines and other circuits in the THz and microwave realms. In [75], the transition from coplanar waveguides to SSP modes was explored. On the contrary, [76] emphasizes the use of conductor backed microstrips for a more efficient transition compared to those proposed using CPW. Moreover, Zhang et al. [77] achieved improved transmission efficiency by gradually converting the Quasi-TEM mode of the microstrip to the TM mode of SSP. Further advancements include Pan et al.'s use of metamaterial particles to control SSP wave propagation [78]. Further research has delved into fundamental and higher-order SSP modes within multi-layer composites, as outlined in [79]. Researchers have developed filtering structures in both THz frequency regimes and microwave frequencies [80–82].

In [69], the authors devised a wideband filter utilizing dumbbell grooves at both THz and microwave frequencies. However, this design did not address spurious mode suppression in the stop band, thereby limiting its performance. In [82], a series of split ring resonators (SRRs) and complementary split ring resonators (CSRRs) within the developed SSP transmission lines to achieve filtering characteristics was proposed. Furthermore, [83] proposed a manually tuneable and switchable filtering configurations to showcase the reconfigurable capability of the SSP wave propagation. Another study reported manual dispersion tunability using a different set of lumped SMD capacitors [84]. Nevertheless, the manual tuning feature of these designs restricts their practical usage since it necessitates the realization and fabrication of the complete circuit with different parameters designated for the particular tuning state. In [85], a simulation study was conducted for a passband reconfigurable SSP filter using a SPICE model of varactor diodes that were incorporated in the designed transmission lines. However, the design exhibited inefficiency in terms of fabrication and power management, requiring 24 variable capacitive elements as tuning elements. In [86], a reconfigurable dual-band band-pass filter was proposed, leveraging SSP through a planar ring resonator. In [66], a versatile SSP-based bandpass filter with a T-shaped resonator loaded with varactor diodes was proposed, designed for operation in both THz and gigahertz frequency ranges.

In [87], the focus was on amplification of propagating SSP modes, utilizing a ring resonator-based filtering structure. In [67], Xiao-Lan et al. pioneer an SSP-based transmission line employing a capacitor loading technique. This configuration provided flexible and reconfigurable dispersion control, coupled with highly selective filtering performance. In [83], bandwidth tunability was demonstrated in a bandstop filter operating at 7 GHz, employing an SSP microstrip line with defected units. In [88], Longfang Yea et al. designed a notch filter based on SSP, employing patterned conductive surfaces for frequency tuning. The dispersion characteristics are systematically engineered by adjusting parameters of periodically arranged spiral-shaped units. In [89], a dual-band rejection filter was introduced, featuring an SSP-based transmission line with six coupled circular ring resonators etched among ground planes of the centre corrugated strip. Tuning capabilities were achieved by varying the parameters of the ring resonators. The literature on various SSP-based filter designs is presented in Table 12.2.

12.4.4 Surface Plasmon-Based Sensors

The field of sensors has recently garnered substantial attention across multiple domains. Various sensors were proposed for diverse applications [90], persistently

TABLE 12.2
Literature Survey on SSP-Based Filters

Work Done	Type of the Filter	Tuning Parameter	Frequency	References
Broadband bandpass filter using SSP	BPF	—	7–10 GHz	2015 [80]
Localized SSP-based transmission line with defect units	BSF	Resonant frequency (change in physical dimensions)	7 GHz	2016 [83]
Notch filter based on spiral-shaped units	BSF	—	1.29 THz	2018 [88]
Resonator coupled to SSP transmission line	BPF	Bandwidth (using varactor diode)	0.80 THz, 3.2 GHz	2019 [66]
SSP filter design using coupled circular ring resonator	Dual-band BSF	Resonant frequency (change in physical dimensions)	8.6–11.2, 20–21.8 GHz	2020 [89]
Ring resonator-based filtering structure for SSP amplification	Dual-band BPF	Resonant frequency (change in physical dimensions)	1.78, 3.44 GHz	2020 [80]

exploring new techniques to enhance sensor performance. Particularly, there's a significant demand for miniaturized, label-free, and highly sensitive biosensors. Unlike label-based sensing requiring a secondary molecule, label-free sensors enable real-time detection and monitoring of analyte molecules, eliminating costly laboratory measurements and time consumption [91]. These requirements have led to the proposal of sensors utilizing various technologies–optical, electrochemical, microwave, millimetre-wave, THz, and ultrasonic [92]. Among these technologies, RF-to-THz biosensors have gained attention for their inherent advantages like low power, non-ionizing behaviour, and chemical inertness [93]. Microwave sensing has been a major focus due to its promising development, finding applications in various industries and healthcare sectors such as fluid level measurement [94], soil moisture detection [95], food adulteration detection [96], and healthcare applications [97].

The microwave sensing platform remains appealing not only for its robustness, high reliability, and low manufacturing costs but also for its rapid analysis capabilities. Moreover, THz waves, being non-ionizing and featuring wavelengths comparable to many molecules sizes [98], offer effectiveness in developing non-intrusive sensors, label-free, highly sensitive, positioning them as promising candidates in sensor development. The life sciences and medical therapies consistently drive a significant demand for biosensing in biomedical research. Biosensing primarily aims to detect biomolecules and their interactions. Over the past decade, extensive research has focused on accurately detecting biochemical [99] due to their diverse applications, such as food quality testing, disease detection, drug analysis, and environmental monitoring. Some of the main applications of sensors have been shown in Figure 12.9.

The electromagnetic sensors operating from RF to THz frequencies can be broadly categorized into non-resonant and resonant types. Non-resonant sensors exhibit

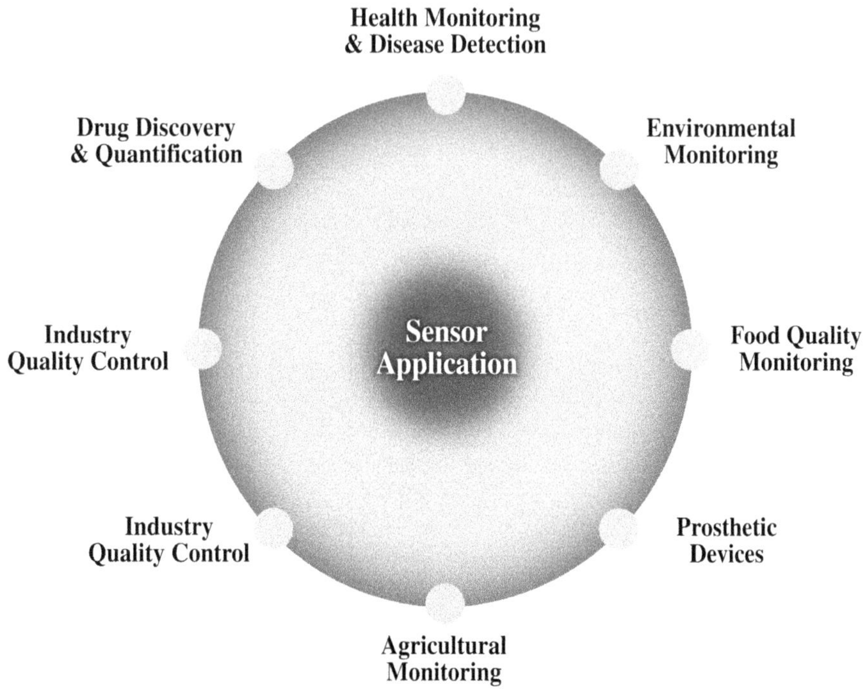

FIGURE 12.9 Applications of THz/microwave sensors.

broadband characteristics and utilize free space and transmission line methods for their implementation [100]. In contrast, resonant sensors showcase a narrow-band response, offering heightened accuracy and sensitivity to even minute changes in the dielectric properties of the substance under test [101]. Assessing the sensor's response involves measuring the magnitude and phase of transmission and reflection coefficients, along with examining their impedance characteristics.

Numerous studies have explored biosensing applications employing the RF-to-THz resonance approach [102]. In [103], waveguide resonators were utilized, while [102] employed dielectric resonators for characterizing biosamples. Achieving precise real-time decision making in lab-on-chip sensor setups hinges on the essential requirements of sensor miniaturization and sensitivity. However, these waveguides and dielectric resonator-based sensors, although effective, are non-planar and relatively large in size. Their substantial dimensions render them unsuitable for integration within lab-on-chip systems. Although these RF-to-THz planar electromagnetic sensors demonstrate optimal sensitivity for solid samples under test, their performance with aqueous samples remains unsatisfactory [104]. Designing sensors for aqueous biosamples poses a challenge due to the high presence of lossy water molecules, absorbing electromagnetic energy and causing attenuation. This attenuation significantly degrades sensor performance [105]. Resonant sensors with high-Q factors that strongly localize and confine the EM-field are required to address this issue, along with techniques to reduce sample size, enabling the sensing of even slight variations in the samples [106]. To achieve highly confined and localized EM waves, the use of SSP polariton structures presents a promising solution [107]. These SSP configurations significantly slow down EM wave propagation, tightly confining it within the resonating structure.

12.4.5 Dielectric Characterization of Materials

Dielectric characterization involves determining the electrical properties of a material, specifically its ability to store electrical energy when exposed to an electric field. This characterization is typically described by two key parameters: permittivity and loss tangent. Permittivity reflects the material's capacity to store electrical energy, while the loss tangent measures its ability to dissipate this energy. These properties play a critical role in designing and optimizing electromagnetic devices like antennas, sensors, and waveguides, especially within specific frequency ranges. To conduct dielectric characterization, researchers employ various techniques such as resonant cavity methods, network analyzers, and impedance analyzers [108]. These methods allow for precise measurement and analysis of how materials interact with electromagnetic fields, providing crucial insights for engineering applications and device performance optimization.

The dielectric constant or permittivity of a material is influenced by various factors including its molecular structure, the types and concentrations of atoms or molecules present, the material's polarity, temperature, and the frequency of the applied electromagnetic field.

Sensors incorporating SP concepts are emerging as promising tools for characterizing materials based on their dielectric properties. These sensors leverage electromagnetic wave interactions with sensing structures, enhancing sensitivity and selectivity for the material under test (MUT). The sensing mechanism relies on changes in resonant frequency and transmission/reflection coefficient of the sensor structure in response to the MUT. By analyzing these changes, SP sensors can determine the dielectric constant and loss tangent of the material.

12.4.6 Sensor Resonator Topologies

The design of a sensor is customized based on specific application requirements, including the selection of resonator topology, positioning of the MUT, and the method of excitation employed. Resonator topology plays a critical role in determining the sensor's sensitivity by influencing the concentration of electric fields and facilitating interaction with the MUT. Sensors can be constructed using either a single isolated unit cell or an array of unit cells [108], with each unit cell potentially containing multiple resonators of varying sizes. Incorporating multiple resonators within a unit cell enables the sensor to achieve multiband resonance, which enhances its accuracy and reliability across different operating conditions [109].

The selection of resonator type and topology in sensors is driven by the specific application requirements, which in turn dictate the overall sensor configuration. Resonators commonly used in these sensors include SRR and their dual, the CSRRs. These offer several advantages as they are self-supporting all-metal structures that can be fabricated by etching the ground plane, enabling compact designs that are easily excited by a transmission line printed on the substrate [109]. SRR within a unit cell can be configured as edge-coupled or broadside-coupled structures, featuring circular or rectangular shapes. Edge-coupled resonators are positioned on the same plane, while broadside-coupled resonators are located on different planes within the substrate. Recent research suggests that placing the MUT between broadside-coupled resonators can optimize the interaction between the MUT and resonators, enhancing the sensor's performance and sensitivity to changes in dielectric properties [110]. In CSRR-based sensors, the sensing area is significantly large because the electric field extends across the ground plane and over the centrally etched ring area. This is in contrast to conventional SRRs, where the electric field is localized primarily in the split region. Rectangular SRRs and CSRRs are straightforward structures that offer high sensitivity in various sensing applications. Asymmetric SRRs, as described in the literature, exhibit greater electric field confinement in the splits compared to symmetrical SRRs, leading to more significant shifts in resonant frequency [110]. Corrugated resonators utilizing SPs, such as ring resonators, represent straightforward yet powerful structures that offer exceptional sensitivity in various sensing applications. These resonators leverage SP polaritons to achieve unique electromagnetic properties, particularly in the THz and infrared spectral ranges [111]. The electrical size of a resonator, proportional to its resonant frequency and dimensions, influences its performance characteristics.

12.4.7 Graphene-Based Surface Plasmon Sensors

The design of the sensor utilizing SP propagation allows for precise control of EM waves. Even minor changes in the refractive index can significantly alter the EM properties of the structure. With this, various THz sensors were developed to achieve efficient detection. Typically, variations in the refractive index of the analyte cause shifts in the sensor's resonant peak, enabling accurate determination of the analyte's refractive index through monitoring these shifts. The performance of THz sensors is assessed using several parameters. Sensor sensitivity, defined as the extent to which the analyte's refractive index affects the resonant peak's position, is a critical measure [112]. Assessing a sensor's sensitivity to a specific substance also involves measuring the extent of resonant peak shifts in response to changes in the analyte's concentration.

In [113], a Vivaldi antenna design was proposed for medical sensing in the THz spectrum, emphasizing high gain and directivity. The sensor covers 0.5–2 THz with over 120% bandwidth. Graphene slabs enhance bandwidth through chemical potential adjustment, impacting gain and return loss. Graphene plasmonic nanoribbons and silicon slabs mitigate these effects, achieving a gain of 11.5 dBi. This design is used for biosensing, distinguishing cancerous from healthy skin tissue in the THz range. Zixuan Du et al. proposed a dynamically tuneable graphene metasurface for multi-band refractive index sensing and a four-state optical switch [114]. Using a graphene ring array, the design achieves high sensitivity (4.25 μm/RIU in mid-infrared, 11.33 μm/RIU in far-infrared) and broad sensing range by tuning graphene's Fermi level. Pinghui Wu et al. proposed a tuneable dual-band perfect absorption structure using cross four split ring resonators (CFSRR) and SiO_2 isolation layers [115].

This structure achieved dual-band perfect absorption, with absorption characteristics adjustable to specific wavelengths by modifying structural parameters. Dynamic tunability was enabled by changing the Fermi level of graphene using voltage gating, reducing manufacturing costs. The structure exhibited high sensitivity for refractive index sensing, making it promising for applications in modulators, sensors, and photoelectric devices. Zhaoxin Geng et al. proposed THz metamaterials for biosensing applications due to their tuneable resonance frequency, matching important biomolecule vibrations like cancer biomarkers [116]. Two metamaterial biosensors integrated with microfluidics are proposed to detect liver cancer biomarkers Alpha fetoprotein (AFP) and Glutamine transferase isozymes II (GGT-II) in early stages. The results showed significant resonance shifts 19 GHz for GGT-II and 14.2 GHz for AFP. This work underscores the power of metamaterial-assisted THz spectroscopy in detecting trace cancer biomarkers, with potential to enhance sensitivity through metamaterial optimization and substrate permittivity reduction for early-stage cancer molecule recognition. In [117], a THz sensor based on a hybrid metal-graphene metasurface was proposed. This structure combines a metal disc with graphene plasmon to enhance light-graphene coupling and achieve multi-band resonance.

A THz metamaterial sensor with surface Pythagorean tree fractal resonators fabricated on flexible polyethylene terephthalate was proposed [118]. It achieved strong absorption peaks exceeding 97% at 0.49 and 0.69 THz, with sensitivity to surrounding refractive index changes of 0.0968 and 0.1182 THz/RIU. Yisong Lv et al. proposed a broadband metamaterial sensor featuring graphene patterns and gold discs, achieving a wide absorption bandwidth of 3.4 THz [119]. Both structures exhibited excellent modulation characteristics, with maximum/minimum modulation depths of 92.7%/69.5% and 91.5%/70.57% when varying the graphene's chemical potential from 0 to 0.5 eV. The proposed design offered a sensitivity of 0.96 THz/RIU. Sevda et al. proposed a metamaterial perfect absorber biosensor for biomedical molecular detection in the near-infrared region [120]. The biosensor, featuring an Au layer with a bracket-shaped cavity, achieved near-unity absorption by exciting localized SPRs. The biosensor demonstrated high sensitivity of 2,120.8 nm/RIU, which increases to 2,500 nm/RIU with the addition of a graphene layer. This compact and sensitive biosensor holds promise for various biomedical applications. The literature on various SP-based sensor designs is presented in Table 12.3.

12.4.8 Spoof Surface Plasmon-Based Sensors

In [121], a SSP-based THz metamaterial sensor featuring square split ring resonators (SSRR) and plus-shaped grooved resonators was demonstrated with sharp resonances. Corrugations on the metasurface enable the excitation of SSP at THz frequencies, resulting in distinctive resonances. The sensor exhibits strong surface current distribution and field confinement, with observed shifts in resonance frequencies corresponding to increased capacitive gap sizes. High sensitivity for ethanol and glucose sensing was achieved, particularly with resonant mode B showing double the sensitivity. A millimetre-wave-based SSP resonator sensor was proposed for measuring glucose concentrations with high sensitivity and a compact chip size [122]. Unlike traditional microwave sensors, this design offered enhanced

TABLE 12.3
State-of-the-Art Graphene-Based SP Sensors

Sensor Structure	Resonant Frequency/ Absorption Peak	Peaks	Sensitivity	Application	References
Circular split-ring resonator	0.55, 0.95 THz	2	150 GHz/RIU	Cancer detection	2017 [116]
Vivaldi antenna	0.5, 2 THz	2	—	Cancer detection	2021 [113]
Periodic unit cell	23.5, 24.3, 27.8 μm	3	5.06 μm/RIU	RI sensing	2021 [117]
Periodic cross split-ring resonator	4.55, 4.16 μm	2	25.77 μm/RIU	RI sensing	2022 [115]
Pythagorean tree fractal resonators	0.49, 0.69 THz	2	0.1182 THz/RIU	Biomedical sensing	2022 [118]
Hybrid graphene-gold pattern	6.5 THz	1	0.96 THz/RIU	Biochemical sensing	2022 [119]
Rectangular SSR	1.42 μm	1	2,500 nm/RIU	Biosensor	2022 [120]
Graphene ring array	10.9, 29.1 μm	2	4.25 μm/RIU	Optics and photonic circuits	2023 [114]

performance and reusability, utilizing a microfluidic channel with minimal sample volume requirements. The sensor operated at 28 GHz and detected glucose concentration changes through the S_{11} parameter, achieving an average sensitivity of 0.015669 dB/(mg/dL) over the range of 0–300 mg/dL. Notably, it could detect concentrations as low as 1 mg/dL with a distinguishable signal of 0.015669 dB from a 3.4 μL sample.

In [123], a sensor was designed for dielectric material testing in microwave and THz frequency regions. The sensor consisted of a main coplanar waveguide (CPW) line integrated with a defected T-shaped SSP line positioned at the substrate's top and bottom, respectively. The design features a defect in the SSP line to create a stop band, localizing transmission energy at specific locations and demonstrated the sensor's ability to distinguish closely spaced dielectric constants, with a high-quality factor ($Q = 165$) and 2.84% sensitivity. Yawei Zhang et al. proposed a Mach–Zehnder Interferometer (MZI) refractive index sensor on a SSP waveguide, demonstrating with dispersion curves and transmission capabilities under varying capacitances, revealing that cut-off frequency increases with reduced groove depth [124]. With specific groove depths, the sensor operated at 8.46 GHz. Experimental results indicated an average sensor error of only 0.77%, highlighting its high accuracy. Ankita

K et al. designed a SP line splitter-based sensor for the THz frequency region [125]. This sensor exhibited improved sensitivity and electric field confinement compared to conventional microstrip designs. Experimental validation using FR-4 samples demonstrated substantial phase differences even with small defects, showcasing the sensor's defect detection capabilities in the microwave frequency range. The proposed sensor offers potential for precise defect detection and phase measurement in THz and microwave applications.

In [126], a SSP transmission line-based THz sensor utilizing a perforated dumbbell-shaped unit cell was proposed. This SSP structure demonstrated enhanced confinement abilities compared to conventional microstrip lines in the THz frequency regime, leading to improved sensor sensitivity. The design incorporated an underlayer ground to achieve efficient impedance matching between conventional microstrip and SSP structures across microwave and THz frequencies. The THz dielectric sensor was developed by creating an appropriate capacitive slot on the dumbbell cells. Xie et al. presented a highly sensitive localized SSP-based sensor for omni-directional inspection of cracks in metal surfaces [127]. The sensor utilized a defected ground structure (DGS) in the form of a conformal multiband split-ring slot (CMSS) etched on the ground of a microstrip line to excite SSPs modes within this structure. The electric SSP mode was leveraged for detecting small cracks on metal surfaces, demonstrating a resonance frequency shift of 400 MHz when inspecting a metal surface with a 0.2-mm-wide and 2-mm-deep crack. Full-wave numerical analysis confirmed the sensor's ability for omnidirectional crack detection.

Xu Chen et al. proposed a THz metamaterial sensor with a metal stripe and three rectangular grooves, achieving sharp resonances for ultra-sensitive sensing [128]. The sensor exhibited a high frequency sensitivity of 1.966 THz per refractive index unit. The study emphasized the impact of metal thickness on sensor performance, suggesting practical applications in the THz regime with planar metamaterial structures. In [129], a microwave planar resonator sensor supporting SSP waves was proposed for wafer dielectric characterization. This low-cost sensor discriminates between wafers based on permittivity and thickness, achieving a sensitivity of 60.22 MHz/ε_r. It effectively measures resonant frequency changes and losses to characterize materials like quartz, borosilicate, BK7 glass, and Silicon wafers. Simulation and measurement confirm its high sensitivity for dielectric material analysis. Yifei Zhang et al. proposed a SP-based biosensor featuring SRRs for ovarian cancer diagnosis at microwave frequencies [130]. This biosensor replaces conventional periodic grooves with SRRs distributed in series on a metal line, maintaining a high cut-off frequency and generating sharp resonances that enhance localized electric fields by a factor of 250. This biosensor effectively detects ovarian clear cell carcinoma (OCCC) and serous ovarian cancer (SOC) tissues by identifying resonant frequency shifts: from 53.990 GHz for normal tissues to ranges like 53.814–53.968 GHz for SOC and 53.698–53.872GHz for OCCC tissues. The biosensor's average resonant point at 53.812 GHz serves as a reference for cancer detection. The literature on various SSP-based sensor designs is presented in Table 12.4.

TABLE 12.4
State of the Art of SSP-Based Sensors

Sensor Structure	Frequency	Peaks	Analysis Techniques	Sensitivity	Application	References
A corrugated metal stripe perforated by three rectangular grooves	0.58, 0.89 THz	2	Linear fitting function	1.966 THz/RIU	Industrial quality control	2017 [128]
SSP Transmission line with dumbbell-shaped unit cell	13.2, 10.6 GHz	1	Numerical analysis	—	Polymer industry	2018 [126]
A defected ground structure based on the CMSS structure	5.83 GHz	1	Numerical analysis	—	Crack detection	2019 [127]
CPW line and defected T-shaped SSP line	15.7 GHz, 0.2 THz	1	Parabolic curve fitting	2.84%	Biological sensing	2020 [123]
Ring-shaped resonator with periodic grooves	28 GHz	1	—	0.015 dB/(mg/dL)	Biochemical sensing	2021 [122]
Square split-ring resonator	0.59, 1.7 THz	2	—	0.40 THz/RIU	Biomedical sciences	2022 [121]
U-shaped metal transmission line	8.26 GHz	1	Parabolic curve fitting	0.77%	Microwave integrated circuits	2022 [124]
Cross-shaped resonator	4.3 GHz	1	Polynomial fit	64.53 GHz/ε_r	Industrial quality control	2022 [129]
Split-ring resonator	53.99 GHz	1	Box plots and spectral curves	3.27 GHz/ε_r	Cancer diagnosis and classification	2022 [130]

12.5 SUMMARY

Designing THz tuneable filters poses significant challenges, notably the necessity for precise tunability and efficient signal propagation with high confinement. Leveraging graphene as a support material for SP propagation offers a promising solution to address these challenges. By varying the chemical potential of graphene, tunability becomes achievable, facilitating dynamic adjustments of the filter's properties. Additionally, integrating SSP propagation structures provides an alternative approach, enabling precise control over signal propagation through engineered metallic patterns. To further enhance tunability, variable capacitor devices can be incorporated into the filter design, offering efficient control over resonant frequencies. By combining these strategies, the challenges associated with achieving precise tunability and efficient signal propagation in THz filters can be effectively mitigated, resulting in enhanced performance and adaptability for diverse THz applications. Achieving highly sensitive, label-free detection with precise molecular specificity poses a significant challenge for traditional biochemical sensors, especially in complex environments. THz sensors based on SP resonance offer a solution by enhancing the interaction between THz radiation and biomolecules. This enhancement enables the detection of minute biomolecule quantities without labels, ensuring high sensitivity and specificity. The accurate and rapid detection of liquid samples in multicomponent mixtures is a complicated challenge. Integrating machine learning-based algorithms with sensors provides a promising remedy to this challenge. These algorithms analyze complex sensor data patterns to accurately identify and quantify alcohol concentrations. By learning from large datasets, ML algorithms enhance sensor performance, sensitivity, and provide real-time monitoring capabilities, overcoming the limitations of traditional sensors in multicomponent liquid analysis.

REFERENCES

1. S. Cherry, Edholms law of bandwidth, *IEEE Spectrum* 41 (7) (2004) 58–60. https://doi.org/10.1109/MSPEC.2004.1309810
2. M. Jacob, S. Priebe, T. Kurner, C. Jastrow, T. Kleine-Ostmann, T. Schrader, An overview of ongoing activities in the field of channel modeling, spectrum allocation and standardization for mm-wave and THz indoor communications, in: *2009 IEEE Globecom Workshops*, Honolulu, HI, 2009, pp. 1–6. https://doi.org/10.1109/GLOCOMW.2009.5360685
3. D. O'Brien, G. Faulkner, K. Jim, E. Zyambo, D. Edwards, M. Whitehead, P. Stavrinou, G. Parry, J. Bellon, M. Sibley, V. Lalithambika, V. Joyner, R. Samsudin, D. Holburn, R. Mears, High-speed integrated transceivers for optical wireless, *IEEE Communications Magazine* 41 (3) (2003) 58–62. https://doi.org/10.1109/MCOM.2003.1186546
4. P. Siegel, Terahertz technology, *IEEE Transactions on Microwave Theory and Techniques* 50 (3) (2002) 910–928. https://doi.org/10.1109/22.989974
5. M. Tonouchi, Cutting-edge terahertz technology, *Nature Photonics* 1 (2) (2007) 97–105.
6. J. Federici, L. Moeller, Review of terahertz and subterahertz wireless communications, *Journal of Applied Physics* 107 (11) (2010) 111101. https://doi.org/10.1063/1.3386413
7. H.-J. Song, T. Nagatsuma, Present and future of terahertz communications, *IEEE Transactions on Terahertz Science and Technology* 1 (1) (2011) 256–263. https://doi.org/10.1109/TTHZ.2011.2159552
8. M. Nagel, P. H. Bolivar, M. Brucherseifer, H. Kurz, Integrated THz technology for label-free genetic diagnostics, *Applied Physics Letters* 80 (1) (2002) 154–156.
9. P. Bolivar, M. Brucherseifer, J. Rivas, R. Gonzalo, I. Ederra, A. Reynolds, M. Holker, P. de Maagt, Measurement of the dielectric constant and loss tangent of high dielectric-constant materials at terahertz frequencies, *IEEE Transactions on Microwave Theory and Techniques* 51 (4) (2003) 1062–1066. https://doi.org/10.1109/TMTT.2003.809693
10. K.-C. Huang, Z. Wang, Terahertz terabit wireless communication, *IEEE Microwave Magazine* 12 (4) (2011) 108–116. https://doi.org/10.1109/MMM.2011.940596
11. T. G. Phillips, J. Keene, Sub-millimeter astronomy (heterodyne spectroscopy), *Proceedings of the IEEE* 80 (1992) 1662–1678.

12. J. A. Zeitler, P. F. Taday, D. A. Newnham, M. Pepper, K. C. Gordon, T. Rades, Terahertz pulsed spectroscopy and imaging in the pharmaceutical setting: A review, *Journal of Pharmacy and Pharmacology* 59 (2) (2007) 209–223.
13. V. Anitha, A. Beohar, A. Nella, THz imaging technology trends and wide variety of applications: A detailed survey, *Plasmonics* 18 (2) (2023) 441–483. https://doi.org/10.1007/s11468-022-01775-9
14. F. Shi, H. Ning, W. Huangfu, F. Zhang, D. Wei, T. Hong, M. Daneshmand, Recent progress on the convergence of the internet of things and artificial intelligence, *IEEE Network* 34 (5) (2020) 8–15. https://doi.org/10.1109/MNET.011.2000009
15. M. A. Hamdani, A. Nasri, H. Zairi, Design of a microwave biosensor using a defected csrr for cancer cells characterization, in: *2018 IEEE 4th Middle East Conference on Biomedical Engineering (MECBME)*, Tunis, Tunisia, 2018, pp. 7–10. https://doi.org/10.1109/MECBME.2018.8402395
16. P. R. West, S. Ishii, G. V. Naik, N. K. Emani, V. M. Shalaev, A. Boltasseva, Searching for better plasmonic materials, *Laser and Photonics Reviews* 4 (6) (2010) 795–808.
17. R. Zia, J. A. Schuller, A. Chandran, M. L. Brongersma, Plasmonics: The next chip-scale technology, *Materials Today* 9 (7) (2006) 20–27.
18. W. Park, Optical interactions in plasmonic nanostructures, *Nano Convergence* 1 (2) (2014) 2.
19. W. M. Albers, I. Vikholm-Lundin, *Surface Plasmon Resonance on Nanoscale Organic Films*, Springer, New York, 2011, pp. 83–125.
20. D. K. Gramotnev, S. I. Bozhevolnyi, Plasmonics beyond the diffraction limit, *Nature Photonics* 4 (2). https://doi.org/10.1038/nphoton.2009.282
21. G. Grosso, G. P. Parravicini (Eds.), Electrons in one-dimensional periodic potentials, in: Solid State Physics, Academic Press, London, 2000, pp. 1–36. https://doi.org/10.1016/B978-012304460-0/50001-3
22. A. Ghahremani, G. Moradi, Planar tunable graphene based low-pass filter in the terahertz band, *Applied Optics* 57 (27) (2018) 7823–7829. https://doi.org/10.1364/AO.57.007823
23. M. Kong, Y. Wu, Z. Zhuang, W. Wang, Y. Liu, Graphene-based THz tunable bandstop filter with constant absolute bandwidth, *Progress in Electromagnetics Research Letters* 71 (2017) 141–147. https://doi.org/10.2528/PIERL17101301
24. B. Schiffman, G. Matthaei, Exact design of band-stop microwave filters, in: *PTGMTT National Symposium Digest*, Santa Monica, CA, Vol. 63, 1963, pp. 69–74. https://doi.org/10.1109/PTGMTT.1963.1123223
25. A. A. Maradudin, Surface electromagnetic waves on structured perfectly conducting surfaces, in: *Modern Plasmonics*, Vol. 4 of Handbook of Surface Science, North-Holland, 2014, pp. 223–251. https://doi.org/10.1016/B978-0-444-59526-3.00007-0
26. J. B. Pendry, L. Martin-Moreno, F. J. Garcia-Vidal, Mimicking surface plasmons with structured surfaces, *Science* 305 (5685) (2004) 847–848.
27. A. P. Hibbins, B. R. Evans, J. R. Sambles, Experimental verification of designer surface plasmons, *Science Mag.* 308 (5722) (2005) 670–672.
28. C.-H. Kim, K. Chang, Ring resonator bandpass filter with switchable bandwidth using stepped-impedance stubs, *IEEE Transactions on Microwave Theory and Techniques* 58 (12) (2010) 3936–3944.
29. D. Martin-Cano, M. L. Nesterov, A. I. Fernandez-Dominguez, F. J. Garcia-Vidal, L. Martin-Moreno, E. Moreno, Domino plasmons for subwavelength terahertz circuitry, *Optics Express* 18 (2) (2010) 754–764.
30. X. Shen, T. J. Cui, D. Martin-Cano, F. J. Garcia-Vidal, Conformal surface plasmons propagating on ultrathin and flexible films, *Proceedings of the National Academy of Sciences of the United States of America* 110 (1) (2013) 40–45.

31. M. Navarro-Ca, M. Beruete, S. Agrafiotis, F. Falcone, M. Sorolla, S. A. Maier, Broadband spoof plasmons and subwavelength electromagnetic energy confinement on ultrathin meta films, *Optics Express* 17 (20) (2009) 18184–18195.
32. M. Kong, Y. Wu, Z. Zhuang, W. Wang, Y. Liu, Graphene-based thz tunable bandstop filter with constant absolute bandwidth, *Progress in Electromagnetics Research Letter* 71 (2) (2017) 141–147. https://doi.org/10.2528/pierl17101301
33. V. Dmitriev, G. Tavares, C. Nascimento, Graphene terahertz filter, in: *2015 SBMO/IEEE MTT-S International Microwave and Optoelectronics Conference (IMOC)*, Porto de Galinhas, Brazil, 2015, pp. 1–5. https://doi.org/10.1109/IMOC.2015.7369136
34. W. L. Barnes, A. Dereux, T. W. Ebbesen, Surface plasmon subwavelength optics, *Nature* 424 (6950) (2003) 824–830. https://doi.org/10.1038/nature01937
35. V. W. Brar, M. S. Jang, M. Sherrott, J. J. Lopez, H. A. Atwater, Highly confined tunable mid-infrared plasmonics in graphene nanoresonators, *Nano Letters* 13 (6) (2013) 2541–2547. https://doi.org/10.1021/nl400601c
36. O. Paul, R. Beigang, M. Rahm, Highly selective terahertz bandpass filters based on trapped mode excitation, *Optics Express* 17 (21) (2009) 18590–18595. https://doi.org/10.1364/oe.17.018590
37. F. Tavakoli, F. B. Zarrabi, H. Saghaei, Modeling and analysis of high sensitivity refractive index sensors based on plasmonic absorbers with fano response in the near-infrared spectral region, *Applied Optics* 58 (20) (2019) 5404–5414.
38. D. Li, R. B. Kaner, Graphene-based materials, *Science* 320 (5880) (2008) 1170–1171.
39. A. K. Geim, K. S. Novoselov, The rise of graphene, *Nature Materials* 6 (2007) 183–191.
40. Z. Fei, M. D. Goldflam, J. S. Wu, S. Dai, M. Wagner, A. S. McLeod, M. K. Liu, K. W. Post, S. Zhu, G. C. A. M. Janssen, M. M. Fogler, D. N. Basov, Edge and surface plasmons in graphene nanoribbons, *Nano Letters* 15 (12) (2015) 8271–8276. https://doi.org/10.1021/acs.nanolett.5b03834
41. Z. Fang, Z. Liu, Y. Wang, P. M. Ajayan, P. Nordlander, N. J. Halas, Graphene-antenna sandwich photodetector, *Nano Letters* 12 (7) (2012) 3808–3813. https://doi.org/10.1021/nl301774e
42. M. Liu, X. Yin, X. Zhang, Double-layer graphene optical modulator, *Nano Letters* 12 (3) (2012) 1482–1485. https://doi.org/10.1021/nl204202k
43. M. R. J. Azizpour, M. Soroosh, N. Dalvand, Y. Seifi-Kavian, All-optical ultra-fast graphene-photonic crystal switch, *Crystals* 9 (9) (2009) 461.
44. F. Haddadan, M. Soroosh, N. Alaei-Sheini, Designing an electro-optical encoder based on photonic crystals using the graphene–Al_2O_3 stacks, *Applied Optics* 59 (7) (2020) 2179–2185.
45. M. H. Sani, A. A. Tabrizi, H. Saghaei, R. Karimzadeh, An ultrafast all-optical half adder using nonlinear ring resonators in photonic crystal microstructure, *Optical and Quantum Electronics* 52 (107). https://doi.org/10.1007/s11082-020-2233-x
46. F. Mehdizadeh, M. Soroosh, H. Alipour-Banaei, E. Farshidi, Ultra-fast analog-to-digital converter based on a nonlinear triplexer and an optical coder with a photonic crystal structure, *Applied Optics* 56 (7) (2017) 1799–1806. https://doi.org/10.1364/AO.56.001799
47. J. T. Kim, C.-G. Choi, Graphene-based polymer waveguide polarizer, *Optics Express* 20 (4) (2012) 3556–3562. https://doi.org/10.1364/oe.20.003556
48. A. K. Varshney, N. P. Pathak, D. Sircar, Design of graphene-based thz antennas, in: B. Iyer, S. L. Nalbalwar, N. P. Pathak (Eds.), *Computing, Communication and Signal Processing*, Springer, Singapore, 2019, pp. 29–35.
49. H. Saghaei, M. K. Moravvej-Farshi, M. Ebnali-Heidari, M. N. Moghadasi, Ultra-wide mid-infrared supercontinuum generation in $AS_{40}Se_{60}$ chalcogenide fibers: Solid core PCF versus SIF, *IEEE Journal of Selected Topics in Quantum Electronics* 22 (2) (2016) 279–286. https://doi.org/10.1109/JSTQE.2015.2477048

50. R.-P. Pan, C.-Y. Chen, C.-F. Hsieh, C.-L. Pan, A liquid-crystal-based terahertz tunable Lyot filter, in: *2005 IEEE LEOS Annual Meeting Conference Proceedings*, Sydney, NSW, 2005, pp. 981–982. https://doi.org/10.1109/LEOS.2005.1548341
51. M. MacDonald, A. Alexanian, R. York, Z. Popovic, E. Grossman, Spectral transmittance of lossy printed resonant-grid terahertz bandpass filters, *IEEE Transactions on Microwave Theory and Techniques* 48 (4) (2000) 712–718. https://doi.org/10.1109/22.841963
52. H. Saghaei, Supercontinuum source for dense wavelength division multiplexing in square photonic crystal fiber via fluidic infiltration approach, *Journal of Radioengineering* 26 (1) (2017) 16–22. https://doi.org/10.13164/re.2017.0016
53. Y. Kim, S. Lin, H. Wu, R. Pan, A tunable terahertz filter and its switching properties in terahertz region based on a defect mode of a metallic photonic crystal, *Journal of Applied Physics* 109 (12) (2011) 123111. https://doi.org/ 10.1063/1.3603009
54. H. Chu, C. H. Gan, Active plasmonic switching at mid-infrared wavelengths with graphene ribbon arrays, *Applied Physics Letters* 102 (23) (2013) 231107. https://doi.org/10.1063/1.4810003
55. W. Su, B. Chen, Graphene-based tunable terahertz filter with rectangular ring resonator containing double narrow gaps, *Pramana: Journal of Physics* 89 (3) (2017) 1–5. https://doi.org/10.1007/s12043-017-1436-6
56. M. Azimbeik, N. Shoghi Badr, S. Ghasem Zadeh, G. Moradi, Graphene based high pass filter in terahertz band, *Optik* 198 (2019) 163246. https://doi.org/10.1016/j.ijleo.2019.163246
57. K. Yang, S. Liu, S. Arezoomandan, A. Nahata, B. Sensale-Rodriguez, Graphene-based tunable metamaterial terahertz filters, *Applied Physics Letters* 105 (9) (2014) 093105. https://doi.org/10.1063/1.4894807
58. A. Ghahremani, G. Moradi, Planar tunable graphene based low-pass filter in the terahertz band, *Applied Optics* 57 (27) (2018) 7823. https://doi.org/10.1364/ao.57.007823
59. W. Su, A four-port ultra-compact terahertz splitting filter based on graphene nanoribbon, *IEEE Photonics Technology Letter* 31 (1) (2019) 86–89. https://doi.org/10.1109/LPT.2018.2884048
60. Y. Cai, K. D. Xu, R. Guo, J. Zhu, Q. H. Liu, Graphene-based plasmonic tunable dual-band bandstop filter in the far-infrared region, *IEEE Photonics Journal* 10 (6) (2018) 1–9. https://doi.org/10.1109/JPHOT.2018.2876681
61. A. Moazami, M. Hashemi, N. C. Shirazi, High efficiency tunable graphene based plasmonic filter in the THz frequency range, *Plasmonics* 14 (2) (2019) 359–363. https://doi.org/10.1007/s11468-018-0812-5
62. S. Naghizade, H. Saghaei, Tunable graphene-on-insulator band-stop filter at the mid-infrared region, *Optical and Quantum Electronics* 52 (4) (2020) 1–13. https://doi.org/10.1007/s11082-020-02350-4
63. Y. Wang, H. Liu, S. Wang, M. Cai, Wide-range tunable narrow bandstop filter based on bilayer graphene in the mid-infrared region, *IEEE Photonics Journal* 12 (4) (2020) 2–8. https://doi.org/10.1109/JPHOT.2020.3004005
64. Y. Cai, K. D. Xu, R. Guo, J. Zhu, Q. H. Liu, Graphene-based plasmonic tunable dual-band bandstop filter in the far-infrared region, *IEEE Photonics Journal* 10 (6) (2018) 1–9. https://doi.org/10.1109/JPHOT.2018.2876681
65. D. Correas-Serrano, J. S. Gomez-Diaz, J. Perruisseau-Carrier, A. Alvarez-Melcon, Graphene-based plasmonic tunable low-pass filters in the terahertz band, *IEEE Transactions on Nanotechnology* 13 (6) (2014) 1145–1153. https://doi.org/10.1109/TNANO.2014.2344973
66. R. K. Jaiswal, N. Pandit, N. P. Pathak, Spoof surface plasmon polaritons based reconfigurable band-pass filter, *IEEE Photonics Technology Letters* 31 (3) (2019) 218–221. https://doi.org/10.1109/LPT.2018.2889007

67. X.-L. Tang, Q. Zhang, S. Hu, A. Kandwal, T. Guo, Y. Chen, Capacitor loaded spoof surface plasmon for flexible dispersion control and high selectivity filtering, *IEEE Microwave and Wireless Components Letters* 27 (9) (2017) 806–808. https://doi.org/10.1109/LMWC.2017.2734738
68. Y. Liu, J. Yan, Y. Shao, J. Pan, C. Zhang, Y. Hao, G. Han, Spoof surface plasmon polaritons based on ultrathin corrugated metallic grooves at terahertz frequency, *Applied Optics* 55 (7) (2016) 1720–1724.
69. Y. J. Zhou, B. J. Yang, Planar spoof plasmonic ultra-wideband filter based on low-loss and compact terahertz waveguide corrugated with dumbbell grooves, *Applied Optics* 54 (14) (2015) 4529–4533.
70. X. Liu, L. Zhu, Q. Wu, Y. Feng, Highly-confined and low-loss spoof surface plasmon polaritons structure with periodic loading of trapezoidal grooves, *AIP Advances* 5 (7) (2015) 077123.
71. G. Kumar, S. Pandey, A. Cui, A. Nahata, Planar plasmonic terahertz waveguides based on periodically corrugated metal films, *New Journal of Physics* 13 (3) (2011) 033024.
72. A. Kianinejad, Z. N. Chen, C.-W. Qiu, Design and modeling of spoof surface plasmon modes-based microwave slow-wave transmission line, *IEEE Transactions on Microwave Theory and Techniques* 63 (6) (2015) 1817–1825.
73. A. Kianinejad, Z. N. Chen, C. Qiu, Full modeling, loss reduction, and mutual coupling control of spoof surface plasmon-based meander slow wave transmission lines, *IEEE Transactions on Microwave Theory and Techniques* 66 (8) (2018) 3764–3772. https://doi.org/10.1109/TMTT.2018.2841857
74. A. Kianinejad, Z. N. Chen, C.-W. Qiu, Low-loss spoof surface plasmon slow wave transmission lines with compact transition and high isolation, *IEEE Transactions on Microwave Theory and Techniques* 64 (10) (2016) 3078–3086. https://doi.org/10.1109/TMTT.2016.2604807
75. H. F. Ma, X. Shen, Q. Cheng, W. X. Jiang, T. J. Cui, Broadband and high-efficiency conversion from guided waves to spoof surface plasmon polaritons, *Laser & Photonics Reviews* 8 (1) (2014) 146–151.
76. Z. Li, L. Liu, B. Xu, P. Ning, C. Chen, J. Xu, X. Chen, C. Gu, Q. Qing, High-contrast gratings-based spoof surface plasmons, *Scientific Reports* 6 (21199). https://doi.org/10.1038/srep21199
77. W. Zhang, G. Zhu, L. Sun, F. Lin, Trapping of surface plasmon wave through gradient corrugated strip with underlayer ground and manipulating its propagation, *Journal of Applied Physics* 106 (2) (2015) 021104.
78. B. Pan, Z. Liao, J. Zhao, T. Cui, Controlling rejections of spoof surface plasmon polaritons using metamaterial particles, *Optics Express* 22 (11) (2014) 13940–13950.
79. B. C. Pan, J. Zhao, Z. Liao, H. C. Zhang, T. J. Cui, Multi-layer topological transmissions of spoof surface plasmon polaritons, *Scientific Reports* 6 (22702). https://doi.org/10.1038/srep22702
80. L. Zhao, X. Zhang, J. Wang, W. Yu, J. Li, H. Su, X. Shen, A novel broadband band-pass filter based on spoof surface plasmon polaritons, *Scientific Reports* 6 (1) (2015) 36069. https://doi.org/10.1038/srep36069
81. X. Gao, L. Zhou, Z. Liao, H. Ma, T. Cui, An ultra-wideband surface plasmonic filter in microwave frequency, *Applied Physics Letters* 104 (19) (2014) 191603.
82. Q. Zhang, H. C. Zhang, J. Y. Yin, B. C. Pan, T. J. Cui, A series of compact rejection filters based on the interaction between spoof spps and csrrs, *Scientific Reports* 6 (1) (2016) 28256. https://doi.org/10.1038/srep28256
83. B. Xu, Z. Li, L. Liu, J. Xu, C. Chen, C. Gu, Bandwidth tunable microstrip band-stop filters based on localized spoof surface plasmons, *Journal of the Optical Society of America B* 33 (7) (2016) 1388–1391.

84. X.-L. Tang, Q. Zhang, S. Hu, A. Kandwal, T. Guo, Y. Chen, Capacitor loaded spoof surface plasmon for flexible dispersion control and high selectivity filtering, *IEEE Microwave and Wireless Components Letters* 27 (9) (2017) 806–808. https://doi.org/10.1109/LMWC.2017.2734738
85. H. C. Zhang, P. He, X. Gao, W. X. Tang, T. J. Cui, Pass-band reconfigurable spoof surface plasmon polaritons, *Journal of Physics: Condensed Matter* 30 (13) (2018) 134004. https://doi.org/10.1088/1361–648X/aaab85
86. R. Jaiswal, N. Pandit, N. Pathak, Spoof surface plasmon polariton-based reconfigurable band-pass filter using planar ring resonator, *Plasmonics* 14 (2019) 631–646. https://doi.org/10.1007/s11468-018-0841-0
87. R. K. Jaiswal, N. Pandit, N. Pathak, Amplification of propagating spoof surface plasmon polaritons in ring resonator-based filtering structure, *IEEE Transactions on Plasma Science* 48 (9) (2020) 3253–3260. https://doi.org/10.1109/TPS.2020.3014856
88. L. Ye, W. Zhang, B. K. Ofori-Okai, W. Li, J. Zhuo, G. Cai, Q. H. Liu, Super subwavelength guiding and rejecting of terahertz spoof SPPs enabled by planar plasmonic waveguides and notch filters based on spiral-shaped units, *Journal of Lightwave Technology* 36 (20) (2018) 4988–4994. https://doi.org/10.1109/JLT.2018.2868129
89. E. Farokhipour, M. Mehrabi, N. Komjani, C. Ding, A spoof surface plasmon polaritons (SSPPs) based dual-band-rejection filter with wide rejection bandwidth, *Sensors* 20 (24) (2020) 7311. https://doi.org/10.3390/s20247311
90. X. Zhao, B. Chen, G. Wei, J. M. Wu, W. Han, Y. Yang, Polyimide graphene nanocomposite foam-based wind-driven triboelectric nanogenerator for self-powered pressure sensor, *Advanced Materials Technologies* 4 (5) (2019) 1800723.
91. W. Jiang, X. Wang, Z. Yang, P. Peng, Y. Lu, H. Ren, The present situation and application of biosensor, in: *Proceedings of 2011 International Conference on Electronic & Mechanical Engineering and Information Technology*, Harbin, China, 2011, pp. 1743–1746.
92. A. B. Socorro, E. Santamaría, J. Fernández-Irigoyen, I. D. Villar, J. M. Corres, F. J. Arregui, Fiber-optic immunosensor based on an etched SMS structure, *IEEE Journal of Selected Topics in Quantum Electronics* 23 (2) (2017) 314–321.
93. P. Mehrotra, B. Chatterjee, S. Sen, EM-wave biosensors: A review of RF, microwave, mm-wave and optical sensing, *Sensors* 19 (5) (2019) 1013.
94. M. A. Karimi, M. Arsalan, A. Shamim, Multi-channel, microwave based, compact printed sensor for simultaneous and independent level measurement of eight liquids, *IEEE Sensors Journal* 19 (14) (2019) 5611–5620.
95. S. Birgermajer, G. Kitic, V. Crnojevic-Bengin, Microwave soil moisture sensor based on phase shift detection of transmission coefficient in bandstop structures, in: *Proceedings of 19th Telecommunications Forum (TELFOR) Papers*, Belgrade, Serbia, 2011, pp. 27–30.
96. N. K. Tiwari, S. P. Singh, M. J. Akhtar, Novel improved sensitivity planar microwave probe for adulteration detection in edible oils, *IEEE Microwave and Wireless Components Letters* 29 (2) (2019) 164–166.
97. W. Yu, S. Y. Huang, T-shaped patterned microstrip line for non-invasive continuous glucose sensing, *IEEE Microwave and Wireless Components Letters* 28 (10) (2018) 942–944.
98. P. H. Siegel, Terahertz technology in biology and medicine, *IEEE Transactions on Microwave Theory and Techniques* 52 (10) (2004) 2438–2447.
99. C. Chen, Y. Guo, P. Chen, H. Peng, Recent advances of tissue interfaced chemical biosensors, *Journal of Materials Chemistry B* 8 (2020) 3371–3381.
100. N. M. N. Haase, G. Fuge, H. K. Trieu, A.-P. Zeng, A. F. Jacob, Miniaturized transmission-line sensor for broadband dielectric characterization of biological liquids and cell suspensions, *IEEE Transactions on Microwave Theory and Techniques* 63 (10) (2015) 3026–3033.

101. H. Choi, J. Naylon, S. Luzio, J. Beutler, J. Birchall, C. Martin, Design and in vitro interference test of microwave non-invasive blood glucose monitoring sensor, *IEEE Transactions on Microwave Theory and Techniques* 63 (10) (2015) 3016–3025.
102. M. Zhang, J. T. W. Yeow, Nanotechnology-based terahertz biological sensing: A review of its current state and things to come, *IEEE Nanotechnology Magazine* 10 (3) (2016) 30–38.
103. B. Kapilevich, B. Litvak, Optimized microwave sensor for online concentration measurements of binary liquid mixtures, *IEEE Sensors Journal* 11 (10) (2011) 2611–2616.
104. C. Watts, S. M. Hanham, J. P. K. Armstrong, M. M. Ahmad, M. M. Stevens, N. Klein, Microwave dielectric sensing of free-flowing, single, living cells in aqueous suspension, *IEEE Journal of Electromagnetics, RF and Microwaves in Medicine and Biology* 4 (2) (2020) 97–108.
105. D. Jahn, A. Soltani, J. C. Balzer, M. Koch, W. Withayachumnankul, Fabry-Perot interferometer for sensing polar liquids at terahertz frequencies, *Journal of Applied Physics* 121 (2017) 204502.
106. I. Al-Naib, Biomedical sensing with conductively coupled terahertz metamaterial resonators, *IEEE Journal of Selected Topics in Quantum Electronics* 23 (4) (2017) 1–5.
107. Z. Xu, P. Mazumder, Bio-sensing by Mach–Zehnder interferometer comprising doubly-corrugated spoofed surface plasmon polariton (DC-SSPP) waveguide, *IEEE Transactions on Terahertz Science and Technology* 2 (4) (2012) 460–466.
108. J. Baker-Jarvis, M. D. Janezic, D. C. Degroot, High-frequency dielectric measurements, *IEEE Instrumentation and Measurement Magazine* 13 (2) (2010) 24–31. https://doi.org/10.1109/MIM.2010.5438334
109. T. Pechrkool, P. Sangmahamad, P. Thiamsinsangwon, T. Sutham, B. Kumkhet, V. Pirajnanchai, High-sensitivity contactless microwave sensor based on rectangular complementary split ring resonator for glucose concentration characterization, in: *2022 37th International Technical Conference on Circuits/Systems, Computers and Communications (ITC-CSCC)*, Phuket, Thailand, 2022, pp. 971–974. https://doi.org/10.1109/ITC-CSCC55581.2022.9895084
110. A. M. Albishi, M. K. E. Badawe, V. Nayyeri, O. M. Ramahi, Enhancing the sensitivity of dielectric sensors with multiple coupled complementary split-ring resonators, *IEEE Transactions on Microwave Theory and Techniques* 68 (10) (2020) 4340–4347. https://doi.org/10.1109/TMTT.2020.3002996
111. A. Vivek, K. Shambavi, Z. C. Alex, A review: Metamaterial sensors for material characterization, *Sensor Review* 39 (3) (2019) 417–432.
112. A. S. Saadeldin, M. F. O. Hameed, E. M. A. Elkaramany, S. S. A. Obayya, Highly sensitive terahertz metamaterial sensor, *IEEE Sensors Journal* 19 (18) (2019) 7993–7999. https://doi.org/10.1109/JSEN.2019.2918214
113. S. Poorgholam-Khanjari, F. B. Zarrabi, Reconfigurable Vivaldi THz antenna based on graphene load as hyperbolic metamaterial for skin cancer spectroscopy, *Optics Communications* 480 (2021) 126482. https://doi.org/10.1016/j.optcom.2020.126482
114. Z. Du, R. Zhou, S. Luo, D. Zhao, W. Long, Q. Ling, Z. Yu, D. Chen, Tunable multi-band absorbers based on graphene metasurfaces for infrared sensing and switching, *Optics Communications* 534 (2023) 129320. https://doi.org/10.1016/j.optcom.2023.129320
115. P. Wu, X. Zeng, N. Su, M. Chen, Y. Zeng, Y. Yu, A graphene perfect absorber with tunable, dual band, high sensitivity characteristics, *Diamond and Related Materials* 125 (2022) 109002. https://doi.org/10.1016/j.diamond.2022.109002
116. Z. Geng, X. Zhang, Z. Fan, X. Lv, H. Chen, A route to terahertz metamaterial biosensor integrated with microfluidics for liver cancer biomarker testing in early stage, *Scientific Reports* 7 (2017) 16378. https://doi.org/10.1038/s41598-017-16762-y
117. H. Shen, C. Liu, F. Liu, Y. Jin, B. Guo, Z. Wei, F. Wang, C. Tan, X. Huang, H. Meng, Multi-band plasmonic absorber based on hybrid metal-graphene metasurface for refractive index sensing application, *Results in Physics* 23 (2021) 104020. https://doi.org/10.1016/j.rinp.2021.104020

118. Y. I. Abdulkarim, O. Altintas, A. S. Karim, H. N. Awl, F. F. Muhammadsharif, F. Ö. Alkurt, M. Bakir, B. Appasani, M. Karaaslan, J. Dong, Highly sensitive dual-band terahertz metamaterial absorber for biomedical applications: Simulation and experiment, *ACS Omega* 7 (42) (2022) 38094–38104. https://doi.org/10.1021/acsomega.2c06118
119. Y. Lv, W. Liu, J. Tian, R. Yang, Broadband terahertz metamaterial absorber and modulator based on hybrid graphene-gold pattern, *Physica E: Low-dimensional Systems and Nanostructures* 140 (2022) 115142. https://doi.org/10.1016/j.physe.2022.115142
120. S. Seyyedmasoumian, A. Attariabad, A. Pourziad, M. Bemani, Refractive index biosensor using metamaterial perfect absorber based on graphene in near-infrared for disease diagnosis, *IEEE Sensors Journal* 22 (15) (2022) 14870–14877. https://doi.org/10.1109/JSEN.2022.3187011
121. R. Bhati, M. Jewariya, A. K. Malik, Spoof surface plasmon-based terahertz metasensor for glucose and ethanol, *Applied Physics A* 128 (2022) 840. https://doi.org/10.1007/s00339-022-05943-w
122. Y. Kim, A. Salim, S. Lim, Millimeter-wave-based spoof localized surface plasmonic resonator for sensing glucose concentration, *Bio-sensors* 11 (2021) 358. https://doi.org/10.3390/bios11100358
123. S. P. Singh, N. K. Tiwari, M. J. Akhtar, Spoof surface plasmon-based coplanar waveguide sensor for dielectric sensing applications, *IEEE Sensors Journal* 20 (1) (2020) 193–201. https://doi.org/10.1109/JSEN.2019.2942709
124. Y. Zhang, Y. Liu, H. Xi, T. Meng, G. Zhao, A Mach–Zehnder interferometer refractive index sensor on a spoof surface plasmon polariton waveguide, *Electronics* 11 (2022) 3944. https://doi.org/10.3390/electronics11233944
125. A. Kumari, S. P. Singh, N. K. Tiwari, M. J. Akhtar, Design of a differential spoof surface plasmon sensor for dielectric sensing and defect detection, *IEEE Sensors Journal* 22 (4) (2022) 3188–3195. https://doi.org/10.1109/JSEN.2022.3142067
126. M. J. Akhtar, N. K. Tiwari, S. P. Singh, Spoof surface plasmon based planar THz sensor system using dumbbell shaped unit cell, in: *2018 International Symposium on Antennas and Propagation (ISAP)*, Busan, Korea (South), 2018, pp. 1–2.
127. Z. Xie, G. Wang, L. Sun, Y. Li, R. Cao, Localised spoof surface plasmon based sensor for omni-directional cracks detection in metal surfaces, *IET Microwaves, Antennas & Propagation* 13 (2019) 2061–2066. https://doi.org/10.1049/iet-map.2018.5749
128. X. Chen, W. Fan, Ultrasensitive terahertz metamaterial sensor based on spoof surface plasmon, *Scientific Reports* 7 (2017) 2092. https://doi.org/10.1038/s41598-017-01781-6
129. I. Llamas-Garro, Z. Brito-Brito, F. Mira, M. T. de Melo, J.-M. Kim, Microwave spoof surface plasmon sensor for dielectric material characterization, *IEEE Sensors Letters* 6 (5) (2022) 1–4. https://doi.org/10.1109/LSENS.2022.3165215
130. Y. Zhang, Y. Xia, H. Ling, J. Zhang, K. Li, C. Yuan, H. Ma, W. Huang, Q. Wang, X. Su, Label-free diagnosis of ovarian cancer using spoof surface plasmon polariton resonant biosensor, *Sensors and Actuators B: Chemical* 352 (2022) 130996. https://doi.org/10.1016/j.snb.2021.130996

13 Exploring NOMA in Satellite and mmWave Photonic Communication for Future Networks

Sk Thaherbasha, SD Nageena Parveen, Aluvoju Vivek, Munjala Praharsha, Pradeep Dhara, and Vineetha Paul Srungarapati

13.1 INTRODUCTION

This chapter reviews the integration of Non-Orthogonal Multiple Access (NOMA) with satellite and millimetre-wave (mmWave) photonic communication technologies that are foreseen to revolutionize the future wireless networks. It explains in depth the principles and benefits of NOMA, while discussing its synergy with satellite and mmWave systems, and the role of photonic technologies in enhancing performance. Solutions have been proposed, thus, by highlighting issues and impediments such as interference in a signal, its propagation effect, resource sharing, among others; consequently, the book will help scholars and researchers grasp insight on the basis of which it will assist real-life application studies with the prospect of 6G networks, IoT connection, and deep space applications, discussing current challenges to advance it and suggest areas for possible research in order to determine how it is able to lead future advancements.

13.1.1 Overview of Future Wireless Communication Needs

The journey of communication technology has started many years ago. The main significance came when Alexander Graham Bell has invented the telephone in early 1876. A big revolution has taken place while moving from wired telephone to wireless in 1940. In wireless technology, mobile radio telephone-0G was the pre-cellular generation. In the 1940s, Motorola and Bell systems are pioneers who gave birth to 0G. Communication between users exists as pressing a button to speak and leaving that button to listen to other side, and it was a manual technology popularly known as push to talk. This technology has limited operation and cannot be used by everybody. The overview of various generations (1G to 5G) of wireless communication technology is shown in Table 13.1.

DOI: 10.1201/9781003599111-13

TABLE 13.1
Overview of 1G to 6G

Generation	Speed	Technology	Features
1G (1970–1980s)	14.4 kbps	AMPS, NMT, TACS	Voice only
2G (1990–2000)	9.6/14.4 kbps	TDMA, CDMA	Voice and data
2.5, 2.75G	171.2 kbps 20–40 kbps	GPRS EDGE	Voice, data, and mobile internet email
3G (2004–2005)	3.1 Mbps 500–700 kbps	CDMA2000 UMTS and EDGE	Voice, data, multimedia, video calling, and live TV
3.5G (2006–2010)	14.4 Mbps 1–3 Mbps	HSPA HSPA+	All the services from 3G with high speed and more mobility
4G (2010 to till)	100–300 Mbps 3–5 Mbps	WiMax LTE and WiFi	High speed, voice over IP, HD multimedia
5G (2020)	1–10 Gbps	LTE-A, OMA and NOMA	Low latency, IoT, high speed data, smart health care, autonomous driving, etc.
6G (2030 and beyond)	100 Gbps to 1 Tbps	AI-based networking, Terahertz (THz) communication	Ultra-low latency, AI integration, holographic communication, smart cities, real-time sensing, global connectivity, and extended reality applications

The switch from 5G to 6G unlocks all new horizons in wireless communication, with present access through components such as terahertz (THz) band communication, pervasive artificial intelligence, and intelligent communication environments. It also incorporates terrestrial and space networks into the "Internet of Space Things" by integrating CubeSats (Cube-Satellite) and UAV (Unmanned Aerial Vehicle) technologies [1], and it closes connectivity gaps in remote areas in addition to ultra-dense networks serving real-time applications such as autonomous systems and remote healthcare. With data rates of terabits per second, ultra-low latency, and designed for energy efficiency, it is able to cater to developing needs such as intelligent industrial automation and precision agriculture. Intelligent communication environments will dynamically adapt the wireless propagation medium within them to minimize interference and improve reliability. Artificial Intelligence (AI) and automation will be the key drivers for all 6G network operations, optimizing resource allocation, spectrum use, and scalability to give seamless, global, and sustainable connectivity.

AI has brought a change in wireless communications, from cognitive radios to remote sensing, computer vision, network management, and many more. AI uses channel estimation and symbol detection and eliminates the constraints of conventional techniques in massive MIMO (Multiple-Input Multiple-Output) communications because of machine learning (ML) speed in parallel processing. Data-driven techniques are mainly substituted in direct approaches instead of the model-based counterparts while rendering more robust task-related schemes, like the ones required for symbol detection and channel estimation, for a pretty disparate scenario observed in the real-life analysis. For instance, an optimum setup of the configuration for predicting traffic flow resource allocation would be performed for reinforcement learning. The corresponding application of Edge Ai facilitates making quicker real-time decisions. For example, AI contributions to network orchestration include supervised, reinforcement, and unsupervised learning techniques that realize disadvantageous functions such as traffic prediction, resource optimization, and even security improvements. However, the research community is very active in overcoming the challenges arising from lack of standardized algorithms and good quality datasets and the requirement for effective comparison methods considering possible scenarios of future wireless networks such as 6G [2].

This has been enabled by 5G and 6G advancements while integrating further into smart device communications beyond wired connections. Thus, IoT systems use wireless sensor networks and devices to embrace existing computing further beyond industrial applications into home applications. For instance, such technologies could deliver very low latencies, the highest reliabilities, and utilization of energy for applications ranging from mMTC (massive Machine-Type Communications) to eMBB (enhanced Mobile Broadband) and URLLC (ultra-reliable low-latency communications). Examples of such innovation enabled by the combination of IoT with 5G are edge computing and fog computing, which involve moving all data processing activities as close as possible to the sources. Future 6G is probably going to realize global coverage by satellite integration and advanced spectrum usage, so that its users can go on freely without the threat of interruption. The multi-layered structure of IoT also possesses critical security provisions to cover applicability, network, data, and perception layers to help relieve evolving threats. Future deployments are expected to focus more on energy-efficient packaging, optimal spectrum usage, and software-defined radio for the independence of the kind of environment within which reliability of wireless communication is expected. The promises made between IoT and next-generation technologies include delivering further automation towards applications like smart cities and mission-critical systems. Research is still ongoing, despite the developments, thus improving interoperability issues, path loss, and propagation dynamics, all in the effort of producing strong and healthy IoT ecosystems for future networks [3].

Recently, mmWave has started to play a very important role in future generation networks, since it would provide ultra-high data rates, massive bandwidth, and low latency for the increasing demands of advanced applications. Working within the frequency range of 30–300 GHz, mmWave provides much larger spectrum availability compared to traditional bands, which enables high increases in capacity and faster communication. This makes it perfect for applications that require high data, such as augmented reality, virtual reality, and ultra-high-definition video streaming. Besides, mmWave plays a vital role in enabling technologies like 5G and 6G, powering smart cities, autonomous vehicles, and industrial automation with high-speed real-time connectivity. While limited range and susceptibility to blockage are some of the challenges with using mmWave, innovation in beamforming and antenna design is overcoming these barriers. Innovation in beamforming and antenna design has been surmounting such barriers and will be crucial to robust, scalable, efficient future generation networks.

This is the necessity for mmWave photonic communication due to the increasing inability of regular mmWave communication systems, which will not be able to meet the future prospects of a network like 6G and beyond, whereas mmWave communication can give excellent speeds and frequencies; most of the time, this sort of communication is constrained. It has bandwidth, limited range, and scalability barriers. The new mmWave photonic communication solves issues by incorporating optical technologies on communication, thus offering exceptional performance and advanced applications beyond what is currently achievable within radio frequency (RF) technology.

While both technologies operate on the mmWave spectrum, the mmWave photonic communication merges the advantages of photonics and mmWave frequencies to overcome bandwidth and speed limitations of traditional mmWave communication, thus making it the cornerstone for future high-performance networks. In order to meet the requirements of future wireless communication systems, the research mainly focuses on: (1) Improving spectral efficiency through the use of (a) higher-order modulation methods and (b) multiple access schemes. (2) Examining the unlicensed spectrum's potential through the use of (a) the visible light spectrum (430–790 THz) and (b) infrared (300 GHz to 430 THz).

To establish communication between BSs (Base Station) and users, the geographical area is partitioned into several number of cells. The designed cellular system should provide service to the maximum number of users in a given cell. The allocated bandwidth for a cell is limited and the maximum number of users need to access the channel with better quality of service. Hence, in cellular systems, various MASs (Mobile Access System) are employed at BS for effective utilization of frequency resources. To accommodate multiple users, many MASs are proposed for 1G to 5G. While designing MASs, the main requirements should keep in mind such as:

1. It should drive more users without interference.
2. It can provide maximum spectrum efficiency.
3. The handover to adjacent cells cannot be hard.

Various multiple access techniques, which are used for 1G to 6G along with future generation communication systems, are classified and described in Figure 13.1.

13.1.2 Role of NOMA in Next-Generation Networks

Next-generation wireless communication network system, in particular, the sixth generation (6G), and the technology of NOMA for networking have gradually evolved as a revolution. As compared to the OMA (Orthogonal Multiple Access) systems, NOMA allows a certain number of users to access the channel using frequency, time, or code division; NOMA also uses power division and the SIC (Successive Interference Cancellation) technique. This innovation pinpoints the increasing call for massive connectivity, low latency, and superior spectral efficiency. Since 6G networks are anticipated to support various technologies and applications, including IoT, autonomous systems, and immersive virtually based systems, it is obvious that

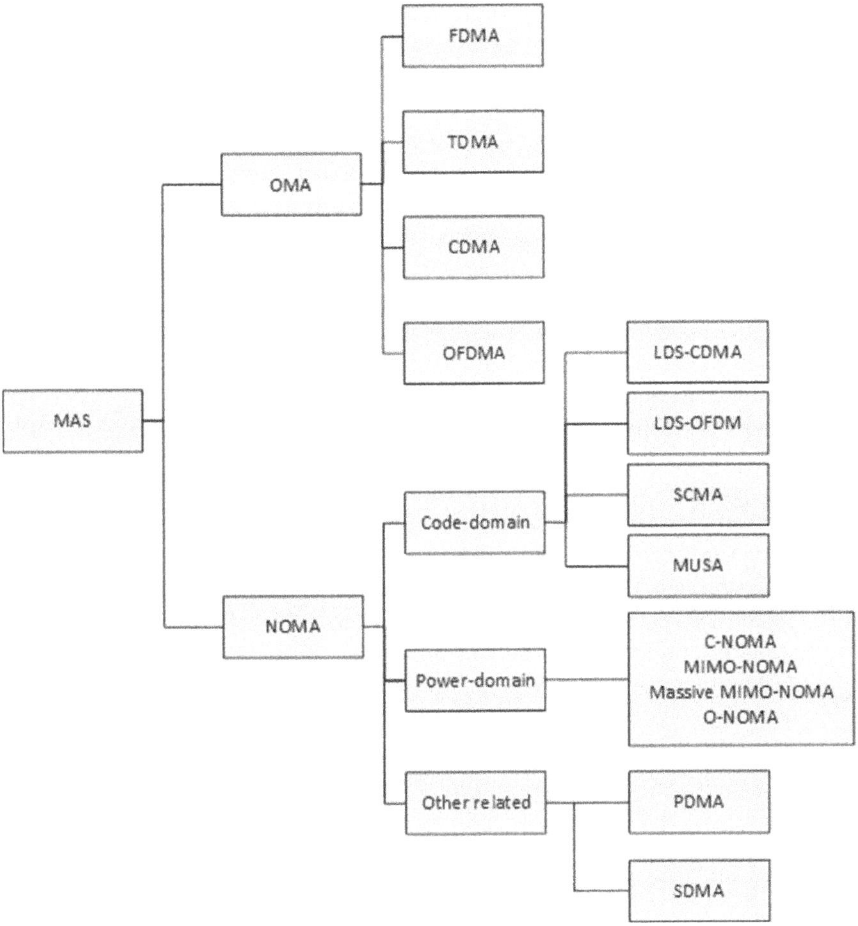

FIGURE 13.1 Classification of multiple access schemes.

NOMA is a fundamental tool that needs to replace the conventional access schemes and methods. Moreover, integrating with advanced technologies, such as ML, reconfigurable intelligent surfaces, and THz communication, established NOMA as a fundamental technology for supporting a variety of heterogeneous scenarios with high demands in the context of the enabling digital world. NOMA's integration with the latest trends and innovative technologies like THz (Terahertz) communications, OTFS (Orthogonal-Time-Frequency-Space) modulation, and RIS (Reconfigurable Intelligent Surfaces) is steering a paradigm shift in the wireless communication. By reducing the inter-symbol interference of the arriving signals in high-mobility scenarios, the OTFS-NOMA can support the demands of the stationary as well as the fast-moving user [4]. Similarly, THz-NOMA fully exhausts the bandwidth available in the THz band to deliver ultra-high data rates satisfying beyond 5G requirements. Reconfigurable intelligent surfaces complement NOMA networks by actively modifying the physical characteristic of the wireless channel to improve reception and quality. These integrations are crucial for the subject of combating the obstacles of compact urban environments, rapidly integrative technologies, and the upcoming sixth generation of networks. Various aspects of NOMA with future wireless communication systems are explained as follows:

i. **Connectivity in 6G Networks:** This approach gives massive boosts in spectral efficiency and allows 6G networks to meet the massive connectivity requirements of smart cities and the Internet of Things, where billions of devices must communicate reliably in real time. NOMA is ensured through dynamic resource allocation in order to provide diversity QoS (quality-of-service) levels, ensuring the requirements for both high data rate and ultra-low latency. It is thereby well-suited for meeting all the demands of next-generation wireless communication.
ii. **Resources Management with Machine Learning:** Resource management is improved through ML in NOMA networks through online user clustering and power control. For example, reinforcement learning takes into account changes in the network environment, increases the efficiency of using the spectral resources, and decreases latency in the high traffic area. Also, through ML techniques, new approaches assist in handling multiple objectives; concerning the QoS requirements of 6G networks and complex scenarios, resources are smoothly allocated.
iii. **Autonomous Systems:** An important role of NOMA in enhancing the automaticity of systems is the presence of reliable communication and decrease of latency in conditions of various dynamicity. It enables the forming of groups that contain devices with different mobility rates, improving signal quality, and increasing spectral efficiency. This capability guarantees medium and unrestricted communication between autonomous robots, UAV, and vehicles in logistics, agriculture, and defence industries. Due to the potential of establishing high performance, even while operating within the context of very dynamic environments requiring real-time functionality, NOMA is vital to the safety, synchronization, and integration of self-controlled systems [5].

13.2 FUNDAMENTALS OF NOMA

Non-orthogonal multi-access is an advanced communication technology that works by allowing more than one user to access the same frequency at the same time so that better spectral efficiency could be achieved. From Figure 13.2, the base station transmits to User1 and User2 using some amplified power level difference where the power domain multiplexing is employed. User2 has a more powerful signal that allows decoding of its data directly, while U1 does a SIC scheme by first decoding and subtracting User2 signal then proceeding for its own signal. The layering thus ensures separating signals and recovering data sharing the same frequency. This is one of the most crucial advantages of NOMA in employing advantageously more users in 5G networks and beyond by managing resource efficiency [6].

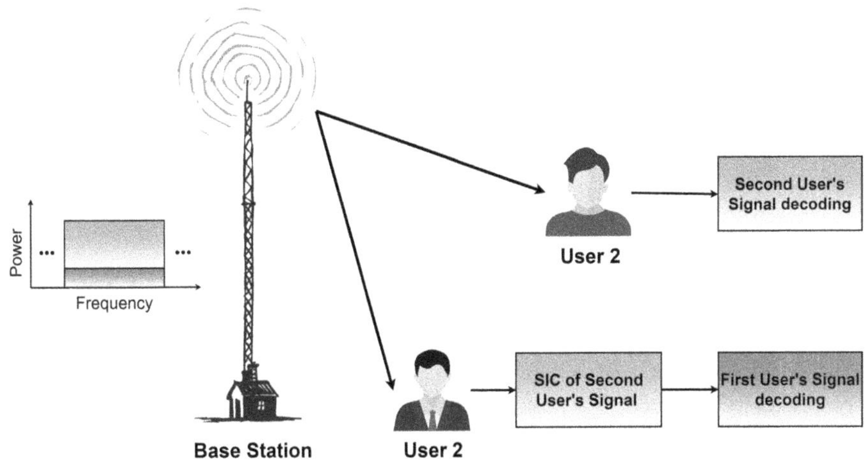

FIGURE 13.2 NOMA wireless communication with two users.

13.2.1 Principles of Non-orthogonal Multiple Access

1. **Power Domain Multiplexing (PDM):** PDM is an important method in NOMA where different signals sent through the same frequency band are distinguished by their power levels. Therefore, users of PDM have different power levels to transmit at the same time without interference, which increases both spectral efficiency and the number of users supported. This technique can be used well in utilizing the given bandwidth and at the same time maintain the latency and increase the overall throughput in the wireless systems.
2. **Superposition Coding:** In NOMA, the multiple signals are encoded at different power levels and together the signals are broadcast in the same frequency band. This is said to be superposition. This technique enables different users who have different channel quality to operate in the same resource block, while the signals in this block are added and then decoded by successively interring them using the SIC at the receiving end, thus increasing spectral efficiency and the system capacity. The superposition coding is best suited for many users and diverse requirements environment including heterogeneous network and high traffic density scenarios.
3. **Successive Interference Cancellation (SIC):** In NOMA, SIC is a decoding operation used in order to handle interference produced by multiple users present in the network. In SIC, the base station initially demodulates and subtracts the strongest user signal and considers it noise for the next user, then demodulates the weaker user's signal. This decision-making procedure improves the spectral efficiency of the system and allows for the possibility of multiple user conversation even at different power levels.
4. **User Differentiation via Power Allocation:** In NOMA, user differentiation is done through the allocation of power levels corresponding to the status

of the users' channel conditions. Due to benefits of channel quality, strong users are assigned less power to interfere with weak users. On the other hand, the weaker users with unfavourable channel conditions are assigned a higher transmission power in order to have their signals properly detected. This makes the strategy not only accommodate the trade-off between the spectral efficiency and the fairness but also guarantee a successful SIC at the receiver side since the signal from the stronger user can still be separated after detecting and decoding that from the weaker user.

5. **Fairness and Quality of Service (QoS):** Fairness and QoS in NOMA are achieved through careful power allocation and resource management. Power levels in the NOMA are divided depending upon the user channel conditions; poor channel quality are assigned with more power and QoS to meet the needed quality; on the other hand, powerful users channel quality provides less power to help the entire efficiency of the system. It helps solve a major problem where more reserved and less experienced users are left out while more skilled users continue to dominate the system. Moreover, the power of NOMA to grant fungible power to individual users can also be applied in a dynamic basis towards certain users, e.g., emergency response personnel in crash-sensitive conditions, or IoT devices, which require comparatively low latency, enabling solution to heterogeneous multi-user cases.

13.2.2 Comparison with Traditional Access Techniques with NOMA

Table 13.2 provides a detailed comparison of traditional multiple access techniques (FDMA, TDMA, CDMA, and OMA) with NOMA. It clearly states how advanced NOMA is in terms of spectral efficiency, support for massive device connectivity, making it crucial for next-generation communication systems.

13.2.3 Benefits of NOMA in Wireless Communication

The key benefits of NOMA in wireless communication are:

1. **Efficient Spectrum Utilization:** NOMA implements power-domain multiplexing under which different users can transmit on the same frequency at the same time. It greatly enhances the efficiency of the used frequency spectrum while eliminating the major disadvantage of the conventional orthogonal multiple access methods like TDMA or FDMA where resources are shared among users. For instance, NOMA can achieve a spectral overall efficiency rate of up to 7 bits/Hz, while traditional OMA systems would only give a maximum of 4–5 bits/Hz under similar conditions.
2. **Increased Connectivity:** On arrival of many users, more can be supported in using NOMA on a specific resource block, making it useful for scenarios such as 5G and IoT where maximum connectivity is required. For instance, both NOMA and OMA can be compared to show that in a single cell, NOMA can accommodate 10 or 20 folds more numbers of devices than the OMA.

TABLE 13.2
Comparison with Traditional Access Techniques with NOMA

Feature	FDMA (Frequency Division Multiple Access)	TDMA (Time Division Multiple Access)	TDMA (Time Division Multiple Access)	OMA (Orthogonal Multiple Access)	NOMA (Non-orthogonal Multiple Access)
Access method	Users are allocated distinct frequency bands	Users are allocated distinct time slots	Users share the same frequency and time resources but are distinguished by unique codes	Combines features of FDMA, TDMA, and CDMA to allocate orthogonal resources.	Users share the same frequency and time resources, differentiated by power levels.
Spectral efficiency	Limited by the number of frequency bands available	Limited by the number of time slots and overhead of switching	Higher than FDMA and TDMA, but interference management is complex	Moderate, constrained by strict orthogonality in resource allocation	High, as multiple users share the same resources simultaneously
Scalability	Poor scalability due to limited frequency spectrum	Limited scalability; adding more users requires splitting existing slots	Good scalability, as multiple users can share the same spectrum	Moderate scalability, with strict limits on orthogonality.	Excellent scalability; supports massive connectivity by efficiently using available resources
Latency	Low latency, as users can transmit simultaneously	Higher latency due to waiting for time slots.	Moderate latency, depending on decoding delays	Moderate latency, similar to TDMA or CDMA depending on implementation	Low latency, as users transmit concurrently on the same resources
Resource allocation	Fixed frequency bands	Fixed time slots.	Fixed spreading codes.	Static orthogonal resources (frequency, time, or code).	Dynamic power allocation based on user channel conditions

(*Continued*)

TABLE 13.2 (*Continued*)
Comparison with Traditional Access Techniques with NOMA

Feature	FDMA (Frequency Division Multiple Access)	TDMA (Time Division Multiple Access)	TDMA (Time Division Multiple Access)	OMA (Orthogonal Multiple Access)	NOMA (Non-orthogonal Multiple Access)
Fairness	Fair for users with similar frequency conditions	Fair for users with similar time-slot conditions.	Fairness depends on code allocation and interference management.	Fair under orthogonal resource sharing but inefficient for diverse user demands	Highly fair; ensures weaker users receive higher power for better performance
Complexity	Simple to implement and manage	Moderate complexity for slot synchronization	High complexity due to code management and decoding requirements	Moderate complexity, combining features of traditional methods	High complexity due to SIC and real-time power allocation
Energy efficiency	Relatively low, as fixed frequency resources may be underutilized	Moderate, but switching between time slots adds overhead	High energy consumption due to complex decoding.	Energy efficiency varies, but orthogonality constraints lead to underutilized resources in some cases.	High, as shared resources reduce overall energy consumption and optimize usage.
Suitability for IoT	Limited, not suitable for massive device connectivity	Poor, as scalability issues arise with increased users	Better than FDMA and TDMA, but decoding complexity limits suitability	Suitable only for small-scale IoT systems due to scalability constraints	Ideal for IoT, supporting massive connectivity and diverse QoS requirements

3. **Enhanced Throughput:** Additional user signals imposed upon the existing signals to enable high priority users, those closer to the base station, to transmit at the same time with low priority users, those who are farther from the base station. With this, the system will have least time that will be utilized hence maximizing the total throughput of the system. Several simulations reveal that NOMA enhances the system's throughput to about 30%–50% more than OMA under user distance disparity environments.
4. **Reduced Latency:** In this case, users with NOMA can transmit data at once and do not experience delays, a feature unlike in OMA where a user has to wait for the other one to finish communicating before he or she can proceed. Especially in case of real-time applications such as video streaming and augmented reality, it becomes all the more crucial.

13.3 OVERVIEW OF SATELLITE TECHNOLOGIES FOR WIRELESS NETWORKS

Satellite communication systems are the actual backbones of wireless modern networks, providing complete, reliable worldwide coverage, as well as flexible connectivity in places where many people and very few people exist. Other assets like space and the free propagation of the transmitting and receiving ends put satellite communications at a different level—a level unattainable by terrestrial networks and their concrete infrastructures. They are dependent upon their ejection points between terrestrial-based phone networks, hence they are quite suited for areas that do not fall under the scopes of coverage by traditional cellular technologies. Examples would be oceans, seas, deserts, polar regions, etc. Thus, satellite technologies form a major backbone for key applications such as broadcasting and navigation, broadband Internet, and other emerging ones like 6G.

Satellites in wireless networks are mainly categorized into three types. They are Geostationary Earth Orbit (GEO), Medium Earth Orbit (MEO), and Low Earth Orbit (LEO). This categorization is done on the basis of altitude and operational features. GEO satellites are placed about 35,786 km above the earth and provide their broad coverage and consistent connectivity. This makes them ideal for broadcasting and also for monitoring weather. MEO satellites are about 8,000–20,000 km in the atmosphere above the earth; applications of these satellites are mainly in navigation systems like GPS since they offer a good blend between latency and coverage. LEO satellites are in between 500 and 2,000 km from the earth; they provide low-latency, very-high-speed, broadband services and applications for IoT. Figure 13.3 illustrates the paths of the all three orbits. Hence, every orbital category provides different potentialities and accuses impediments that form in the wireless communication network [7].

The operational backbone of satellite systems includes transponders, antennas, and ground stations. On the onboard equipment of a satellite, it receives signals from the earth, amplifies them, and retransmits them over to designated specific regions. Antennas installed on satellites and ground stations are designed for specific frequency bands such as the C-band, Ku-band, Ka-band, and V-band; each of these bands has

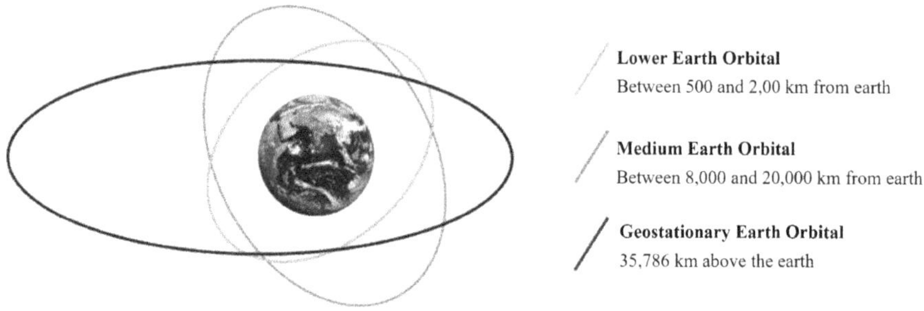

FIGURE 13.3 Satellite categories in wireless networks.

its unique advantages in terms of data rate, coverage area, and interference control. Ground station Earth stations are the ones that track satellites by using parabolic antennas and work with uplink and downlink of data within a terrestrial network. With advances in technology, including phased array antennas and reconfigurable transponders, there have further been improvements in beam flexibility and spectral efficiency.

Satellites will be an inherent part of NTN (Non-Terrestrial Networks) in the future wireless ecosystems, even including the 6G scenario. Satellites have increasingly become part of terrestrial infrastructure to provide seamless coverage within hybrid systems. The deployment of mmWave frequency bands for satellite links is promising ultra-high-speed data transfer, while optical satellite communications provide much higher bandwidth and better resistance against interference. Such advancements will not only ripen further possibilities but also open doors for applications like augmented reality (AR), autonomous transportation, and real-time remote sensing [8].

However, like other systems, they are limited by high deployment costs, attenuation of signals due to atmospheric interference, and latency. Fortunately, the development of reusable launch vehicles, advanced satellite manufacture techniques, and new innovations such as optical satellite communication promise to mitigate these challenges. Initiatives such as mega-constellations (e.g., SpaceX's Starlink, OneWeb, and ISRO's satellite programs) further served to extend the reach of satellite airborne wireless networks. These promise to deliver cost-effective ubiquitous connectivity and applications from smart cities and self-driving cars to deep space.

13.3.1 Challenges in Satellite-Based Communication

Satellite-based communication is characterized by several constraints, which affect the system performance, reliability, and efficiency. These challenges range from the basics, such as signal loss and delay, to policy and security questions and compatibility with the emerging technology platforms, the latter of which needs to be figured out to maintain sound and perfect communication. Below are some challenges mentioned in detailed fashion [9]:

1. **Signal Attenuation and Propagation Loss:** Satellite communication experiences signal weakening and transmission loss as main hurdles because of the trouble links and diverse interference occurrences contributing to signal strength. Free-space path loss (FSPL) is directly proportional to the square

of the distance of satellite to receiver and the square of the signal frequency; the power is reduced a lot. However, rain fade, fog, and cloud absorption considerably affect high-frequency bands such as Ka-band and V-band, resulting in signal distortions and reduced dependability due to multipath, in which the signals reflect back and interfere.

2. **Latency and Delay:** Latency and delay issues in satellite communication arise primarily from the vast distances and signal processing requirements. Propagation delay arises from the fact that signals travel at the speed of light $c \approx 3 \times 10^8$ m/s; geostationary satellites are situated at about 35,786 km above the earth's surface, and hence, a signal in transit would take about $2 \times \frac{35,786}{c}$ m/s ≈ 240 ms for round trip; at least a part of the processing delay is caused by the time needed for such processes as signal encoding, which accumulates in both satellite onboard systems and ground station infrastructure, further increasing total communication latency.
3. **Interference and Spectrum Management:** Satellite communication involves challenges like interference and spectrum management. Adjacent satellite interference takes place when signals from adjacent satellites intersect since the angle of separation is small. Spectral efficiency is improved with frequency reuse, but at the same time, co-channel interference takes place between the overlapping beams. From experience, terrestrial interference occurs when ground-based systems operate on similar or neighbouring frequencies, thus interfering with satellite communication links. Such matters need sophisticated methods of interference cancelling and synchronization of the signals.
4. **Doppler Shift:** Doppler shift on the other hand is the tendency of the frequency of a signal to change when a satellite is moving and in regard to the ground station. In Low-Earth Orbit (LEO) systems, where satellites move at high velocities (typically ~7.8 km/s), the received signal frequency f_r is shifted from the transmitted frequency f_t by an amount given by:

$$\Delta f = f\frac{v}{c}$$

where υ is the relative velocity between the satellite and receiver, and c is the speed of light. This change can result in time-scale synchronization and a decrease in performance for systems that share information unless it is corrected using Doppler shift [10].
5. **Orbital Dynamics and Mobility:** Orbital dynamics and mobility pose significant challenges in satellite-based communication, particularly for maintaining precise positioning and seamless connectivity. Satellite drift is caused by perturbing forces such as lunar and solar gravity, atmospheric drag in cases of low earth orbit satellites, and solar radiation pressures that distort the required position of the satellite from its planned path. In LEO and MEO systems, the position changes quickly and requires handover within a short period, which introduces latency, synchronization problem, and signal fading at the time of handover.
6. **Beam-forming and Coverage Challenges:** There are several beam-forming and coverage challenges in satellite communication, mainly due to the need to deliver signals to different geographical regions and various

user densities. Non-uniform coverage happens because although the satellite beams are created with fixed designs of footprints, the whole area that is mapped does not relay strong signals at different terrains, climate, and population densities. To this end, dynamic beam steering through phased array antennas is adopted so that the satellite can flexibly point its beams to certain mobile users or areas, which have high need for communication resources while minimizing crosstalk interference.

7. **Energy Efficiency Constraints:** In satellite communication, capacity for energy efficient transmission is limited due to low power generated from solar panels and batteries covering the satellite on-board, and this limits both the power used in transmission and the degree of signal processing that is feasible. Due to the existence of these power limitations, controlling and deploying communication power, computational loads, and heat generated by hardware elements inside the space craft becomes challenging but crucial to avoid overheating of systems while at the same time providing reliable information transfer.
8. **Integration with Emerging Technologies:** In the context of NOMA, one of the major issues relates to user clustering, where users are clubbed based on their channel quality, and the second is the interference management because the time frequency and code resources that the multiple users are sharing leads to intra-cluster interference, which has to be handled wisely with the help of power control techniques. One of the primary issues in incorporating photonic communication into satellites is the phasing of photonic beams; this must be precise, as it affects the signal power over ranges, and also in controlling conversion losses during optoelectronic and electro-optical conversions to decrease overall system efficiency. These integration issues require proper design for efficient integration of NOMA with the photonic communication and eradication of interference, alignment, and conversion efficiency [11].
9. **Security and Reliability:** Satellite communication systems are susceptible to jamming in which interference has capability to overpower the satellite signal by providing a strong noise or a fake signal, which is a problem concerning the reliability of data. The primary threats to satellite networks and communication protocol are intrusions on satellite networks, spoofing, whereby the attacker mimics the ground stations, and data interception, whereby attackers intercept information being transmitted thus compromising its secrecy. The current threats require higher levels of security incorporating encryption, authentication, and interfering signals denial to realize continuity and reliability in the communication channels.

13.3.2 Integration of NOMA in Satellite Networks

Satellite communication networks can provide coverage where the networks cannot deploy with conventional architecture like terrestrial. Recently, PD-NOMA is

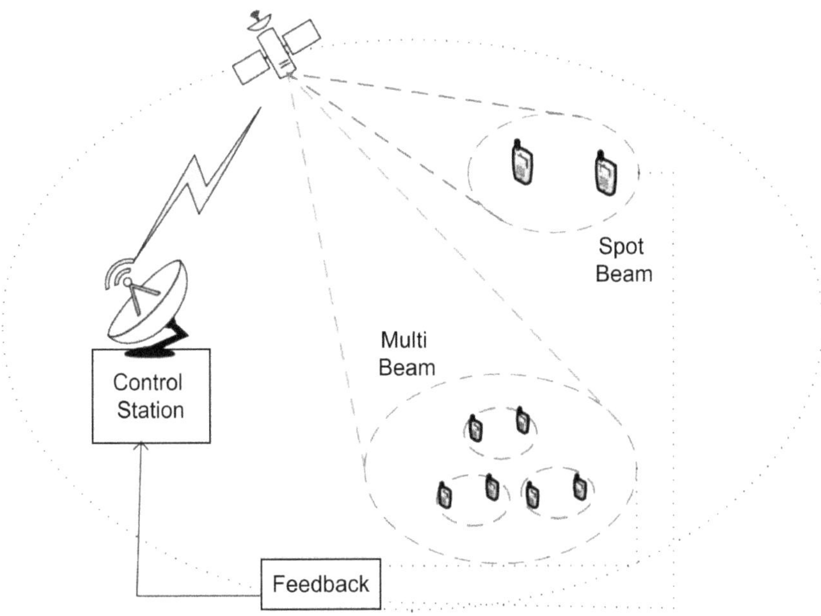

FIGURE 13.4 Integration of NOMA in satellite networks.

applied to satellite communication networks and proven that it meets the requirements of future wireless communication systems. Downlink is based on channel quality, which can be known at satellite from users. The users whose channel quality is low or low antenna gain can be allocated with more power. Similarly, the high antenna gain (high channel quality) user is allocated with low power where the SIC is applied to decode its original signal. The satellite transmits the data by forming spot beam for two-user NOMA and a multi-beam for multi-user NOMA. NOMA-based satellite communication networks can be extended to other application areas such as cooperative and IoT. The users served with a multiple satellite works as BS and the remaining worked as relay. In the first time slot, the main satellite transmits the data to the user as well as other satellite (relays). In the second time slot, user will get amplified signal from the satellite (relay) (Figure 13.4).

13.3.3 Fundamentals of NOMA in Satellite Networks

In the conventional Orthogonal Multiple Access (OMA) systems, users are assigned orthogonal resources, such as time, frequency, or code, and this leads to underutilization of available spectral particularly in large arrays of users. On the other hand, NOMA uses power-domain multiple access, which even makes it possible for users with poor channel quality to access the same resource block. This is by classification whereby all the users split according to channel gain where the users with better channel gains are given lower power and the users with poor channel gains are assigned high power (Figure 13.5).

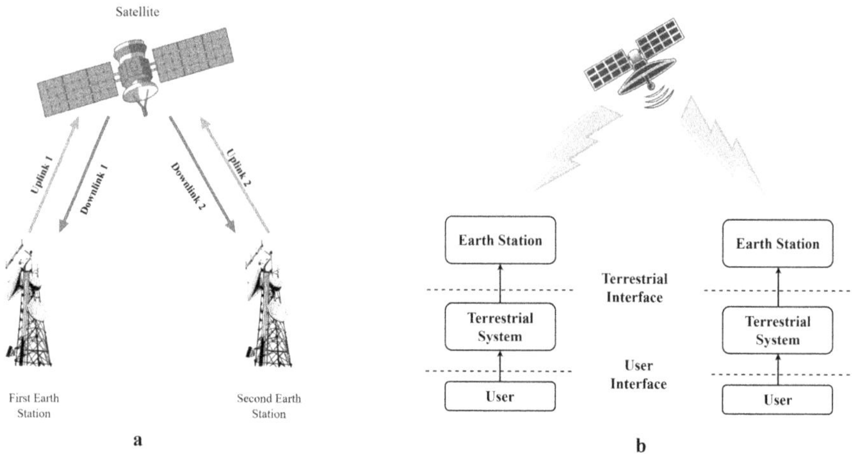

FIGURE 13.5 (a) Two-way satellite communication. (b) General architecture of satellite communications.

In satellite networks, primarily two areas have major concern in integration of NOMA. They are:

1. **Power Allocation:** The received power is shared more dynamically across the users depending on the channel condition, thus improving the design for both coverage and system capacity.
2. **Signal Decoding:** In weak channel conditions, users demodulate their signals first before passing them through SIC and then pass them to strong users thereby enhancing cancellation of interferences.

13.3.4 Challenges in Integrating NOMA with Satellite Networks

While NOMA has significant potential, integrating it with satellite communication networks presents several technical challenges. Some of them are:

1. **Channel Impairments and Fading Effects:** Satellite communication paths are mostly affected by factors such as the free space path loss, atmospheric attenuation, and multi-path fading. These factors have a direct impact on the performance of NOMA as proper Channel State Information (CSI) is vital for power control decision-making. Satellite links entail a huge coverage area that causes spatial dynamics in the channel that affects power distribution. These variations form a significant problem due to the fact that one cannot easily determine the best way to distribute power among the users in the various regions with different signal intensities.
2. **User Differentiation and Power Allocation:** Power control in NOMA-based satellite networks has emphasized calls for the determination of channel gains to users in large coverage areas. In LEO, MEO, and GEO satellite

systems, the relative motion of the users and their distance from the satellite make estimation of the channel conditions challenging. Power control policies have to be adaptive because of the variable position of the user and the variability of the signal transmission path [12].

3. **Successive Interference Cancellation (SIC):** In NOMA, SIC requires a high SINR (Signal-to-Interference-plus-Noise Ratio) to decode the superposed signals effectively and improve interference management. Unlike conventional cellular networks, user distance and propagation delay are random in satellite networks and cause a varying SINR, which makes it challenging to implement SIC users. Moreover, since the number of users of SIC increases with system complexity, implementation of SIC using higher complexity processing elements at the satellite or ground stations will be disadvantageous in terms of implementation time and computational load.
4. **Interference Management:** Interference management is a significant issue in NOMA-based satellite networks because inter-satellite and inter-cell interference results from synchronous transmission of several users over the same frequency resources. In GEO and MEO systems, due to the large area of coverage, users are often in different areas of the earth and therefore interfere spatially with the quality of the signal as received. Applications of efficient interface control methods like beam forming, power control, and use of other higher order codes are vital for health system in a more congested environment.

13.3.5 Synergies between NOMA and Satellite Communication Technologies

Despite challenges mentioned, integrating NOMA with satellite networks can provide several benefits as:

1. **Enhanced Spectral Efficiency:** This reduces the need for additional spectrum allocation, making more efficient use of limited and expensive spectrum resources. As a result, NOMA improves the throughput and capacity of satellite communication systems, especially in resource-constrained environments.
2. **Improved Coverage and Capacity:** Compared to other approaches, NOMA improves satellite network capacity as the system provides a higher power level to users with low signal strength to enable them join the network well. This power distribution helps the users at the edge of the satellite coverage, which always receive low signals for communication. In contrast to the compared traditional OMA that grants the orthogonal resources and faces the coverage to the network fringes problem, the NOMA decreases the general service area and increases the capacity.
3. **Support for Massive Connectivity:** When combined with the satellite networks, NOMA improves the mMTC and IoT support by allowing the uplink of many low power devices at once. NOMA does this through what is

known as "power domain multiplexing," whereby multiple devices can use the same frequency resources through different amounts of transmit power. NOMA is therefore a perfect fit for IoT networks to seamlessly accommodate the high-traffic, low-density sporadic data requirements.

4. **Energy Efficiency:** NOMA improves energy efficiency in satellite networks by dynamically allocating power based on channel conditions and data requirements of each user. This makes the resource distribution more efficient in power and hence decreasing the total power consumed, especially in the uplink from the ground terminals to satellites. In the long-haul mission or even deep space transmission, this is important since it introduces a means of stretching the life of satellite network by reducing power consumption.

13.4 mmWAVE COMMUNICATION

Next-generation wireless communication systems, including 5G, 6G, and advanced satellite communications, are incorporated with a crucial enable technology called mmWave technology. It operates at the frequency band of 30–300 GHz and the Wavelengths vary between 1 and 10 mm. Figure 13.6 gives the alignment of mmWave in frequency spectrum [13].

Key features of mmWave technology are:

1. **High Frequency and Short Wavelength:** mmWave exists on the operating frequency band of 30–300 GHz, which is between the microwave and the infrared region of the electromagnetic spectrum. This high-frequency range offers the prospect of a wide bandwidth and therefore high-speed transmission. In the spectrum ranging from 1 to 10 mm, signals in the mmWave spectrum are different from traditional radio waves in that they possess significantly shorter wavelengths. This makes the implementation of antennas small and compact with mmWave adaptive into the modern density network infrastructure.
2. **Large Bandwidth:** The wide bandwidth of mmWave technology achieves data rates on tens of Gigabits per second. For some bandwidth-intensive applications such as 4k/8k video and, of course, Augmented Reality/Virtual Reality, these high rates can be a prerequisite for wide bandwidths. Hence,

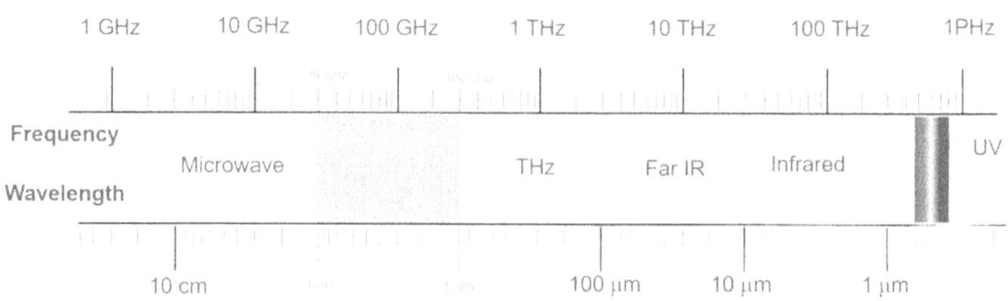

FIGURE 13.6 mmWave frequency band.

wide bandwidth makes the high throughputs possible in mmWave systems communication capabilities, which are free from latency and time delays. The extension of spectrum in mmWave bands frees up traffic load in low frequency bands, hence making efficient usage of overall networks. Therefore, mmWave contributes towards satisfying the ever-increasing need of high throughputs in areas with high traffic demand and data congestion.

3. **Propagation Characteristics:** mmWave signals are required to have a line of sight for communication since their experience with loss is considerable in the presence of an obstruction, hence making diffraction and scattering difficult. Because of their high frequency, mmWave signals witness increased free-space path loss, which is defined as a loss in strength of the signal rapidly over distance. The atmospheric components, such as oxygen and water vapor, add attenuation to the signal, which further reduces its range. Bad weather has a tendency of even adding to the effects of attenuation from precipitation, such as rain, snow, and fog, which further reduces the effectiveness of mmWave systems under such environments.
4. **Directional Beamforming:** mmWave technology uses directional beamforming to focus narrow beams of signals, thereby increasing the strength and eliminating interference. It brings much better spatial reuse with maximum efficiency for many users sharing the same spectrum. Multiple antenna arrays are manipulated in mmWave systems towards particular users, thereby enhancing communication quality. Such systems create what are known as massive MIMO systems that adopt the technology of having too many antennas, thus improving spatial diversity and spectral efficiency and increasing the capacity of the system as a whole.
5. **High-Density Connectivity:** Because of the high loss characteristics of mmWave signals, these signals cannot be used effectively to transmit signals over long distances. To address this, mmWave systems utilize small cell networks as base stations, which occupy limited space due to their size and are located in crowded areas such as city centres and stadia. These small cells offer localized coverage and high data rates for short range. The ability to incorporate multiple pencil beams also makes the mmWave efficient in many users targeting within these crowded spaces.
6. **Low Latency:** mmWave technology means ultra-low latency because it has high data throughput that is very important for latency-sensitive applications such as self-driving vehicles and robotic surgery. High frequency means mmWave completes signal processing in a short time, and the delay in transmission and reception of data is reduced. Subsequently, efficient beamforming minimizes communication time by directing the signal more accurately towards a particular receiver. This capability is critical to support the desired latency range in the 5G and beyond networks, where communication delays must be on the order of milliseconds.
7. **Multipath Fading and Interference:** Due to their extreme reflectance and scattering properties, these wavelengths are more likely to suffer multipath fading caused by reflections, scattering, or diffraction when intersecting with obstacles like trees, buildings, and human bodies, impairing

the quality of signals. Because of its nature of becoming weak on encountering such obstacles, signals transmitted over the mmWave face greater losses. Beamforming helps to focus mmWave signals in specific directions and targets precise areas, thereby reducing interference. However, in dense urban areas, it is still very hard to manage interference due to the natural overlap of signals generated by various small cells that could interfere with each other.

Applications of mmWave in High-Speed Networks:

1. **Vehicular Networks:** The low latency at the frequency range of 30–300 GHz and ultra-high-speed data transfer are used in the millimetre-wave technology, which has benefited the vehicular networks. Gigabit-level data rates at high-frequency spectrum enable these types of applications for the vehicle to everything (V2X) including vehicle to vehicle (V2V), vehicle to infrastructure (V2I), and vehicle to system (V2S). The mmWave technology is essential to autonomous driving as it enables the real-time sharing of sensor data, high-definition maps, and videos between cars and their environment. This will help in making correct decisions in situations that concern collision, taking of lanes, and taking right decisions on traffic flow. Incorporating notch beamforming and multi-antenna systems ensures that possible problem areas, such as signal weakening caused by barriers or unfavourable climate, are addressed to ensure effective connectivity within rapidly ever-changing urban landscapes. This ensures that vehicular infotainment also receives mmWave guarantees that promise seamless high throughputs needed for AR navigation, cloud-based gaming, and Union5 streaming services to the passengers. This is through offloading, whereby complex computational tasks are transferred by the vehicle to proximal RSUs or edge servers for efficiency through reduced latency. Building on these features, NOMA enhances mmWave for multiple user access to the same frequency resources, thereby increasing spectral efficiency for the increasing number of connected vehicles. According to recent estimates, by 2030, the mmWave-enabled vehicular networks will reduce traffic accidents by about 25% and pave the way for the highly efficient next-generation ITS that will amplify sophisticated safety features and entertainment features in conjunction with each other [14].
2. **5G/B5G/6G Cellular Networks:** In the context of the future 5G and beyond (B5G/6G) cellular networks, the technology of mmWave has garnered much attention in order to meet increasing demands for higher data rates, massive connectivity, and even less latency. Millimetre-Wave (mmWave), which is used in 5G frequency bands including 28 and 39 GHz and is planned in 6G higher than 100 GHz, offers a multi-Gbps data rate and thus suits highly populated city environments and industrial use. Massive MIMO techniques utilize large numbers of arrays of antennas—hundreds, sometimes—in small areas due to the short lengths of mmWave. In addition, with high levels of beamforming and beam-steering, attenuation and blockages are mitigated

such that mmWave has superior performance even in very unfavourable conditions. Such features place mmWave as an essential technology for eMBB and mMTC, two of the basic building blocks of the new-generation networks. In the context of 6G, terahertz communication or also known as mmWave is expected to pave ways to applications including but not limited to holographic telepresence, ultra immersion virtual reality (VR), and mission critical ultra-reliable low latency communication (URLLC). As latencies are expected to go down below 1 ms, 6G networks would hence enable real-time command and control of autonomous systems including robot controlled surgery and industrial automation. Further, mmWave will be expanded all the way to the THz band (~300 GHz) to achieve up to 1 Tbps data transfer rates for real-time smart city twins and highly connected IoT environments. To complement these innovations, new concepts such as RIS and AI-based resource management will be integrated, thereby augmenting mmWave's capacity to address propagation issues while enhancing network efficiency. These innovations position mmWave at the centre of the evolution of cellular networks and future concept applications.

3. **UAV-Assisted Networks:** These have incorporated mmWave communication, low latency communication, and broadband as it serves to enable multifarious purposes such as a detailed disaster response, distant object detection, and surveillance in UAVs. Because UAVs are targeted for operation within the frequency bands of mmWave, data rates higher than 10 Gbps are more easily provided that fits to be transmitted to the vast dataset. These high data rates allow for almost real-time downlink of critical information, such as imagery of disasters, topographical maps, and sensors data back to ground control centres. Because of its wave shape, mmWave signals are directional, and together with the performance of beamforming, the communication occurs even in complex environments. This directional transmission also reduces interference, so UAVs can form a mesh network where there is no man-made one, in disaster areas or remote places. For example, using UAVs in earthquake-prone areas can be helpful for demonstrating how those UAVs equipped with mmWave technology can facilitate data transfer from the operational teams to headquarters, thus improving overall cooperation efficiency and reducing the time of response. Besides emergency applications, the effectiveness of mmWave communication expands the functionality of UAVs as imaging and monitoring devices. It is used in transmitting of high-resolution video feeds and detailed sensor data needed in precision agriculture, whereby real-time imaging assists in crop health and resource utilization. Likewise in infrastructure inspection, UAVs can relay topographic pictures of bridges, power lines, or pipe lines to diagnose the presence of possible flaws. Nevertheless, communication using mmWave technology in UAVs has faced certain drawbacks such as Doppler shift due to UAV mobility and rain attenuation. The above limitations of this technique can be eliminated with adaptive beam tracking, advanced modulation schemes, and error correction algorithms for suitable communication even under adverse conditions. mmWave technology, with integration

with UAVs, would be the significant technology required for the growth of the future wireless communication and automobile systems.

4. **Satellite Communications:** Mobile satellite communications cannot be complete without the use of mmWave technology with a focus on Ka-band 26.5–40 GHz and Q/V bands 33–75 GHz. These frequency ranges allow HTS (High Throughput Satellite) vessels to provide broadband services in geographically challenging areas such as rural areas, cruise ships, and airplanes. Specifically, mmWave satellites can provide up to 100 Mbp+ per user and therefore support the mass of data-consuming applications such as video streaming or telemedicine. Available modulation techniques such as 64-QAM with H-PCA, ACM, and automatic power control guarantee reliable data transmission through rain-affected paths. Because of the small wavelength of mmWave frequencies, satellite antennas can be very small, which leads to a reduction in the cost of satellite apparatus and increased adaptability of satellites. Besides the broadband internet, mmWave is central in enabling satellite networks with the terrestrial 5G and the evolving 6G networks. This convergence enables proper connectivity for P2P and CP2P, such as IoT global networks, drone shipping, and data exchange in deep-space operations. For example, in mmWave bands, phased array antennas are able to steer a beam precisely and offer high directivity such that satellite to the ground link is stable over vast distances. Next-generation advances will be used for frequencies such as THz frequencies (~300 GHz), which aim to provide data rate more than 1 Tbps including newly emerging applications like creation of real-time digital twin, interplanetary communications, etc. The mmWave satellite communication market is poised to surge above $75 billion by 2030, creating emerging technology developments in international connectivity and advanced networks.

13.4.1 Combining NOMA with mmWave for Enhanced Performance

mmWave -NOMA is one of the latest paradigms in wireless communication, leveraging the high-frequency spectrum of mmWave bands in addition to spectral efficiency introduced by NOMA. The mmWave spectrum normally involves a wide range from 30 to 300 GHz and can offer ultra-wide bandwidths, facilitating ultra-high data rates and dense user connectivity-alien concepts to most current wireless systems and quite essential for applications in next-generation networks like 5G and 6G. As we know, NOMA is an advanced multiple access technique that enables simultaneous transmission of multiple user signals in the same frequency resources by exploiting power-domain multiplexing and SIC at the receiver. This will supplement mmWave technology in overcoming some of its inherent challenges, such as highly directional propagation and sensitivity to blockages, by intelligent user pairing and power allocation strategies. Integration of mmWave and NOMA can achieve much better capacity and coverage in wireless systems, taking full advantage of the peculiarities of both technologies. For example, mmWave directional beamforming, combined with the NOMA ability to serve multiple users within a single beam, optimizes resource usage and reduces interference. Besides, mmWave-NOMA systems are particularly suited for high-density scenarios, such as those occurring in urban areas or IoT applications, where traditional access techniques have scalability limitations.

The basic points to be considered in mmWave-NOMA relate to the propagation losses and beam misalignment inherent to the mmWave frequencies, aside from performing optimal SIC with a view to handling large power level differences among distinct users for effective separation. Some advanced solutions, including hybrid preceding and dynamic beam allocation, are crucial to address those challenges with a view to ensuring smooth integration. Therefore, mmWave-NOMA is poised to act as a key enabler to realize high-speed, low latency, and energy-efficient communication targets imposed for the future wireless networks.

As the demand of data traffic continuously increases, BS cannot keep up due to the limited number of resources. Hence, the future generation network requires a huge frequency band in order to provide resources to every user. The mmWave band from 3 to 300 GHz is planned to utilize for communication. In mmWave communications, providing service to huge number of users is a big challenge because in mmWave band the propagation losses are high and it leads to high cost in hardware development. Hence, providing the service for multiple users in one resource block is the key concept to overcome the challenges. PD-NOMA serves multiple users in one resource block, hence it meets the requirements of mmWave communication. The conventional OMA approaches cannot provide efficient service to users when compared to NOMA. NOMA has an extra domain to access the channel. With conventional OMA approach, each user is allocated a frequency resource with beam forming technique. Similarly, with NOMA, a group of users will get a single frequency resource with beam forming, increasing the spectral efficiency of mmWave communication. In mmWave-NOMA, the beam-forming is employed such that the users near to BS have a wider beam and a smaller gain, and the users away from the BS have a narrow beam and a higher gain. Figure 13.7 illustrates the trans-reception phenomenon in NOMA-based mmWave communication. The RF chain is utilized by a group of users by employing superposition coding at BS and SIC performed at the user to decode their message signal.

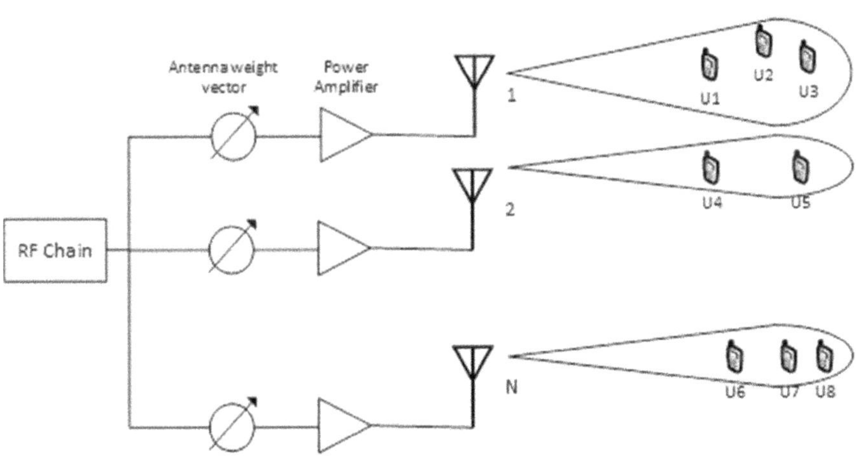

FIGURE 13.7 NOMA-based mmWave communication.

13.5 PHOTONIC TECHNOLOGIES IN COMMUNICATION

Microwave Photonics (MWP) is a multidisciplinary research area that combines microwave engineering and photonics to advance the generation, modulation, and distribution of microwave signals. This integration takes advantage of the high frequency characteristics of microwave systems and low loss broadband nature of optical signals, meeting present day requirement of high-speed communication, radar, and sensing. Normally, MWP utilizes the act of demodulation of microwave signals carried on light waves and the utilization of photonic devices including laser, modulator, and photodetector, as shown in Figure 13.8. This has led to improvements in the signal processing ability, the use of the frequency spectrum, and the performance of the networks. For example, optical systems offer more than 100 GHz bandwidth compared to approximately 1 GHz with conventional electronic systems; for this reason, MWP is an important innovation for next-generation networks, including 5G, 6G, and quantum networks.

These MWP systems, constituted by photonic oscillators, mixers, and filters, have more performance than their electronic peers. For example, the photonic oscillators provide highly stable microwave signals, suited for applications requiring high accuracy. The photonic mixer has good bandwidth capabilities as far as the conversion of matching frequency from the modulated signal to the LO (Local Oscillator) is concerned up to a ratio larger than 40 GHz. Coherent filters help in the regulation of signal bandwidth and minimization of signal interference with high precision. Such characteristics make MWP systems compact, energy-saving, and less costly but with enhanced performance. This makes them suitable for use in such things as high-definition radar and state-of-the-art defence communication.

Figure 13.8 shows a photonic-based mmWave communication system that would use optical components to process mmWave signals for high-speed wireless communication. Starting from an RF input from an antenna in mmWave range, the signals would be modulated on an optical carrier with an optical modulator driven by a laser.

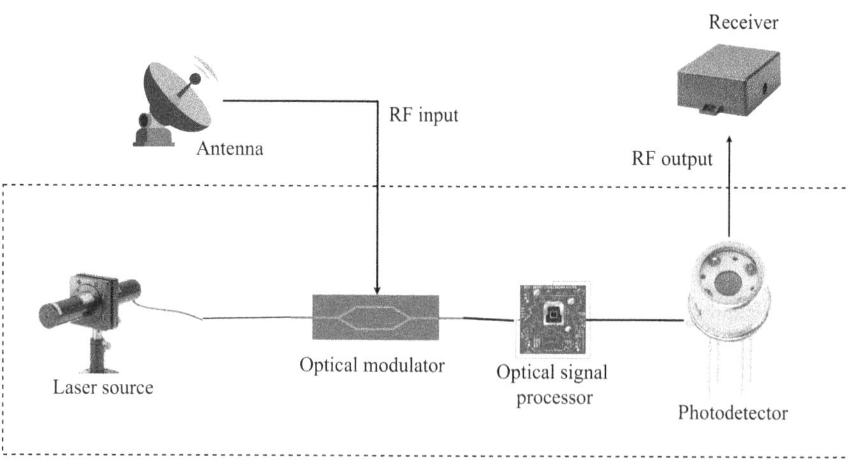

FIGURE 13.8 Photonic-based mmWave signal processing system.

After modifying that optical signal made by the photodetector as an mmWave RF signal to input this signal to a receiver, this whole arrangement can enable transmission to be high-bandwidth, low-latency, and so forth. It would make such a setup very useful to eliminate challenges such as signal attenuation and limited range when used in conjunction with the advantages of photonic technologies [15].

Microwave Photonics or MWP can greatly alter the signal transmission of satellites and space at considerable distance with attenuation of signals. This is achieved by employing the use of optical fibres that can tolerate signal loss of 0.2 dB/km. In the case of radar and sensing systems, MWP facilitates minimizing distortion of signal quality and lead time by offering highly effective signal delivery. Some of the most important benefits for future technologies that even can deliver a much faster data rate than over 100 Gbps and can operate well at various signal frequencies. These make it highly recommended to be used in all sectors such as IoT, self-driving cars, space travels, solving major problems in speed, storage, and reliability [16].

13.5.1 Role of Photonics in mmWave and Satellite Systems

Photonics plays an important role in efficiency-enhancing high-frequency data transmission in mmWave communication systems. For instance, it is possible to operate photonic integrated circuits at frequencies up to 100 GHz with minimal attenuation and without significant distortion of the signal during mmWave signal modulation and detection. Optical amplifiers and modulators that can have up to 40 dB signal gain are used to achieve a significantly improved link budget, thus allowing longer distances for the transmitter before degradation of the signal occurs. Low loss optical fibres with losses as low as 0.2 dB/km help in minimizing signal attenuation over large path lengths, which are suited for high-frequency mmWave links. High-capacity data streams can be transmitted at data rates of up to 100 Gbps in mmWave photonic systems, essential for supporting future applications such as 5G and beyond, with the aid of photonics. The integration of photonic technology with mmWave systems enables low latency (i.e. less than 1 ms) and high throughput needed for 6G (and the IoT) to support the stringent requirements of high-rate data and ubiquitous connectivity.

For satellite systems, photonics acts as an important activator to the communication algorithms for high-speed data transmission with a high capacity. This is made possible by integrating photonic technologies, including optical fibres and free space optical (FSO) communication, which allow data transmission at speeds greater than 100 Gbps, several orders of magnitude faster than traditional RF systems. Photonics makes it possible for satellites to have payloads that are smaller and lighter than they might otherwise be, as well as optical components that are lighter than and require less space to house than electronic counterparts. Besides, wafers that work at higher frequencies with wavelengths between 0.8 and 1.6 μm are widely used in space communication only because these photonic devices, such as laser transmitter and laser receiver, can work without signal degradation and high bandwidth efficiency. One way to meet the challenge above is through utilization of photonics in satellite systems that would reduce interference caused by the atmosphere from adversely affecting signal strength and reliability of signal transmission under adverse weather conditions. Moving forward, as satellite constellations for both global communications and Earth observation are foreseen to increase, next-generation satellites will be powered by photonics that can have data rates reaching up to 1 Tbps in inter-satellite links.

13.5.2 Advantages of Photonic Signal Processing

1. **Highly Bandwidth Capacity:** Using photonic systems, signals could be processed with even astonishing rates up to THz, while electronics-based systems could go up to a maximum of Gigahertz. Due to this bandwidth, the photonic signal processors can accommodate much larger data than the electronic signal processors that are suitable for high-speed communication systems such as 5G and 6G [17].
2. **Low Power Consumption for High Data Rates:** Lasers and modulators in the optical circuits consume less power per operational bit rate as compared to electronic circuits, more so at large data rates. For instance, the photonic signal processors can provide efficiency level in picojoules per bit (pJ/bit), which is far better compared to electronic devices in similar operations.
3. **Reduced Signal Loss over Long Distances:** Signal quality in photonic systems is high where light signals passing through optical fibres undergo negligible loss of about 0.2 dB/km at wavelength of 1,550 nm. However, the attenuation of electronic signals through copper wire can be as high as 10 dB/km, and therefore, photonic systems are appropriate for use in satellite and undersea cables.
4. **Immunity to Electromagnetic Interference (EMI):** In optical signals transmission, signals are transmitted by photons that are immune to electromagnetic fields. This immunity guarantees steadfast signal conditioning in areas with potential EMI interference, such as industrial complexes, medical diagnostic work where EMI threatens data quality, or deep interstellar communication where interference would ruin data content.

13.5.3 NOMA with Photonic Communication

Wireless communication in deep space MWP is important in unravelling the high-capacity long-haul data transmission for interplanetary missions and astronomical research. The use of MWP technology to overcome the constraints of conventional RF systems that operate at reliable though comparatively narrow bandwidths and limited data rates required for transmitting massive flows of scientific data acquired by orbiting probes and telescopes. It ensures the accurate functioning of the signal while at the same time providing a solution to the problem of noise for quality click of the signal over millions of kilometres of links. Using photonic filter components and modulators, MWP systems operate in very high spectral efficiency while having the potential to deliver connectivity in real time for deep space missions and scientific research studies. These developments are critical for missions like sample return missions, asteroid sample return, and all other missions to far-off celestial bodies.

Further, deep-space communication employing MWP also includes command and control of spacecraft immersed in extreme space environment. The application of photonic technologies reduces the problem of signal attenuation due to large distance and other related space interferences that may hamper the communication process. It also supports applications such as high-resolution images and video transmission of planetary surfaces, the tracking of environmental parameters across other planets, and even in connecting various space-based observatories for research. To develop future missions such as manned Mars missions, lunar colonies, and long-distance space telescopes, MWP's high degrees of accuracy, low latency, and reliability are necessary in the space science and exploration field. This integration opens new opportunities for creating and fulfilling great interstellar dreams in the future.

Underwater Communication: MWP has emerged to be crucial in data transfers through some of the key underwater applications, as seen in Figure 13.9. It is one of the essential components of oceanographic exploration, because it enables continuous monitoring of environmental conditions such as climate change and salinity, which are significant in determining the impacts of climate change and conservation of marine ecosystems. This technology is used in defence missions to assist secured and high-speed exchanges of data between underwater vehicles and submarines to surface stations for surveillance and reconnaissance purposes. It connects other sensors to offshore industries such as oil and gas, where service and inspection are done on offshore platforms and ROVs (Remotely Operated Vehicle). Reliable

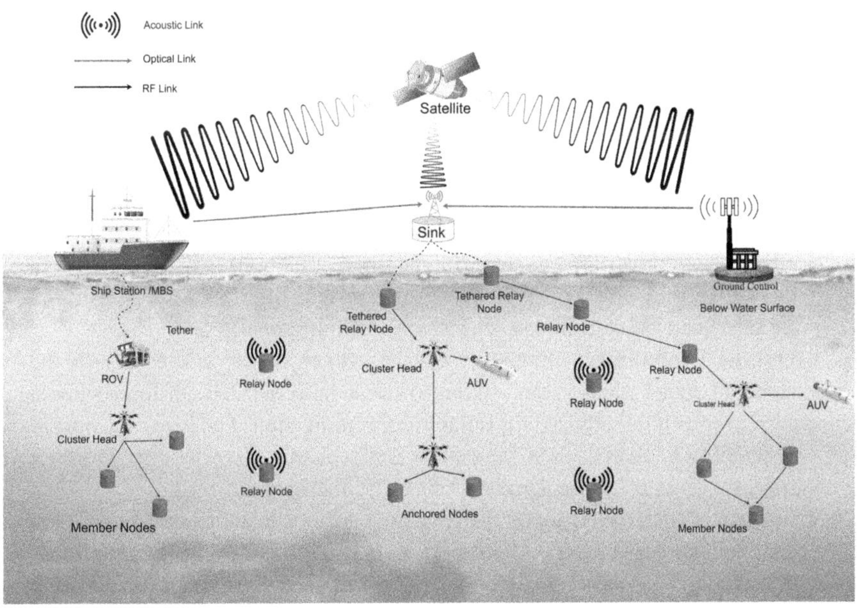

FIGURE 13.9 Under water communication.

wireless links are shown in Figure 13.9. It is possible to overcome the problems such as signal loss and spreading in water.

Other use cases of the technology include the "underwater IoT," which can include sensor clusters and devices for pollution and early earthquake detection in the ocean. Microwave-photonics communication is a low-latency, high-throughput means of transmitting real-time high-resolution video. The technology can be applied to map cultural sites by underwater archaeologists much better. It also enables the underwater turbine performance sensors and the surface control units to interact with each other for their operation in marine renewable energy scenarios. These scientific and industrial applications form the crux of the MWP importance in the discovery aspirations of science, in the conservation of ecology, and in industrial growth. System overview of an underwater network integrates different technologies to ensure reliable connectivity using acoustic, optical, and RF links [18,19].

13.6 FUTURE DIRECTIONS AND OPEN RESEARCH AREAS

In NOMA-based communication systems, performance is commonly evaluated using the following metrics:

1. **Spectral Efficiency (SE):** This provides information on general efficiency regarding the utilization of available spectrum. It gives a good view of how much information may be generated in the concerned bandwidth. It is represented by bit second −1 per Hz, bps/Hz. The higher the spectral efficiency, the more the utilization of bandwidth, and it can be considered as one important concern nowadays in modern communications, wherein spectrum resources are quite scarce.
2. **Energy Efficiency (EE):** The metric quantifies the number of transmitted data bits by per-unit energy consumption in bits per joule. It becomes vital for centres where energy consumption is of primary concern, for instance, IoT devices and 5G networks.
3. **Throughput:** Throughput represents the achieved data rate that could be effectively attained by a system for specified conditions, often measured in bits per second, bps. This definition brings in real-life effects on transmitted signals, such as interference or practical coding schemes.
4. **Outage Probability:** Outage probability refers to the probability of the received signal quality, say, signal-to-noise ratio or SNR, falling below a certain threshold required for reliable communication. The lower the outage probability is, the better is the system performance and reliability in that it reflects communication is most likely to be successful.
5. **Ergodic Capacity:** It gives the average capacity that the communication channel can deliver over a long period. It is the measure of the maximum achievable data rate that can be supported by the system over time, taking into consideration all channel variations. Higher Ergodic capacity shows that the system is more effective and capable of supporting high data rates.

6. **Bit Error Rate (BER):** BER is the fraction of bits received in error due to noise, interference, or distortion in the channel. It includes one of the most vital parameters to assess the error performance of a communication system. Thus, a low BER means very high reliability and accuracy during transmission.

In this chapter, evaluation of the outage performance is more explored, and the remaining metrics can be followed from Table 13.3.

Performance analysis of NOMA-based communication systems must therefore be carried out considering various factors affecting the signal received. Because of fading, the amplitude of the signal fluctuates with time and thus affects its strength, which is essentially the strength with which the system is able to decode the original signal back. The event of outage occurs when signal strength falls below a threshold value or protection ratio. This threshold is such that if the received signal falls below it, it is said that an outage has occurred, and this event has a certain probability of occurrence, known as the outage probability. With regard to the assessment of system performance, there are three possible scenarios where the definition of outage probability is applied:

TABLE 13.3
Performance Metrics of NOMA-Based Communication Systems

Metric Name	References	Comments
Outage probability	[20–22]	Outage probability evaluates the system's reliability by measuring the likelihood of the signal failing to meet a certain threshold. It plays a key role in determining the robustness of NOMA in multi-user environments
Ergodic capacity	[23–25]	Ergodic capacity indicates the maximum achievable rate of the system over time, providing insights into the system's ability to support high data rates under ideal conditions
BER	[26–28]	BER provides an essential measure of system performance, indicating the number of received bits incorrectly decoded due to noise and interference, crucial for determining transmission quality
Spectral efficiency	[29–31]	Spectral efficiency measures the data transmission rate per unit bandwidth and is essential for determining how efficiently the system utilizes the available spectrum
Energy efficiency	[32–34]	Energy efficiency evaluates how well the system conserves power while achieving high performance, playing a crucial role in green communication
Fairness index	[35–37]	Fairness in NOMA ensures that all users receive an equitable share of resources, important for achieving user satisfaction and improving overall network performance
Throughput	[38–40]	Throughput measures the rate at which data is successfully transmitted across the network, which directly impacts the user experience in high-demand environments

SNR-Limited Systems: In a noise-limited environment, SNR is the main determinant of outage probability. Low values of SNR yield high outage probabilities since the signal is overpowered by the noise and hence it can never be decoded.

SIR-Limited Systems: In interference-limited channels, the interference levels determine the performance in terms of outage probability. For large interference, the quality of the received signal will be degraded, leading to an outage when the SIR of the signal falls below a threshold value.

SINR Systems: A system in which both interference and noise are important, the outage probability is controlled by the SINR. For lower SINR, the outage probability is higher because both interference and noise limit the capacity of the system to decode the signal successfully. The calculation of an outage probability for these cases contributes to evaluating the main aspects of the NOMA system performance—reliability and efficiency. Besides, further important metrics, including Ergodic capacity, BER, and energy efficiency, will be the determining factors of throughput, the quality of service, and, correspondingly, the power consumptions of the system under various interference/noise/channel conditions.

The outage probability, Ergodic capacity, BER, and energy efficiency in different performance metrics are derived under the conditions of different fading scenarios in order to effectively demonstrate NOMA system performance. Among them, several typical channel fading scenarios involve Rayleigh, Rician, and Nakagami-*m*, all related to some specific real propagation environments. For example, Rayleigh fading models scenarios without a dominant line-of-sight—a common scenario in urban areas—while Rician fading incorporates a strong line-of-sight path, typical in rural or satellite communications. These testing metrics under these conditions will help in the design of the system to be able to handle challenges such as multipath propagation, shadowing, and interference. This comprehensive testing helps in finding out the limitations of the system and optimizing parameters such as power allocation and user pairing for improved robustness and efficiency.

In designing adaptive and resilient communication systems, consideration of multiple fading scenarios is very important. In fact, most of the time, the environment is not static; it varies dynamically owing to the channel conditions. Testing of systems under such diverse scenarios provides insight into the behaviour of the system, both for favourable as well as adverse conditions, and hence allows performance benchmarking with high accuracy. For example, outage probability and BER are directly impacted by fading severity, while Ergodic capacity and energy efficiency show the system's ability to maintain reliable communication and utilization of resources. By considering these variations, engineers can ensure that NOMA systems perform consistently across different deployment contexts, enhancing their reliability, quality of service, and user experience. The various faded channels are summarized in Table 13.4.

NOMA systems, which are targeted to improve spectral efficiency and allow a number of users to acquire simultaneous accessibility, have several disadvantages. It includes the issues of imperfect CSI, imperfect SIC, inter-cell interference, and many such factors arising in the worst scenarios that may lead to performance deteriorations.

TABLE 13.4

Fading Channel	Proposed by	Other Fading Models Included
η-μ	[41]	One-sided Gaussian, Hoyt, Rayleigh, and Nakagami-m
κ-μ	[41]	One-sided Gaussian, Rayleigh, Rician, and Nakagami-m
α-μ	[42]	One-sided Gaussian, Rayleigh, Nakagami-m, exponential, gamma, and Weibull
κ-μ shadowed faded	[43,44]	Rayleigh, Nakagami
α-κ-μ	[45]	Rayleigh, Rice, Nakagami-m, κ-μ, Weibull, and One-Sided Gaussian
α-η-μ	[46]	Rayleigh, Hoyt, One-Sided Gaussian, Weibull, and Nakagami—m, α-μ, and η-μ
α-κ-μ shadowed faded	[47]	One-sided Gaussian, Rayleigh, Nakagami-m, Hoyt Rician, α-μ, η-μ, κ-μ, α-κ-μ, Weibull, α-η-μ, κ-μ shadowed, Rician Shadowing

13.6.1 Imperfect Channel State Information (CSI)

NOMA requires reasonably accurate CSI in resource optimization, power control, and user pairing. Basically, CSI is the perception regarding the channel condition between different transmitters and receivers. Imperfect CSI denotes poor estimation of the channel condition owing to noise, delayed feedback, or quantization error. This further leads to the worst power allocation, affecting SINR and subsequently increasing the outage probability. Furthermore, inaccurate CSI could result in ineffective pairing of users in which the expected performance benefits of NOMA regarding higher throughput and fairness are not effectively gained.

13.6.2 Imperfect SIC

SIC plays a very important role in NOMA, which is used at the receiver for signal separation among multiple users. Imperfect SIC refers to that condition wherein interferences by the previously decoded signal are not absolutely cancelled. It can be due to hardware constraints, estimation errors, or decoding inaccuracy. Under imperfect SIC conditions, the SINR values in subsequent users, especially those with much weaker signals, are reduced, leading to higher BERs and lower throughputs. It also reduces the system capability for fairness among users, as higher-priority users can still cause interference to the lower-priority users.

13.6.3 Inter-cell Interference

In multi-cell networks, due to this interference, the signals from neighbouring cells interfere with those in the target cell and, therefore, result in inter-cell interference. The impairment is significantly worse for edge users that are closer to neighbouring cells and receive higher interference levels. Inter-cell interference reduces SINR and causes degraded QoS and increased outage probability and, therefore, higher BER. This often requires effective coordination techniques such as interference alignment or cooperative multi-cell processing.

All these impairments together affect the reliability, fairness, and efficiency of NOMA systems. The remedies for these involve sophisticated techniques: improved CSI estimation, advanced SIC algorithms, and inter-cell interference mitigation strategies are all essential for the full realization of NOMA in modern wireless networks.

The metrics such as outage probability, Ergodic capacity, BER, and energy efficiency can be obtained more precisely if the above impairments—imperfect CSI, imperfect SIC, and inter-cell interference—are accurately modelled and accounted for during the system's performance evaluation.

Recently, NOMA has been one of the important technologies that attract increasing attention in modern wireless communication research, promising a solution to improve spectral efficiency, energy efficiency, and fairness, among users. Unlike other orthogonal multiple access techniques, NOMA allows more than one user to utilize the same time-frequency resources based on power domain multiplexing and SIC. This grants NOMA significantly superior system capacity and user connectivity, making the technology of paramount importance for 5G and further applications: IoT networks, satellite communications, and dense urban environments. Optimizing power allocation in NOMA, Development of NOMA: Practical problems, Further advanced NOMA Deployments—some other important models: Cooperative NOMA, Massive MIMO-NOMA systems, and Hybrid NOMA schemes remain in focus during ongoing research by scientific and academic circles within state-of-the-art NOMA studies. These studies aim at enhancing the robustness, scalability, and efficiency of NOMA systems in diverse dynamically changing wireless environments.

To conduct research in NOMA, follow the outlined methodology to systematically address key aspects of system design and performance evaluation.

Step 1: NOMA System Selection—Application Domain

Identify the application domain in which the NOMA system has to be deployed. The various application domains can be:

1. **Underwater Communication:** It is characterized by severe attenuation and multipath propagation; hence, it requires high robustness.
2. **Satellite Transmission:** It involves dealing with long distances and Doppler effects; line-of-sight propagation is the primary concern.
3. **IoT Networks:** Energy consumption is at a very low value and efficiency at a high value due to the energy-constrained devices.
4. **Cellular Networks in an Urban Environment:** These address issues including dense user environments and multipath fading.

Step 2: Select Suitable Channel Scenarios

Choose the scenario that best depicts which limitation is dominant in the communication channel:

1. **SNR-Limited Systems:** These are the systems in which noise predominates, for instance, underwater or satellite communication.
2. **SIR-Limited Systems:** When interference is the main challenge, for example, dense urban networks.
3. **SINR Systems:** The scenarios where noise and interference both play a significant role in performance degradation—for example, in heterogeneous networks.

Step 3: Choosing Appropriate Fading Scenarios

The channel fading model should characterize the real propagation environment:

1. **Rayleigh Fading:** It is suitable for NLOS conditions usual in urban or underwater communications.
2. **Rician Fading:** It refers to dominant line-of-sight path, like in satellite communication.
3. **Nakagami-*m* Fading:** It refers to generalized conditions when either LOS or NLOS is variable in nature.
4. **Other Models:** These are advanced models for special fades, like κ-μ, η-μ, or shadowed κ-μ.

Step 4: Model System Impairments

The model system should include real-world impairments for accurate performance evaluation:

Imperfect CSI:

1. Introduce channel estimation errors with statistical models.
2. Evaluate the effect of this on power allocation, SINR, and fairness.

Imperfect SIC:

1. Model partial interference cancellation.
2. Analyze the influence of residual interference on subsequent users' SINR and BER.

Hardware Noise:

1. Include hardware impairments caused by phase noise or quantization errors.
2. Assess the cumulative effect this might have on system reliability and efficiency.

13.7 CONCLUSION

As one of the major multiple access techniques, NOMA significantly improves spectral efficiency and user connectivity by allowing simultaneous transmission of multiple users over the same frequency band. This capability positions NOMA as a fundamental technology for next-generation wireless systems, including 6G networks, the Internet of Things, and deep-space communication.

Satellite communication systems are especially suited for wide-area and global coverage, and as such, aptly positioned for addressing connectivity issues in remote or under-served areas. By including NOMA in a satellite system, perspectives towards capacity improvement open, while various natures of use cases can also be supported. This poses very unique challenges such as signal interference, propagation delays, and problems related to resource allocation. In this chapter, such challenges were analyzed while proposing possible strategies that may overcome these challenges. The combination of mmWave with NOMA enhances the network capacity for high-density user environments while addressing some issues like propagation loss and channel fading.

The key role of photonics for such an ecosystem is related to ultra-fast signal processing and low-propagation loss for both mmWave and satellite systems. Photonic technologies further provide advanced solutions in the framework of signal generation, routing, and processing, boosting overall general performance and energy efficiency of such systems. The integration of NOMA into Photonics Communication Systems: This merge marks the next generation towards ultra-reliable, high-speed communications networks.

The huge possibility of using NOMA combined with satellite, mmWave, and photonic technologies—from 6G networks and IoT to deep-space communication—has been discussed in this chapter. In fact, the integration of these technologies forms not just an achievement, but a paradigm shift that might alter the very design and deployment strategies of wireless networks. Future research will have to continue into the optimization of these integrations, find new use cases, and solve the remaining technical challenges before the vision of a seamless, high-performance, universally accessible communication infrastructure can be realized.

REFERENCES

1. Sainath, B., and S. K. Tadinada. Environment-aware green UAV-assisted, CubeSat communication network energy efficiency and outage probability analysis. *IEEE Journal on Miniaturization for Air and Space Systems* (2024). doi:10.1109/JMASS.2024.3451011
2. Akyildiz, I. F., A. Kak, and S. Nie. 6G and beyond: The future of wireless communications systems. *IEEE Access* 8 (2020): 133995–134030.
3. Venu, N., A. Arun Kumar, and K. K. Vaigandla. Review of Internet of Things (IoT) for future generation wireless communications. *International Journal for Modern Trends in Science and Technology* 8(3) (2022): 1–8.
4. Ding, Z., R. Schober, P. Fan, and H. Vincent Poor. OTFS-NOMA: An efficient approach for exploiting heterogenous user mobility profiles. *IEEE Transactions on Communications* 67(11) (2019): 7950–7965.
5. Liu, Y., W. Yi, Z. Ding, X. Liu, O. A. Dobre and N. Al-Dhahir. Developing NOMA to next generation multiple access: Future vision and research opportunities. *IEEE Wireless Communications* 29(6) (2022): 120–127. doi:10.1109/MWC.007.2100553
6. Shanmugavel, G., and M. S. Vasanthi. Performance of outage probability and ergodic sum rate for two-user paired NOMA downlink system. *Journal of Physics: Conference Series* 2335(1) (2022): 012035. IOP Publishing.
7. Vatalaro, F., G. E. Corazza, C. Caini, and C. Ferrarelli. Analysis of LEO, MEO, and GEO global mobile satellite systems in the presence of interference and fading. *IEEE Journal on Selected Areas in Communications* 13(2) (1995): 291–300. doi: 10.1109/49.345873
8. Ammar, S., C. Pong Lau, and B. Shihada, An in-depth survey on virtualization technologies in 6G integrated terrestrial and non-terrestrial networks. *IEEE Open Journal of the Communications Society* 5 (2024): 3690–3734. doi:10.1109/OJCOMS.2024.3414622
9. Tedeschi, P., S. Sciancalepore, and R. Di Pietro. Satellite-based communications security: A survey of threats, solutions, and research challenges. *Computer Networks* 216 (2022): 109246.
10. Zhang, C., G. Wang, M. Jia, R. He, L. Zhou, and B. Ai. Doppler shift estimation for millimeter-wave communication systems on high-speed railways. *IEEE Access* 7(2019): 40454–40462. doi:10.1109/ACCESS.2018.2861889
11. Akbar, A., S. Jangsher, and F. A. Bhatti. NOMA and 5G emerging technologies: A survey on issues and solution techniques. *Computer Networks* 190 (2021): 107950.
12. Yang, Z., Z. Ding, P. Fan, and N. Al-Dhahir. A general power allocation scheme to guarantee quality of service in downlink and uplink NOMA systems. *IEEE Transactions on Wireless Communications* 15(11) (2016): 7244–7257.

13. Wei, Z., L. Zhao, J. Guo, D. W. K. Ng, and J. Yuan. Multi-beam NOMA for hybrid mmWave systems. *IEEE Transactions on Communications* 67(2) (2018): 1705–1719.
14. Redondi, A. E. C., C. Innamorati, S. Gallucci, S. Fiocchi, and F. Matera. A survey on future millimeter-wave communication applications. *IEEE Access* 12 (2024): 133165–133182. doi:10.1109/ACCESS.2024.3438625
15. Marpaung, D., J. Yao, and J. Capmany. Integrated microwave photonics. *Nature Photonics* 13(2) (2019): 80–90.
16. Minasian, R. A. Photonic signal processing of microwave signals. *IEEE Transactions on Microwave Theory and Techniques* 54(2) (2006): 832–846.
17. Li, E., and A. X. Wang. Theoretical analysis of energy efficiency and bandwidth limit of silicon photonic modulators. *Journal of Lightwave Technology* 37(23) (2019): 5801–5813.
18. Nomikos, N., A. Giannopoulos, P. Trakadas, and G. K. Karagiannidis, Uplink NOMA for UAV-aided maritime Internet-of-Things, *2023 19th International Conference on the Design of Reliable Communication Networks (DRCN)*, Vilanova i la Geltru, Spain, 2023, pp. 1–6. doi:10.1109/DRCN57075.2023.10108290
19. Islam, K. Y., I. Ahmad, D. Habibi, M. Ishtiaque, A. Zahed, and J. Kamruzzaman. Green underwater wireless communications using hybrid optical-acoustic technologies. *IEEE Access* 9 (2021): 85109–85123.
20. Zhang, H., Z. Zhang, and H. Dai. Error rate analysis of NOMA: Principles, survey and future directions. *IEEE Access* 7 (2019): 40432–40450.
21. Kumar, K., R. Vishwakarma, and A. Jain. Cooperative power-domain NOMA systems: An overview. *PMC Open Access*, Dec. 2023.
22. A. Tiwari, S. Patel, and M. K. Yadav. Performance of cooperative relay NOMA with large antenna arrays. *Electronics* 11(21) (2022): 3482–3498.
23. Li, S., R. Z. Zhang, and H. Dai. Ergodic capacity and energy efficiency of NOMA systems with SIC. *IEEE Transactions on Communications* 67(7) (2019): 4685–4699.
24. Xu, H., F. R. Yu, and J. Zhang. Ergodic capacity of NOMA with multiple antennas in the presence of imperfect SIC. *IEEE Transactions on Wireless Communications* 18(8) (2019): 3787–3799.
25. Zhang, D., Z. Li, and Z. Wei. Ergodic capacity of NOMA system in multi-user environment with channel state information at the transmitter. *Wireless Networks* 27(4) (2021): 2071–2081.
26. Li, H., Z. Shi, and Z. Zhang. Bit error rate analysis of NOMA systems with SIC in downlink and uplink scenarios. *IEEE Transactions on Communications* 67(11) (2019): 7736–7748.
27. Zhou, Z., W. Zhuang, and K. K. Wong. Bit error rate performance of NOMA systems with channel state information imperfection. *IEEE Transactions on Vehicular Technology* 69(7) (2020): 8061–8075.
28. Zheng, D., X. Chen, and M. Zhao. BER analysis and optimization for NOMA systems. *IEEE Transactions on Communications* 68(2) (2020): 1153–1166.
29. Zhang, H., H. Dai, and Z. Zhang. Spectral efficiency of NOMA systems in massive MIMO networks. *IEEE Transactions on Wireless Communications* 18(6) (2019): 3257–3269.
30. Wu, D. J., Y. Li, and J. Z. Zhang. Spectral efficiency of downlink NOMA systems with non-linear SIC decoding. *IEEE Transactions on Communications* 68(7) (2020): 4230–4243.
31. Zhang, S., Y. Zhang, and J. Zhang. Spectral efficiency in NOMA with reconfigurable intelligent surfaces. *IEEE Access* 9 (2021): 108–119.
32. Zhang, K. Z., Y. Zou, and Y. Wu. Energy-efficient NOMA-based communications in 5G and beyond. *IEEE Transactions on Communications* 68(4) (2020): 2252–2266.

33. Garcia-Ramos, A., M. Lopez-Benitez, and F. K. Jondral. Energy efficiency optimization for NOMA-based systems. *IEEE Transactions on Wireless Communications* 18(7) (2019): 3421–3434.
34. Wang, X., H. Zhang, and M. Zhao. Energy efficiency and resource allocation for NOMA systems in cloud-RAN. *IEEE Transactions on Wireless Communications* 19(5) (2020): 3161–3173.
35. Mohammed, A., O. Al-Hanbali, and H. S. H. Aghvami. Fairness in non-orthogonal multiple access systems: Challenges and solutions. *IEEE Access* 7 (2019): 149304–149317.
36. Alouini, M., F. B. V. Oliveira, and Y. Zou. Fair resource allocation in NOMA for 5G and beyond. *IEEE Transactions on Communications* 68(9) (2020): 5571–5584.
37. Karakaya, D., S. Z. Zawawi, and M. S. El-Tantawi. A fairness framework for NOMA in 5G systems. *IEEE Transactions on Vehicular Technology* 68(11) (2019): 10653–10664.
38. Yu, F. R., X. Zhang, and M. Li. Throughput optimization of NOMA in 5G networks. *IEEE Transactions on Communications* 67(9) (2019): 6249–6259.
39. Liang, S., W. Chen, and Q. Li. Throughput performance of NOMA-based wireless networks with imperfect SIC. *IEEE Transactions on Wireless Communications* 68(2) (2020): 1584–1598.
40. Ghosh, A. M. B., V. Y. Tsiatsis, and M. T. M. Abadi. Throughput enhancement in NOMA-based communications. *Wireless Networks* 28 (2022): 1057–1070.
41. Yacoub, M. D. The κ-μ distribution and the η-μ distribution. *IEEE Antennas and Propagation Magazine* 49(1) (2007): 68–81.
42. Yacoub, M. D. The α-μ distribution: A physical fading model for the Stacy distribution. *IEEE Transactions on Vehicular Technology* 56(1) (2007): 27–34.
43. Nguyen, T.-L., C.-B. Le, and D.-T. Do. Performance analysis of multi-user NOMA over κ-μ shadowed fading. *Electronics Letters* 56(15) (2020): 771–773.
44. Paris, J. F. Statistical characterization of κ-μ shadowed fading. *IEEE Transactions on Vehicular Technology* 63(2) (2013): 518–526.
45. Fraidenraich, G., and M. D. Yacoub. The α-η-μ and α-κ-μ fading distributions, *2006 IEEE Ninth International Symposium on Spread Spectrum Techniques and Applications*, Manaus, Brazil, 2006. IEEE.
46. Moualeu, J. M., D. B. da Costa, W. Hamouda, U. Dias, and R. A. A. de Souza. Performance analysis of digital communication systems over a-k-μ fading channels. *IEEE Communications Letters* 23(1) (2018): 192–195.
47. Ramirez-Espinosa, P., J. M. Moualeu, D. B. da Costa, and F. J. Lopez-Martinez. The a-k-μ shadowed fading distribution: Statistical characterization and applications, *2019 IEEE Global Communications Conference (GLOBECOM)*, Big Island, Hawaii, 2019, pp. 1–6. IEEE.

14 Modulation and Diversity in Cooperative VLC

Trade-offs and Optimizations for System Design

Prashant Dwivedy, Vipul Dixit, and Atul Kumar

14.1 INTRODUCTION

Traditional radio frequency (RF) solutions face a major issue as a result of the growing spectrum scarcity brought on by the exponential development in consumer demand for wireless networking services and applications. Alternative communication technologies including optical wireless communication (OWC) have gained popularity as a solution to this problem. The use of light-emitting diode (LED) lights for lighting and data transmission distinguishes visible light communication (VLC) from other OWC techniques. As an aid to traditional RF technologies, VLC offers a feasible way to achieve high-speed indoor communication [1,2].

By rapidly varying the light intensity, VLC leverages LEDs to transmit data. The human eye can perceive the average light intensity; consequently, the entire procedure is imperceptible to vision. This enables VLC to seamlessly integrate data transfer and illumination without compromising visual comfort or lighting quality. Furthermore, VLC systems have a high quality of service (QoS) and are reliable and secure due to their immunity to electromagnetic interference [3–5]. By virtue of these attributes, VLC has evolved into a reliable and affordable alternative for traditional RF communication methods.

Nevertheless, the arrangement of VLC across broader sections is constrained by its inherent characteristics, which include a constrained field-of-view (FoV), obstruction concerns, and a reduced range [6,7]. To overcome these obstacles and guarantee sustainable end-to-end communication, cooperative communication (CC) has become an essential method [8,9]. In network systems, CC enables decision-making by authorizing nodes to work together, speeding up response times, and improving overall performance through shared information processing [10]. By resolving uncertainties, a dynamic threshold-based approach can further optimize the performance and achieve prominent gains over conventional fixed-threshold approaches [11]. By using several relay nodes to transfer information from the transmitter to the destination, CC greatly improves system performance by utilizing spatial variety, reducing power consumption, improving area coverage, and increasing dependability [12–14].

DOI: 10.1201/9781003599111-14

In this study, we have employed on–off keying (OOK) and L-ary pulse position modulation (LPPM) as modulation techniques. OOK is a prominent intensity modulation (IM) technique that implies digital data by using the availability or absence of a light pulse. Precisely, binary '1' designates that a light pulse is present, while binary '0' specifies that it is not. OOK is preferred since it is straightforward to comprehend. Nevertheless, by encoding data in one of L different time slots, LPPM is employed for its energy competence and improved security, making it less susceptible to interference and eavesdropping.

This chapter compares and employs diversity approaches, such as equal gain combining (EGC), maximum ratio combining (MRC), and selection-based combining (SBC), to further augment the competence of VLC systems. MRC is a technique that optimizes the output signal-to-noise ratio (SNR) and achieves optimal performance by combining signals from numerous relays with weights proportionate to their SNR. Even while EGC is less complicated than MRC, it nevertheless offers a significant diversity increase by combining signals with identical weights. In contrast to MRC and EGC, SBC offers simplicity at the cost of less diversity gain by choosing the relay with the strongest signal and sending it to the destination.

CC in VLC systems processes signals at relay nodes using amplify-and-forward (AF) and decode-and-forward (DF) procedures. In the DF protocol, the relay sends the source message to downstream nodes after decoding and re-encoding it. Nevertheless, the AF protocol intensifies the received signal before transmitting it. These protocols render VLC systems appropriate for high-speed communication in vast indoor environments by enhancing the network's overall capacity, data throughput, and dependability.

The fundamental concepts of VLC, its benefits over conventional RF systems, and the way CC and modulation techniques can potentially overcome VLC's drawbacks are all covered in this chapter. Furthermore, it examines the diversity strategies (MRC, EGC, and SBC) and contrasts them, giving a thorough rundown of the strategies, applications, and advantages of VLC and emphasizing how it has the potential to entirely transform wireless communication technologies.

Multi-LED phase-shifted OOK (MP-OOK) modulation with multiple LED chips improves bandwidth efficiency and anti-nonlinearity performance over DCO-OFDM, as demonstrated in [15]. The use of silicon photomultipliers (SiPMs) as sensitive VLC receivers is examined in [16], which shows that under 500 lux ambient light conditions, they may reach data rates of over 2 Gbps with a bit error rate (BER) of 10^{-3}. Under 500 lx ambient light, a 20 W VLC system with ON–OFF M-FSK modulation, dimming, and integrated power factor correction achieved 1.11 Mbps over 10 m with a BER of 10^{-3} in the article [17]. The VLC system illustrated in [18] incorporates integrated optics and 680 nm laser diodes for optimal power collection, and a high-sensitivity IC receiver to achieve data speeds of up to 2.5 Gbps across 12 m with a BER$< 10^{-3}$. A power-efficient VLC system using Sub-Carrier Pulse Position Modulation (SC-LPPM) in [19] enhanced illumination and optical power in diffuse, non-line-of-sight environments, achieving a BER of 10^{-9} at 1–2 Mbps. Paper [20] introduces a nested PPM scheme using phosphorescent white LEDs for VLC, enabling dual receiver support with improved bit rates and feasibility through error and SNR analysis. Paper [21] illustrates L-PPM format classification in VLC using machine learning, with 2-PPM outperforming, attaining 97.85% accuracy via Decision Tree at 2.20 m and KNN excelling beyond 2.32 m. Using DF relaying, paper [22] examines the secrecy of a hybrid RF/VLC system and shows that cooperative power saving (CPS) improves security over standalone RF or VLC networks while outperforming non-cooperative power saving (NCPS) in terms of secrecy capacity and outage probability. The use of twin photodetectors in a vehicle-to-vehicle VLC system is examined in [23]. It demonstrates that selection combining enhances signal reception in mobile environments and offers a closed-form calculation for the average BER based on experimental data. Paper [24] offers a generalized ADR structure for indoor multi-cell VLC systems, which reduces SINR fluctuations by up to 32.4 dB with optimum side-detector inclination angles. Three user collaboration and downlink transmission strategies for VLC and RF systems that use non-orthogonal multiple access (NOMA) are presented in paper [25]. The research also suggests an adaptive Qos-oriented approach that addresses connection blockages and mitigates interference to improve system throughput. In [26], the authors look into multi-user multi-cell visible light communication networks that use multiple LED arrays, proposing coordination and cooperation strategies to reduce inter-cell interference through optimal coordinated precoding designs, and providing numerical results that compare the effectiveness of these strategies in maximizing users sum-rate. An orthogonal frequency division multiplexing access (OFDMA)-based indoor VLC and positioning system is presented in paper [27]. Using quadrature phase shift keying (QPSK) mapping, the system chooses three subcarriers with the highest received signal intensity from LEDs for trilateration-based positioning, obtaining a mean positioning error of 1.68 cm and an error vector magnitude greater than 15 dB. An incentive model for relay collaboration in hybrid VLC-RF IoT systems is presented in paper [28]. Through the integration of energy harvesting, agency payments, and improved contract mechanisms, the model maximizes utility while addressing user selfishness. In order to improve user connectivity and sum secrecy rate in multi-user indoor downlink configurations, a combined precoding and user association strategy is proposed in [29], which investigates hybrid RF-VLC systems. This chapter inspects the error investigation of the OOK/L-PPM modulated Co-VLC system. The following are the article's primary contributions:

- In the framework of cooperative VLC (Co-VLC) systems, the chapter provides a thorough examination of OOK and L-PPM modulation techniques along with a performance comparison.
- Diversity combining techniques including EGC, MRC, and SBC are investigated in conjunction with OOK and L-PPM.
- The system's performance is assessed across both perfect and imperfect channel state information (CSI) scenarios.
- It is examined how diversity combining strategies can improve the dependability and effectiveness of Co-VLC systems.
- The study focuses on the way diversity combining strategies might enhance Co-VLC systems' dependability and efficiency.

The remnants of the chapter are structured as follows: Section 14.2 displays the OOK and L-PPM modulation system model, the channel model, and the signal structure for the Co-VLC system. Section 14.3 discusses the BER analysis for the L-PPM/OOK modulated Co-VLC system. Section 14.4 presents the mathematical outcomes and associated interpretations. Section 14.5 signifies the conclusion of the study.

14.2 CO-VLC SYSTEM

14.2.1 System Model

In VLC systems, cooperative transmission is essential, especially when users are disseminated across an extensive section or situated outside the FoV. An indoor VLC configuration has been established to overcome this restriction, with intermediate users to create secure connections by serving as relays. By conserving connectivity even under difficult spatial environments, these intermediary users guarantee the system's resilience. Through the use of competent intermediaries, this strategy not only enhances coverage but also fortifies the communication system's security.

Three user nodes, each with a photodiode (PD) and an LED, and a single source node (T) with a LED are the components of proposed indoor VLC system. The information transmitting node T is located in the center of the ceiling of a $8\times8\times4$ m room, at coordinates $4\times4\times4$ m. The receiving module and the transmitting module are the two key elements that collectively make up the relay node (R). The photodiode in the receiving module, which is primarily responsible for receiving the signal that is transmitted and configuring it for retransmission, is described by its FoV and responsivity ($\mathfrak{R}$). The transmitting module features a LED, which is distinguished by its semi-angle Lambertian pattern ($\alpha_{1/2}$) and optical power (ρ_t). The message must be relayed to the intended recipient or another relay node utilizing this aspect of the system. DF is the predominant relaying strategy employed throughout this investigation.

The Co-VLC system model is shown in Figure 14.1. Assuming there are M users $R_1, R_2, \ldots, R_M$, where the nearest user is represented by R_1 and the farthest user is represented by R_N. As illustrated in the image, all users receive the signal sent by

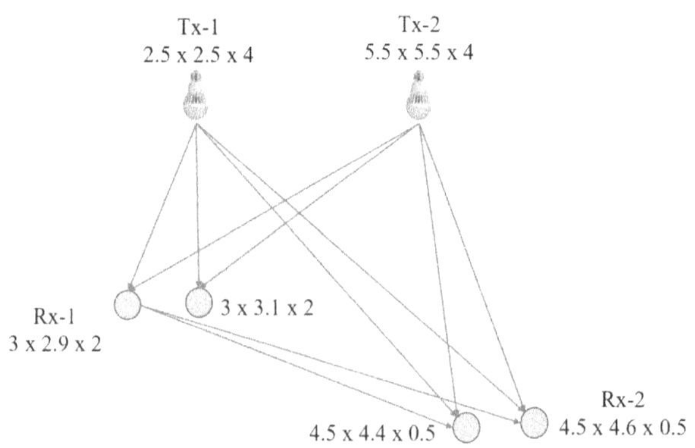

FIGURE 14.1 System model.

the source, and all subsequent users, i.e., $R_{j+1}, R_{j+2}, \ldots, R_M$, receive the signal relayed by R_j $(1 \le j \le M-1)$. The broadcast phase and the cooperation phase are the two distinct stages of the communication process. The source node T uses L-PPM/OOK modulation to send messages to all users during the broadcast phase. The signal is forwarded to later users with the help of prior users during the cooperation phase. Signals from T and all prior users $(R_1, R_2, \ldots, R_{j-1})$ are combined at the jth user node $(2 \le j \le M)$ using methodologies such as EGC, MRC, or SBC [30,31].

14.2.2 Channel Model

In this study, a Lambertian radiation pattern-based line-of-sight (LOS) communication link for data transfer is studied. For the link between the xth transmitting node and the yth receiving node, the LOS channel impulse response (CIR), also known as the channel DC gain, is explained as [32,33]

$$h_{x,y}^{\mathrm{LOS}} = \begin{cases} \dfrac{\mathrm{Ar}}{d_{x,y}^2} G_{\mathrm{OF}}(\beta_{x,y}) G_{\mathrm{OC}}(\beta_{x,y}) R_{\mathrm{LI}}(\gamma_x) \cos(\beta_{x,y}), & 0 \le \beta \le \psi_c \\ 0, & \beta > \psi_c \end{cases}, \tag{14.1}$$

where Ar is the physical area of PD, $d_{x,y}$ is the Euclidean distance between the xth transmitter and the yth receiver, β is the angle of incidence relative to the transmitter plane's normal axis, γ is the irradiance angle concerning the receiver plane's normal axis, ψ_c denotes the FoV of PD, $G_{\mathrm{OF}}(\beta_{x,y})$ is the gain of optical filter, and G_{OC} is the gain of optical concentrator defined as

$$G_{\mathrm{OC}} = \frac{n^2}{\sin^2(\psi_c)}, \quad 0 \le \phi \le \psi_c, \tag{14.2}$$

where n is the refractive index of the optical concentrator, and Lambertian radiant intensity of the xth transmitter is given by

$$R_{\mathrm{LI}}(\gamma_x) = \frac{m+1}{2\pi} \cos^m(\psi_{x,y}), \tag{14.3}$$

where the order of Lambertian emission represented by semi-angle ($\alpha_{1/2}$) at half power of an LED is defined as

$$m = -\frac{\ln(2)}{\ln(\cos(\alpha_{1/2}))}. \tag{14.4}$$

14.2.3 Signal Structure

The signal that the xth transmitting node in OOK modulation emits at any given time is expressed as

$$X = 2\rho_t b_{\mathrm{sym}}, \tag{14.5}$$

Here, ρ_t represents the average transmitted power in watts, and b_{sym} denotes the transmitted symbol measured in bits. Thus, the unit of X_t is also measured in watts. In OOK modulation, no signal is transmitted when $b = 0$. Therefore, the factor of 2 is included to ensure that the average transmitted power remains ρ_t.

L-PPM modulation splits the individual symbol period into L time slots. The signal transmitted by any user in case of L-PPM is represented as [34]

$$X = L\rho_{\text{sym}} b_{\text{slot}}, \tag{14.6}$$

where ρ_{sym} stands for the symbol power and b_{slot} for the transmitted slot data.

Equation (14.7) provides a threshold value that facilitates the detection of symbol transferred from R_1 to R_N:

$$\text{Th}_{x,y} = I_p / 2, \tag{14.7}$$

where I_p is the generated photocurrent. The signal from the xth sending node that is received at the yth receiving node is written as

$$Y = \Re h_{x,y}^{\text{LOS}} X + \eta, \tag{14.8}$$

where $H_{x,y}$ represents the dimensionless LOS channel gain between the xth broadcasting node and the yth receiving node. The detector's responsivity, or $\Re$, is measured in amperes per watt. Consequently, the unit of Y is amperes.

The random noise prevalent in optical communication systems is accurately represented by η. It is measured in amperes and is described as Additive White Gaussian Noise (AWGN) as it includes both shot and thermal noise. The variance of η, represented by σ_η^2, can be determined by

$$\sigma_\eta^2 = \sigma_S^2 + \sigma_T^2, \tag{14.9}$$

where σ_S^2 and σ_T^2 represent the variances of shot noise and thermal noise, respectively, as defined in the following equations [35,36]:

$$\sigma_S^2 = 2qI_b I_2 B + 2q\Re\rho_t B h^{\text{LOS}}, \tag{14.10}$$

$$\sigma_T^2 = \left[8\pi k T_k A_r I_2 B^2 n / G\right] + \left[16\pi^2 k T_k A_r^2 I_3 B^3 n^2 \Gamma / g_m\right]. \tag{14.11}$$

Further, q is the charge of electron, I_b is the background current which is considered for indoor environment, $I_2 = 0.562$ is the noise bandwidth factor, B is the bandwidth of noise which is equal to bandwidth of modulation, k is the Boltzmann's constant, $T_k = 295$ K is the absolute temperature, A_r is the PDs active area, $n = 112$ pF/cm^2 is the fixed capacitance of PD per unit area, $G = 10$ is the open loop voltage gain, $I_3 = 0.0868$, $\Gamma = 1.5$ is the FET channel noise factor, and $g_m = 30$ mS is the transconductance of FET.

14.3 BER ANALYSIS

14.3.1 OOK

14.3.1.1 Perfect CSI

The BER can be assessed as follows, assuming that the recipient has access to CSI knowledge. Both T and $(j-1)$ signals from relays are received directly by the jth user, $1 \leq j \leq M$. The EGC, MRC, and SBC combining algorithms are applied individually and contrasted with each other in order to combine the j signals at the receiving node R_j. It is possible to compute the BER at the jth user as

$$P_j^e = \begin{cases} \left(P_{j-1}^e\left(1-P_{C,j}^e\right)\right)+\left(P_{C,j}^e\left(1-P_{j-1}^e\right)\right), & 2 \leq j \leq M \\ P_1^e, & j=1 \end{cases}, \tag{14.12}$$

where P_1^e is the BER at U_1 defined as

$$P_1^e = Q\left(\frac{2\Re h_{0,1}^{\text{LOS}} \rho_t}{2\sigma_\eta}\right), \tag{14.13}$$

where Q denotes the Q-function and $P(b=0)=P(b=1)=\frac{1}{2}$, i.e., symbols are equiprobable. $P_{C,j}^e$ is the BER at R_j after the implementation of combining technique.

- **EGC:** $P_{C,j}^e$ for EGC is given by

$$P_{C,j}^e = Q\left(\frac{I_p^{\text{EGC}}}{2 \times \sigma_{\text{EGC}}}\right), \tag{14.14}$$

where I_p^{EGC} is the generated photocurrent due to EGC given by

$$I_p^{\text{EGC}} = 2\Re \rho_t h_{\text{EGC}}^{\text{LOS}}, \tag{14.15}$$

while $h_{\text{EGC}}^{\text{LOS}}$ is the EGC channel gain at R_j given by

$$h_{\text{EGC}}^{\text{LOS}} = \sum_{i=0}^{j-1} h_{i,j}^{\text{LOS}}, \tag{14.16}$$

and σ_{EGC} is given by

$$\sigma_{\text{EGC}} = \sqrt{j \times \sigma_\eta^2}. \tag{14.17}$$

- **MRC:** $P_{C,j}^e$ for MRC is given by

$$P_{C,j}^e = Q\left(\frac{I_p^{\text{MRC}}}{2 \times \sigma_\eta}\right), \tag{14.18}$$

where I_p^{MRC} is the generated photocurrent due to MRC given by

$$I_p^{\text{MRC}} = 2\Re\rho_t h_{\text{MRC}}^{\text{LOS}}, \tag{14.19}$$

and $h_{\text{MRC}}^{\text{LOS}}$ is the MRC channel gain at R_j given by

$$h_{\text{MRC}}^{\text{LOS}} = \sqrt{\sum_{i=0}^{j-1} h_{i,j}^2}. \tag{14.20}$$

- **SBC:** $P_{C,j}^e$ for SBC is given by

$$P_{C,j}^e = Q\left(\frac{I_p^{\text{SBC}}}{2\times\sigma_\eta}\right), \tag{14.21}$$

where I_p^{SBC} is the generated photocurrent due to SBC given by

$$I_p^{\text{SBC}} = 2\Re\rho_t h_{\text{SBC}}^{\text{LOS}}, \tag{14.22}$$

and $h_{\text{SBC}}^{\text{LOS}}$ is the SBC channel gain at R_j given by

$$h_{\text{SBC}}^{\text{LOS}} = \max\left[h_{0,j}^{\text{LOS}}, h_{1,j}^{\text{LOS}}, \ldots, h_{(j-1),j}^{\text{LOS}}\right]. \tag{14.23}$$

14.3.1.2 Imperfect CSI

L-PPM modulation splits the individual symbol period into L time slots. Imperfect CSI is the result of noise-induced during the channel anticipation stage. The following is an illustration of the anticipated channel coefficient between the xth transferring node and the yth reception node under noisy CSI:

$$\hat{h}_{x,y}^{\text{LOS}} = h_{x,y}^{\text{LOS}} + \zeta_y, \tag{14.24}$$

where $h_{x,y}^{\text{LOS}}$is the actual channel gain over the link between the xth sending node and the yth reception node, and ζ_y is the error introduced during the estimation of $h_{x,y}^{\text{LOS}}$. It is assumed that ζ_y is Guassian distributed, i.e., $\mathbb{N}\left(0,\sigma_\zeta^2\right)$, and hence, $\hat{h}_{x,y}^{\text{LOS}}$ is a Guassian distributed random variable with $\mathbb{N}\left(h_{x,y}^{\text{LOS}},\sigma_\zeta^2\right)$.

Listed below is the probability density function (PDF) of $\hat{h}_{x,y}^{\text{LOS}}$:

$$f\left(\hat{h}_{x,y}^{\text{LOS}}\right) = \frac{1}{\sqrt{2\pi\sigma_\zeta^2}} e^{-\frac{\left(\hat{h}_{x,y}^{\text{LOS}} - h_{x,y}^{\text{LOS}}\right)^2}{2\sigma_\zeta^2}}, \tag{14.25}$$

It is possible to compute the BER at the jth user as

$$\hat{P}_j^{e,avg} = \begin{cases} \left[\hat{P}_{j-1}^{e,avg}\left(1-\hat{P}_{C,j}^{e,avg}\right)\right] + \left[\hat{P}_{C,j}^{e,avg}\left(1-\hat{P}_{j-1}^{e,avg}\right)\right], & 2 \le j \le M \\ \hat{P}_1^{e,avg}, & j = 1 \end{cases}, \tag{14.26}$$

where $\hat{P}_1^{e,\text{avg}}$ is the BER at R_1 for OOK modulation defined as

$$\hat{P}_1^{e,\text{avg}} = Q\left(\frac{2\Re h_{0,1}^{\text{LOS}}\rho_t}{2\hat{\sigma}_\eta}\right), \tag{14.27}$$

while $\hat{P}_{C,j}^{e,\text{avg}}$ is the BER at R_j after the implementation of combining technique and $\hat{\sigma}_\eta = \sqrt{\sigma_\eta^2 + (\rho_t\Re)^2\sigma_\xi^2}$.

- **EGC:** $\hat{P}_{C,j}^{e,\text{avg}}$ for EGC is given by

$$\hat{P}_{C,j}^{e,\text{avg}} = Q\left(\frac{\hat{I}_p^{\text{EGC}}}{2\times\hat{\sigma}_{\text{EGC}}}\right), \tag{14.28}$$

where $\hat{I}_p^{\text{EGC}}$ is the generated photocurrent due to EGC given by

$$\hat{I}_p^{\text{EGC}} = 2\Re\rho_t\hat{h}_{\text{EGC}}^{\text{LOS}}, \tag{14.29}$$

while $\hat{h}_{\text{EGC}}^{\text{LOS}}$ is the EGC channel gain at R_j given by

$$\hat{h}_{\text{EGC}}^{\text{LOS}} = \sum_{i=0}^{j-1}\hat{h}_{i,j}^{\text{LOS}}, \tag{14.30}$$

and $\hat{\sigma}_{\text{EGC}}$ is given by

$$\hat{\sigma}_{\text{EGC}} = \sqrt{j\times\sigma_\eta^2 + (\rho_t\Re)^2\sigma_\xi^2}. \tag{14.31}$$

- **MRC:** For MRC, $\hat{h}_{\text{MRC}}^{\text{LOS}} = \sum_{i=0}^{j-1}\hat{h}_{i,j}^2$ adheres the non-central chisquare distribution, and thus, $\hat{P}_{C,j}^{e,\text{avg}}$ incapable of being resolved using analytical means [9]. Thus, in this chapter, we quantitatively evaluate BER using the kernel density estimation (KDE) technique. KDE is an invaluable tool for assessing a random variable's PDF in a non-parametric approach so that inferences about the population can be drawn from a sample of finite data.

The signal obtained by the jth user via the xth broadcasting node, assuming that the symbol '1' has been transmitted, can be expressed as

$$Y_{x,j}^1 = 2\Re h_{x,j}^{\text{LOS}}\rho_t + \eta, \tag{14.32}$$

The MRC deployment at the jth node occurs following the acquisition of signals from numerous nodes. The jth node's decision variable can be expressed as

$$D_j^1 = \sum_{i=0}^{j-1}\hat{h}_{i,j}^{\text{LOS}}\times Y_{i,j}^1. \tag{14.33}$$

Therefore, BER in this instance is

$$\hat{P}_j^{e,1} = \left(D_j^1 < \Re\rho_t\sum_{i=0}^{j-1}\hat{h}_{i,j}^2\right). \tag{14.34}$$

The above process is repeated to get an average BER at the jth user, i.e., $\hat{P}_j^{e,\text{avg},1}$.

Additionally, the likelihood of error for sending the symbol '0' is provided by

$$\hat{P}_j^{e,0} = \left(D_j^0 > \Re \rho_t \sum_{i=0}^{j-1} \hat{h}_{i,j}^2 \right), \tag{14.35}$$

where $D_j^0 = \sum_{i=0}^{j-1} \hat{h}_{i,j}^{\text{LOS}} \times Y_{i,j}^0$ and $Y_{i,j}^0 = \eta$.

Further, $\hat{P}_j^{e,\text{avg},0}$ can be obtained as discussed above.

At the jth receiver, the average BER can now be expressed as

$$\hat{P}_{C,j}^{e,\text{avg}} = \frac{1}{2}\left(\hat{P}_j^{e,\text{avg},0} + \hat{P}_j^{e,\text{avg},1} \right). \tag{14.36}$$

- **SBC:** $\hat{P}_{C,j}^{e,\text{avg}}$ for SBC is given by

$$\hat{P}_{C,j}^{e,\text{avg}} = Q\left(\frac{\hat{I}_p^{\text{SBC}}}{2 \times \hat{\sigma}_\eta} \right), \tag{14.37}$$

where $\hat{I}_p^{\text{SBC}}$ is the generated photocurrent due to SBC given by

$$\hat{I}_p^{\text{SBC}} = 2\Re \rho_t \hat{h}_{\text{SBC}}^{\text{LOS}}, \tag{14.38}$$

while $\hat{h}_{\text{SBC}}^{\text{LOS}}$ is the SBC channel gain given by

$$\hat{h}_{\text{SBC}}^{\text{LOS}} = \max\left[\hat{h}_{0,j}^{\text{LOS}}, \hat{h}_{1,j}^{\text{LOS}}, \ldots, \hat{h}_{(j-1),j}^{\text{LOS}} \right], \tag{14.39}$$

and $\hat{\sigma}_\eta$ is defined just below (14.27).

14.3.2 LPPM

14.3.2.1 Perfect CSI

A signal from T and $(j-1)$ signals from relays are received at R_j, $1 \le j \le M$. A combining technique chosen among MRC, EGC, and SBC is employed at R_j. The slot error probability (SLEP) at R_j can be computed by (14.40), assuming that the CSI is known at the receiver.

$$P_j^{e,\text{slot}} = \begin{cases} \left(P_{j-1}^{e,\text{slot}} \left(1 - P_{C,j}^{e,\text{slot}} \right) \right) + \left(P_{C,j}^{e,\text{slot}} \left(1 - P_{j-1}^{e,\text{slot}} \right) \right), & 2 \le j \le M \\ P_1^{e,\text{slot}}, & j = 1 \end{cases}, \tag{14.40}$$

where, for L-PPM modulation the SLEP at R_1 [8], i.e., $P_1^{e,\text{slot}}$ is

$$P_1^{e,\text{slot}} = Q\left(\frac{L \Re \rho_{\text{sym}} h_{0,1}^{\text{LOS}}}{2\sigma_\eta} \right), \tag{14.41}$$

and after the implementation of combining technique, the SLEP at R_j is $P_{C,j}^{e,\text{slot}}$. Where Q denotes the Q-function and $P(b=0)=P(b=1)=\frac{1}{2}$, i.e., symbols are equiprobable.

- **EGC:** After implementing EGC, the SLEP at R_j is provided by the following equation:

$$P_{C,j}^{e,\text{slot}} = Q\left(\frac{I_p^{\text{EGC}}}{2\sigma_{\text{EGC}}}\right), \tag{14.42}$$

where I_p^{EGC} is the generated photocurrent due to EGC given by

$$I_p^{\text{EGC}} = L\Re\rho_{\text{sym}}\sum_{i=0}^{j-1} h_{i,j}^{\text{LOS}}, \tag{14.43}$$

and σ_{EGC} is given by

$$\sigma_{\text{EGC}} = j \times \sigma_\eta. \tag{14.44}$$

- **MRC:** After implementing MRC, the SLEP at R_j is provided by the following equation:

$$P_{C,j}^{e,\text{slot}} = Q\left(\frac{I_p^{\text{MRC}}}{2\sigma_\eta}\right), \tag{14.45}$$

where I_p^{MRC} is the generated photocurrent due to MRC given by

$$I_p^{\text{MRC}} = L\Re\rho_{\text{sym}}\sqrt{\sum_{i=0}^{j-1} h_{i,j}^2}. \tag{14.46}$$

- **SBC:** After implementing SBC, the SLEP at R_j is provided by the following equation:

$$P_{C,j}^{e,\text{slot}} = Q\left(\frac{I_p^{\text{SBC}}}{2\sigma_\eta}\right), \tag{14.47}$$

where I_p^{SBC} is the generated photocurrent due to SBC given by

$$I_p^{\text{SBC}} = L\Re\rho_{\text{sym}} \max\left[h_{0,j}^{\text{LOS}}, h_{1,j}^{\text{LOS}}, \ldots, h_{j-1,j}^{\text{LOS}}\right]. \tag{14.48}$$

Furthermore, the symbol error probability (SEP) of R_j can be acquired by the following equation:

$$P_j^{e,\text{symbol}} = 1-(1-P_j^{e,\text{slot}})^L. \tag{14.49}$$

Finally, the BER of R_j is provided by the following equation:

$$P_j^{e,\text{bit}} = \frac{L/2}{L-1}P_j^{e,\text{symbol}}. \tag{14.50}$$-

14.3.2.2 Imperfect CSI

The channel estimate process introduces noise, which leads to the imperfect CSI case. Considering noisy CSI, the channel coefficient between R_x and R_y can be demonstrated as

$$\hat{h}_{x,y}^{\text{LOS}} = h_{x,y}^{\text{LOS}} + \zeta_y, \tag{14.51}$$

The SLEP for imperfect CSI of R_y can be evaluated by completing a few iterations and averaging the results, hence the average SLEP is given as

$$\hat{P}_j^{e,\text{slot,avg}} = \begin{cases} \left(\hat{P}_{j-1}^{e,\text{slot,avg}}\left(1-\hat{P}_{C,j}^{e,\text{slot,avg}}\right)\right)+\left(\hat{P}_{C,j}^{e,\text{slot,avg}}\left(1-\hat{P}_{j-1}^{e,\text{slot,avg}}\right)\right), & 2 \le j \le M \\ \hat{P}_1^{e,\text{slot,avg}}, & j=1 \end{cases}, \tag{14.52}$$

where $\hat{P}_1^{e,\text{slot,avg}}$ is the average SLEP at R_1 for L-PPM modulation given by

$$\hat{P}_1^{e,\text{slot,avg}} = Q\left(\frac{L\Re\rho_{\text{sym}}h_{0,1}^{\text{LOS}}}{2\widehat{\sigma_\eta}}\right), \tag{14.53}$$

where $\hat{P}_{C,j}^{e,\text{slot,avg}}$ is the average SLEP at R_y after implementation of combining technique and $\widehat{\sigma_\eta} = \sqrt{\sigma_\eta^2 + (0.5L\rho_{\text{sym}}\Re\sigma_\zeta)^2}$.

- **EGC:** After implementing EGC, the average SLEP at R_y is provided by

$$\hat{P}_{C,j}^{e,\text{slot,avg}} = Q\left(\frac{\hat{I}_p^{\text{EGC}}}{2\hat{\sigma}_{\text{EGC}}}\right), \tag{14.54}$$

where $\hat{I}_p^{\text{EGC}}$ is the generated photocurrent due to EGC given by

$$\hat{I}_p^{\text{EGC}} = L\Re\rho_{\text{sym}}\sum_{i=0}^{j-1}h_{i,j}^{\text{LOS}} \tag{14.55}$$

and $\hat{\sigma}_{\text{EGC}} = \sqrt{j\times\sigma_\eta^2 + (\rho_{\text{sym}}\Re)^2\sigma_\xi^2}$.

- **MRC:** When it comes to R_j, $j > 1$, $\hat{h}_j^{\text{MRC}} = \sum_{i=0}^{j-1}\hat{h}_{i,j}^2$ has a non-central chi-square distribution, which means that $\hat{P}_{C,j}^{e,\text{slot,avg}}$ cannot have an analytical solution. Because of this, we use the KDE technique in this investigation to quantitatively determine BER.

  When the symbol '1' is broadcasted, [8] may be used to determine the signal received at R_j from R_x as

$$Y_{x,j}^1 = L\Re h_{x,j}^{\text{LOS}}\rho_{\text{sym}} + \eta. \tag{14.56}$$

Signals from several users are received, and the MRC is implemented at R_j. As a result, R_j decision variable can be defined as

$$D_j^1 = \sum_{i=0}^{j-1} \hat{h}_{i,j}^{\text{LOS}} \times Y_{i,j}^1. \tag{14.57}$$

Consequently, in this case, the SLEP is

$$\hat{P}_j^{e,\text{slot},1} = \left(D_j^1 < \frac{L\Re\rho_{\text{sym}} \sum_{i=0}^{j-1} h_{i,j}^2}{2} \right). \tag{14.58}$$

To obtain an average SLEP at R_j, i.e., $\hat{P}_{j,\text{avg}}^{e,\text{slot},1}$, the aforementioned procedure is performed t times.

Additionally, SLEP for symbol '0' transmission is provided by

$$\hat{P}_j^{e,\text{slot},0} = \left(D_j^0 > L\Re\rho_{\text{sym}} \sum_{i=0}^{j-1} \hat{h}_{i,j}^2 / 2 \right), \tag{14.59}$$

where $D_j^0 = \sum_{i=0}^{j-1} \hat{h}_{i,j}^{\text{LOS}} \times Y_{i,j}^0$ and $R_{i,j}^0 = \eta$. Moreover, the average SLEP at R_j for transmission of symbol '0' is $\hat{P}_{j,\text{avg}}^{e,\text{slot},0}$.

At R_j, the average SLEP can now be expressed as

$$\hat{P}_{C,j}^{e,\text{slot,avg}} = \frac{1}{2}\left(\hat{P}_{j,\text{avg}}^{e,\text{slot},0} + \hat{P}_{j,\text{avg}}^{e,\text{slot},1} \right). \tag{14.60}$$

- **SBC:** After implementing SBC, the SLEP at R_j is provided by the following equation:

$$\hat{P}_{C,j}^{e,\text{slot,avg}} = Q\left(\frac{\hat{I}_p^{\text{SBC}}}{2\hat{\sigma}_\eta} \right), \tag{14.61}$$

where $\hat{I}_p^{\text{SBC}}$ is the generated photocurrent due to SBC given by

$$\hat{I}_p^{\text{SBC}} = L\Re\rho_{\text{sym}} \max\left[h_{0,j}^{\text{LOS}}, h_{1,j}^{\text{LOS}}, \ldots, h_{j-1,j}^{\text{LOS}} \right]. \tag{14.62}$$

14.4 RESULT ANALYSIS

Table 14.1 exhibits the simulation parameters and the room dimensions that were taken into deliberation. The simulation results validate and numerically assess the obtained BER computations for each case. A three-user scenario, i.e., R_1, R_2, and R_3, is taken into consideration for simulation; the coordinates of it are specified in Table 14.1. In particular, R_1 and R_2 serve as relays for R_3, and R_1 serves as a relay for R_2.

TABLE 14.1
Simulation Parameters for Co-VLC System Employing L-PPM Modulation for Users

Description	Value
Dimension of room	8 m×8 m×4 m
Coordinate of source	4 m×4 m×4 m
Coordinate of R_1	3 m×3 m×3 m
Coordinate of R_2	1.8 m×2 m×2 m
Coordinate of R_3	0.5 m×0.5 m×0.5 m
Semi-half angle of LED ($\phi_{1/2}$)	60°
FoV (Ψ_c)	60°
PD area ($\mathcal{A}$)	1 cm^2
Refractive index (N)	1.5
Optical filter gain $\mathcal{G}_\gamma\left(\phi_{k,l}\right)$	1
Transmission rate (r)	50 Mbps
Bandwidth of noise (B)	50 MHz
Background current (i)	202 μA

Figure 14.2 illustrates the BER efficiency of the OOK/L-PPM modulated Co-VLC system when SBC is applied at R_2. At a fixed SNR of 128 dB, the BER curves of R_2 are illustrated against the channel estimate error scenario in the graphical representation. According to the figure, the 8-PPM exhibits the best error performance, followed by 4-PPM, OOK, and 2-PPM within this order. For instance, the BER of R_2 is greater than 10^{-10} for 8-PPM modulation, about 3×10^{-6} for 4-PPM modulation, approximately 10^{-2} for OOK modulation, and approximately 5×10^{-2} for 2-PPM modulation at a channel estimation error of 10^{-12} in Figure 14.2. Additionally, for all mentioned modulation schemes, the system's BER efficiency improves as the channel estimation error decreases. Additionally, 8-PPM has a higher improvement rate in BER against channel estimate error than 4-PPM, OOK, and 2-PPM, in that order.

The BER efficiency of the OOK/L-PPM modulated Co-VLC system when EGC is employed at R_2 is established in Figure 14.3. At a fixed SNR of 128 dB, the BER curves of R_2 are illustrated against the channel estimate error scenario in the graph. Figure 14.3 illustrates that the 8-PPM exhibits the best error efficiency, followed by 4-PPM, OOK, and 2-PPM, in accordance with the study of Figure 14.2. The BER of R_2 is about 2×10^{-9} for 8-PPM modulation, approximately 3×10^{-6} for 4-PPM modulation, about 2×10^{-2} for OOK modulation, and nearly 8×10^{-2} for 2-PPM modulation, for instance, at a channel estimation error of 10^{-12} in Figure 14.2. Additionally, for all mentioned modulation schemes, the system's BER efficiency improves as the channel estimation error decreases. Additionally, 8-PPM has a greater enhancement rate in BER against channel estimate error than 4-PPM, OOK, and 2-PPM, in that order.

The OOK/L-PPM modulated Co-VLC system's BER efficiency when MRC is applied at R_2 is demonstrated in Figure 14.4. At a fixed SNR of 128 dB, the BER

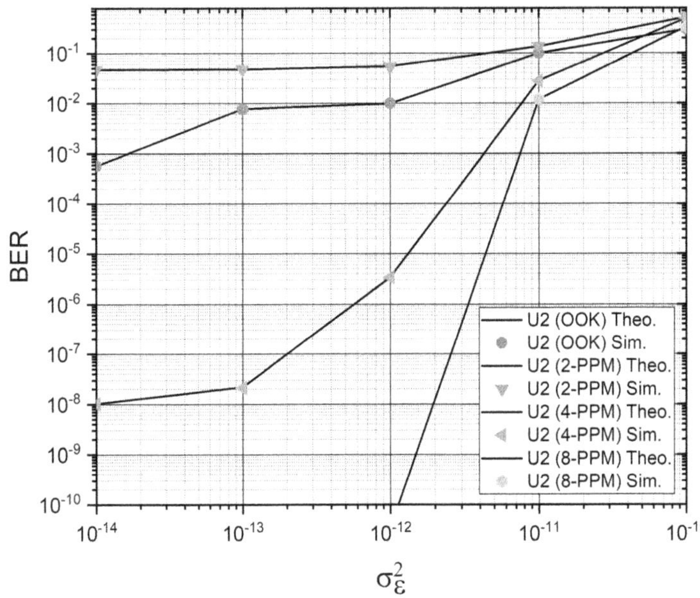

FIGURE 14.2 BER vs. variance for SBC at R_2.

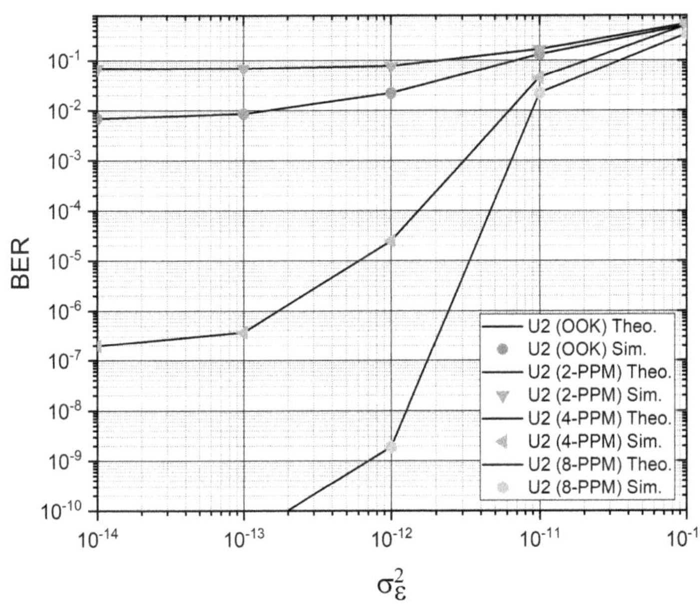

FIGURE 14.3 BER vs. variance for EGC at R_2.

curves of R_2 are illustrated against the channel estimate error scenario in the graph. Figure 14.4 illustrates that the 8-PPM exhibits the best error efficiency, followed by 4-PPM, OOK, and 2-PPM, in accordance with the study of Figures 14.2 and 14.3.

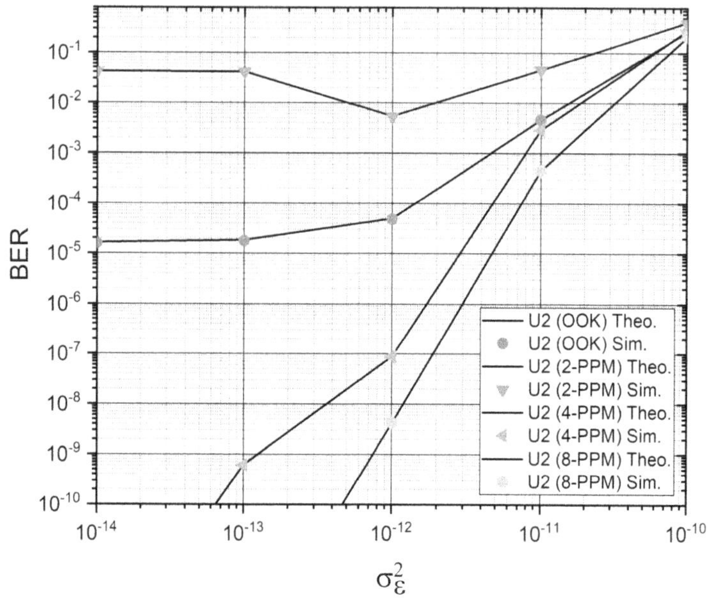

FIGURE 14.4 BER vs. variance for MRC at R_2.

In Figure 14.2, for instance, the BER of R_2 at a channel estimation error of 10^{-12} is about 3×10^{-9} for 8-PPM modulation, approximately 9×10^{-8} for 4-PPM modulation, nearly 5×10^{-5} for OOK modulation, and around 6×10^{-3} for 2-PPM modulation. Additionally, for all aforementioned modulation schemes, the system's BER efficiency improves as the channel estimation error diminishes. Furthermore, the improvement rate in BER contrary to channel estimation error is greater in 8-PPM than 4-PPM, followed by OOK and then 2-PPM, respectively.

The BER efficiency of an 8-PPM modulated VLC system with EGC applied at R_2 and R_3 is illustrated in Figure 14.5. At a given SNR of 128 dB, the BER curves of R_1, R_2, and R_3 are rendered against the channel estimate error scenario in the figure. Additionally, the illustration compares direct transmission and cooperation when the EGC combining technique is implemented. For both R_2 and R_3 users, it is evident that the system works better in a scenario of cooperation than in the case of direct transmission. For instance, in Figure 14.5, at a channel estimation error of 10^{-12}, the BER of R_3 is around 0.6 for direct transmission, about 0.2 for R_2 for direct transmission, almost 2×10^{-4} for R_3 in cooperation case, nearly 2×10^{-8} for R_2 in cooperation case, and approximately 8×10^{-10} for R_1. As the distance from the source LED upsurges, the graph also illustrates that the BER efficiency of R_1 is superior to R_2 and R_3. Furthermore, for every user (i.e., R_1, R_2, and R_3), the system's BER efficiency improves as the channel estimation error decreases. Furthermore, the improvement rate in BER against channel estimation error is greater in R_1 than R_2, followed by R_3 in the cooperation case, and then R_2 and R_3 in the case of direct transmission, respectively.

Figure 14.6 illustrates the BER efficiency of an 8-PPM modulated VLC system when MRC is applied at R_2 and R_3. In the graph, the BER curves of R_1, R_2, and R_3

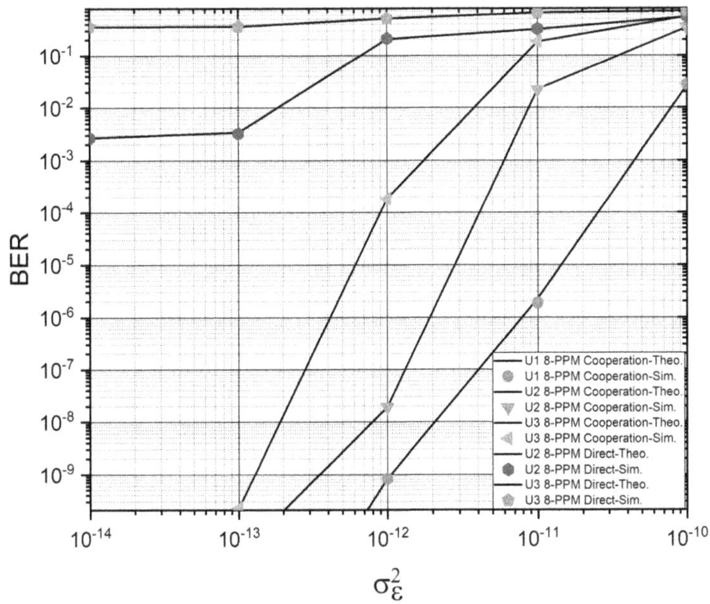

FIGURE 14.5 BER vs. variance for EGC.

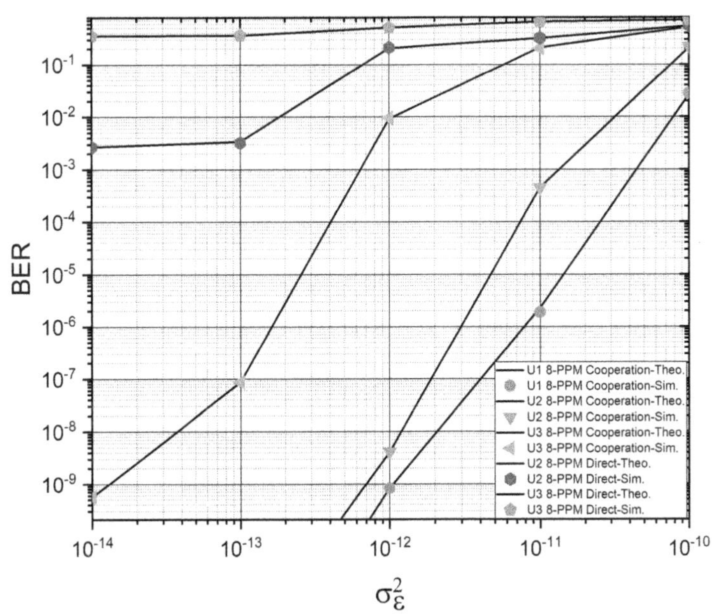

FIGURE 14.6 BER vs. variance for MRC.

are plotted against the channel estimation error scenario at fixed SNR of 128 dB. The figure also shows the comparison of cooperation and direct transmission in case when MRC combining technique is utilized. It clearly shows that the system performs better in the case of cooperation than the direct transmission case for both R_2

and R_3 users. The 8-PPM modulated Co-VLC system's BER efficiency when MRC is employed at R_2 and R_3 is demonstrated in Figure 14.6. At a given SNR of 128 dB, the BER curves of R_1, R_2, and R_3 are illustrated against the channel estimate error scenario in the figure. Additionally, the illustration compares direct transmission and cooperation when the MRC combining approach is utilized. It is evident that for both R_2 and R_3 users, the system performs more effectively when there is cooperation than when there is direct transmission. Furthermore, for all users (i.e., R_1, R_2, and R_3), the system's BER efficiency improves as the channel estimation error declines. Additionally, the improvement rate in BER against channel estimation error is higher in R_1, followed by R_2 and R_3 in the cooperative case, and then R_2 and R_3 in the direct transmission case.

The BER efficiency of an 8-PPM modulated VLC system with SBC applied at R_2 and R_3 is illustrated in Figure 14.7. At a given SNR of 128 dB, the BER curves of R_1, R_2, and R_3 are demonstrated against the channel estimate error scenario in the figure. Additionally, the figure compares direct transmission and cooperation when the SBC combining approach is implemented. It is evident that for both R_2 and R_3 users, the system performs more efficiently when there is cooperation over when there is direct transmission. For instance, in Figure 14.5, at a channel estimation error of 10^{-12}, the BER of R_3 is approximately 0.6 for direct transmission, approximately 0.2 for R_2 for direct transmission, about 5×10^{-6} for R_3 in the cooperation case, nearly 3×10^{-9} for R_2 in the cooperation case, and almost 9×10^{-10} for R_1. Similar to the findings from Figures 14.5 and 14.6, Figure 14.7 also shows that BER efficiency of R_1 is better than R_2, followed by R_3 due to the increasing distance from the source LED. Further, the

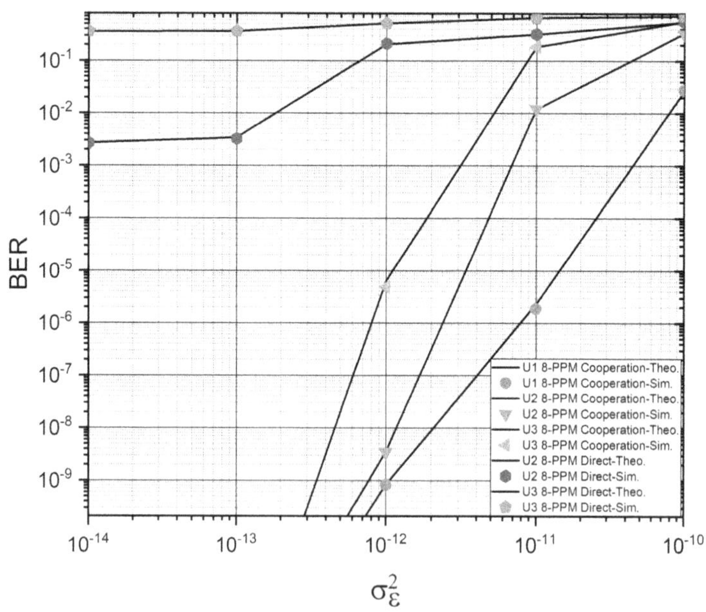

FIGURE 14.7 BER vs. variance for SBC.

BER efficiency of the system ameliorates as the channel estimation error declines for all the users, i.e., R_1, R_2, and R_3. Additionally, the improvement rate in BER against channel estimation error is greater in R_1 than R_2, followed by R_3 in the cooperation case, and then R_2 and R_3 in the case of direct transmission, respectively.

Figure 14.8 demonstrates the BER efficiency of a 4-PPM/OOK modulated Co-VLC system assuming perfect CSI scenarios employing various combining diversity strategies carried out at R_3. The figure illustrates the BER curves of R_3 against the FoV at a certain SNR of 128 dB. The illustration also contrasts the 4-PPM and OOK modulation strategies using EGC, MRC, and SBC. The figure depicts that the BER improves in each scenario when the FoV decreases. The figure additionally demonstrates that, regardless of diversity strategies stated, 4-PPM outperforms OOK. There is barely any variation in the BER efficiency of EGC, MRC, and SBC diversity approaches when OOK modulation is used at high FoV. However, at low FoV, MRC outperforms EGC followed by SBC. In contrast, for all range of FoV MRC outperforms SBC and EGC, respectively, in the 4-PPM modulation scenario.

Figure 14.9 shows the BER efficiency of the OOK/L-PPM modulated Co-VLC system when MRC is applied at R_3. The graph shows the BER curves of R_3 against the FoV at a given SNR of 128 dB. In accordance with Figures 14.2, 14.3, and 14.4, Figure 14.9 shows that the 8-PPM has a superior error performance, followed by 4-PPM, OOK, and 2-PPM. Furthermore, the system's BER efficiency increases as FoV decreases for all of the previously described modulation methods. Additionally, 8-PPM has a higher improvement rate in BER against channel estimate error than 4-PPM, OOK, and 2-PPM in that order.

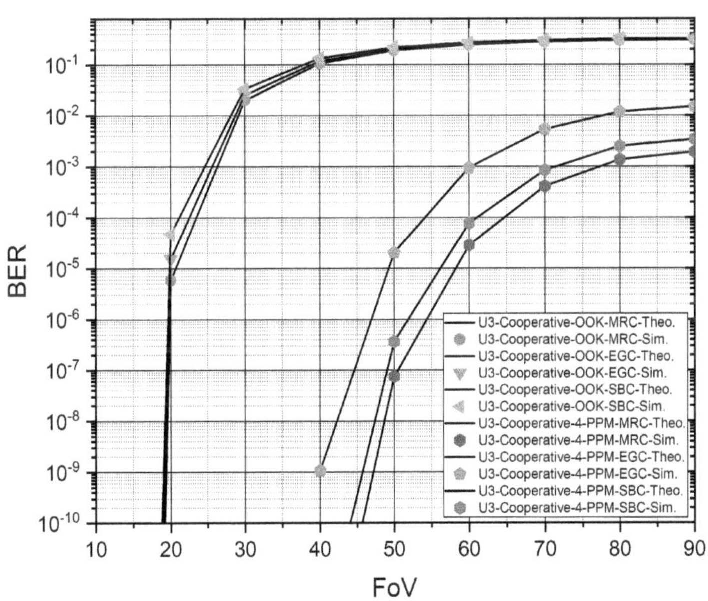

FIGURE 14.8 BER vs. FoV all diversities.

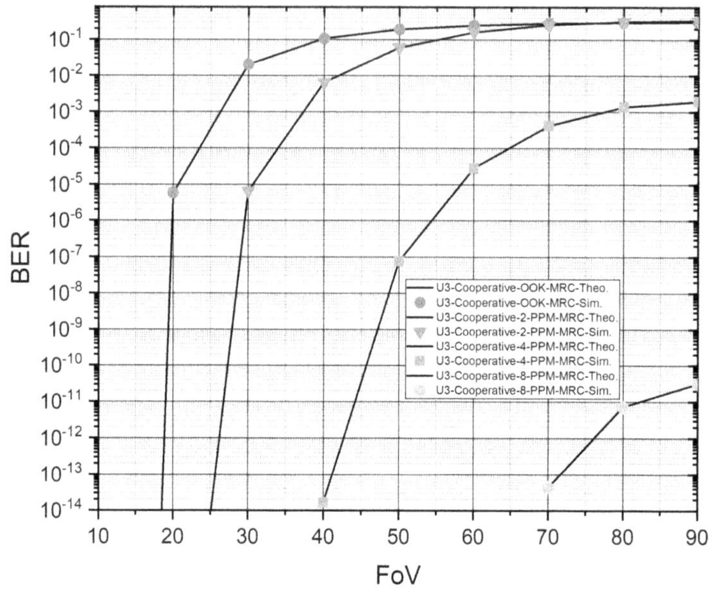

FIGURE 14.9 BER vs. FoV all modulations.

14.5 CONCLUSION

This chapter performs a thorough performance analysis of an L-PPM/OOK modulated Co-VLC system, employing an assortment of diversity combining techniques including EGC, MRC, and SBC, with an emphasis on the impacts of imperfect CSI. Simulation results are used to authenticate the numerical computations for the BER of both modulation schemes, L-PPM and OOK. The outcomes show that 8-PPM accomplishes superior to the other modulation schemes under consideration, followed by 4-PPM, OOK, and 2-PPM. The investigation also highlights the benefits of CC, which accomplishes noticeably better than direct transmission in identical circumstances. The crucial role of CSI in improving system dependability is highlighted by the observation that the BER improves as the channel estimate error reduces. The trade-off between increased reach and error performance is highlighted by the fact that an increase in the number of hops in the communication path results in a decline in BER efficiency. The optimum approach to mitigate the effects of imperfect CSI and improve system performance is MRC, which routinely performs better than SBC and EGC among the diversity combining strategies. These results highlight how Co-VLC systems and sophisticated diversity strategies can be used to overcome the difficulties resulting from actual channel defects. The BER efficiency of EGC, MRC, and SBC for OOK modulation varies barely at high FoV, whereas MRC surpasses EGC and SBC at low FoV. On the other hand, across all FoV ranges, MRC consistently performs better than SBC and EGC for 4-PPM modulation.

REFERENCES

1. S. Gupta, D. Roy, S. Bose, V. Dixit, and A. Kumar, Illuminating the future: A comprehensive review of visible light communication applications, *Optics & Laser Technology*, vol. 177, p. 111182, 2024, https://doi.org/10.1016/j.optlastec.2024.111182.
2. P. Dwivedy, V. Dixit, and A. Kumar, A survey on visible light communication for 6G: Architecture, application and challenges, *2023 International Conference on Computer, Electronics & Electrical Engineering & their Applications (IC2E3)*, Srinagar Garhwal, India, pp. 1–06, 2023, https://doi.org/10.1109/IC2E357697.2023.10262462.
3. A. Kumar, S. K. Ghorai, and A. Chaudhury, BER performance of OFDM based hybrid PLC-VLC system, *2020 International Conference on Computer, Electrical Communication Engineering (ICCECE)*, Kolkata, India, 2020, https://doi.org/10.1109/ICCECE48148.2020.9223079.
4. N. Sharan, S. K. Ghorai, and A. Kumar, PAPR reduction using a Precoder and Compander combination in a NOMA-OFDM VLC system, *2022 2nd International Conference on Artificial Intelligence and Signal Processing (AISP)*, Vijayawada, India, pp. 1–4, 2022, https://doi.org/10.1109/AISP53593.2022.9760659.
5. A. Kumar, and S. K. Ghorai, High-gain and volume-efficient concave–convex lens for multiple-input multiple-output visible light communication system, *Optical Engineering*, vol. 59, no. 3, p. 035106, 28 March 2020.
6. A. Kumar, and S. K. Ghorai, Reduction of channel correlation for multiuser-MIMO system in visible light communications, *2016 International Conference on Signal Processing and Communication (ICSC)*, Noida, India, pp. 22–25, 2016, https://doi.org/10.1109/ICSPCom.2016.7980540.
7. N. Sharan, S. K. Ghorai, and A. Kumar, LED non-linearity mitigation in a NOMA-OFDM VLC system using a union of precoder and companding, *Optical and Quantum Electronics*, vol. 54, no. 10, p. 648, 2022, https://doi.org/10.1007/s11082-022-04043-6.
8. P. Dwivedy, V. Dixit, and A. Kumar, BEP analysis of L-PPM modulated cooperative VLC system with perfect and imperfect CSI, *Optical and Quantum Electronics*, vol. 56, no. 10, p. 1572, 2024, https://doi.org/10.1007/s11082-024-07384-6.
9. P. Dwivedy, V. Dixit, and A. Kumar, On the performance of OOK/L-PPM modulated NOMA cooperative VLC system with diversity combining techniques under imperfect CSI scenario, *IEEE Sensors Journal*, vol. 24, no. 18, pp. 29331–29339, 15 September 2024, https://doi.org/10.1109/JSEN.2024.3435041.
10. C. Bhatnagar, A. Potnis, P. Dwivedy, and S. K. Meena, Performance analysis and optimization schemes for cooperative spectrum sensing and information fusion for cognitive radio: A survey, *2017 1st International Conference on Electronics, Materials Engineering and Nano-Technology (IEMENTech)*, Kolkata, India, https://doi.org/10.1109/IEMENTECH.2017.8077000, pp. 1–6, 2017.
11. S. Soofi, A. Potnis, and P. Diwivedy, Efficient dynamic double threshold energy detection of cooperative spectrum sensing in cognitive radio, *5th International Conference on Emerging Research in Computing, Information, Communication and Applications: ERCICA 2018*, Bangalore, India, vol. 1, 2019.
12. P. Dwivedy, V. Dixit, and A. Kumar, Enhancing performance through diversity schemes in cooperative VLC systems with OOK modulation, *2023 IEEE 7th Conference on Information and Communication Technology (CICT)*, Jabalpur, India, pp. 1–5, 2023, https://doi.org/10.1109/CICT59886.2023.10455319.
13. P. Dwivedy, V. Dixit, and A. Kumar, NOMA cooperative VLC systems: design and performance analysis with OOK and L-PPM modulation, *Applied Optics*, vol. 62, no. 25, pp. 6639–6651, 2023.

14. P. Dwivedy, V. Dixit, and A. Kumar, Cooperative VLC system using OOK modulation with imperfect CSI, *Physica Scripta*, vol. 98, no. 2, pp. 1–10, 2023.
15. D.-F. Zhang, Y.-J. Zhu, and Y.-Y. Zhang, Multi-LED phase-shifted OOK modulation based visible light communication systems, *IEEE Photonics Technology Letters*, vol. 25, no. 23, pp. 2251–2254, 2013.
16. Z. Ahmed, R. Singh, W, Ali, G. Faulkner, D. O'Brien, and S. Collins, A SiPM-based VLC receiver for gigabit communication using OOK modulation, *IEEE Photonics Technology Letters*, vol. 32, no. 6, pp. 317–320, 2020.
17. M. L. G. Salmento, G. M. Soares, J. M. Alonso, and H. A. C. Braga, A dimmable offline LED driver with OOK-M-FSK modulation for VLC applications, *IEEE Transactions on Industrial Electronics*, vol. 66, no. 7, pp. 5220–5230, July 2019.
18. B. Fahs, A. J. Chowdhury, and M. M. Hella, A 12-m 2.5-Gb/s lighting compatible integrated receiver for OOK visible light communication links, *Journal of Lightwave Technology*, vol. 34, no. 16, pp. 3768–3775, 2016.
19. N. A. Mohammed, and K. A. Badawi, Design and performance evaluation for a non-line of sight VLC dimmable system based on SC-LPPM, *IEEE Access*, vol. 6, pp. 52393–52405, 2018.
20. J. Mietzner, and L. Lampe, Nested PPM for visible light communication with heterogeneous optical receivers, *IEEE Photonics Journal*, vol. 16, no. 4, pp. 1–12, 2024.
21. T. T. Ağır, and M. Sönmez, The modulation classification methods in PPM–VLC systems, *Optical and Quantum Electronics*, vol. 55, no. 3, 2023, https://doi.org/10.1007/s11082-022-04520-y.
22. J. Al-Khori, G. Nauryzbayev, M. M. Abdallah, and M. Hamdi, Secrecy performance of decode-and-forward based hybrid RF/VLC relaying systems, *IEEE Access*, vol. 7, pp. 10844–10856, 2019.
23. B. Aly, M. Elamassie, and M. Uysal, Vehicular VLC system with selection combining, *IEEE Transactions on Vehicular Technology*, vol. 71, no. 11, pp. 12350–12355, 2022.
24. C. Chen, W.-D. Zhong, H. Yang, S. Zhang, and P. Du, Reduction of SINR fluctuation in indoor multi-cell VLC systems using optimized angle diversity receiver, *Journal of Lightwave Technology*, vol. 36, no. 17, pp. 3603–3610, 2018.
25. Q. Li, L. Shi, T. Tang, and Z. Xiong, User cooperation and link selection in hybrid VLC/RF systems with NOMA, *IIEEE Wireless Communications Letters*, vol. 13, no. 4, pp. 979–983, 2024.
26. T. V. Pham, and A. T. Pham, Coordination/Cooperation strategies and optimal zero-forcing precoding design for multi-user multi-cell VLC networks, *IEEE Transactions on Communications*, vol. 67, no. 6, pp. 4240–4251, 2019.
27. B. Lin, X. Tang, Z. Ghassemlooy, C. Lin, and Y. Li, Experimental demonstration of an indoor VLC positioning system based on OFDMA, *IEEE Photonics Journal*, vol. 9, no. 2, pp. 1–9, 2017.
28. S. Huang, G. Chuai, W. Gao, and K. Zhang, Agency selling format-based incentive scheme in cooperative hybrid VLC/RF IoT system with SLIPT, *IEEE Internet of Things Journal*, vol. 10, no. 8, pp. 7366–7379, 2023.
29. I. W. G. da Silva, E. E. B. Olivo, M. Katz, and D. P. M. Osorio, Analysis and simulation of precoding and user association for securing hybrid RF/VLC systems, *IEEE Sensors Journal*, vol. 24, no. 20, pp. 33467–33480, 2024.
30. V. Dixit, and A. Kumar, An exact error analysis of multi-user RC/MRC based MIMO-NOMA-VLC system with imperfect SIC, *IEEE Access*, vol. 9, pp. 136710–136720, 2021, https://doi.org/10.1109/ACCESS.2021.3117446.
31. V. Dixit, and A. Kumar, BER performance of MIMO based NOMA VLC system with imperfect SIC, *Transactions on Emerging Telecommunications Technologies*, vol. 33, p. e4422, 2022.

32. V. Dixit, and A. Kumar, Error performance of an L-PPM modulated ADR-based MIMO-VLC system with and without channel estimation error, *Applied Optics*, vol. 62, pp. 2501–2509, 2023.
33. P. Dwivedy, V. Dixit, and A. Kumar, Enhancing performance of image data transmission over cooperative VLC network, *2023 IEEE 15th International Conference on CICN*, Bangkok, Thailand, 2023, https://doi.org/10.1109/CICN59264.2023.10402304.
34. V. Dixit, and A. Kumar, Error analysis of L-PPM modulated ADT based VLC system with perfect and imperfect CSI, *2021 International Conference on Control, Automation, Power and Signal Processing (CAPS)*, Jabalpur, India, https://doi.org/10.1109/CAPS52117.2021.9730708, pp. 1–6, 2021.
35. V. Dixit, and A. Kumar, BER analysis of dynamic FOV based MIMO-NOMA-VLC system, *AEU International Journal of Electronics and Communications*, vol. 142, p. 153989, 2021, https://doi.org/10.1016/j.aeue.2021.153989.
36. V. Dixit, and A. Kumar, Performance analysis of L-PPM modulated NLOS-VLC system with perfect and imperfect CSI, *Journal of Optics*, vol. 23, no. 1, p. 015702, 2020.

Index

L

M

N

O

P

Q

R

S

For Product Safety Concerns and Information please contact our EU representative GPSR@taylorandfrancis.com
Taylor & Francis Verlag GmbH, Kaufingerstraße 24, 80331 München, Germany

www.ingramcontent.com/pod-product-compliance
Ingram Content Group UK Ltd.
Pitfield, Milton Keynes, MK11 3LW, UK
UKHW022217190926
475702UK00003B/409

* 9 7 8 1 0 3 2 9 8 5 2 2 0 *